Fundamentals of O-RAN

IEEE Press
445 Hoes Lane
Piscataway, NJ 08854

Fundamentals of O-RAN

Nishith D. Tripathi
Virginia Tech
Parker, TX, USA

Vijay K. Shah
North Carolina State University and WiSights Lab
Raleigh, NC, USA

IEEE Press
WILEY

Published by John Wiley & Sons, Inc., Hoboken, New Jersey.
Published simultaneously in Canada.

Library of Congress Cataloging-in-Publication Data:

Names: Tripathi, Nishith D., author. | Shah, Vijay K., author.
Title: Fundamentals of O-RAN / Nishith D. Tripathi, Vijay K. Shah.
Description: Hoboken, New Jersey : Wiley, [2025] | Includes index.
Identifiers: LCCN 2024037419 (print) | LCCN 2024037420 (ebook) | ISBN 9781394206803 (hardback) | ISBN 9781394206810 (adobe pdf) | ISBN 9781394206827 (epub)
Subjects: LCSH: Open Radio Access Network.
Classification: LCC TK5103.2 .T733 2025 (print) | LCC TK5103.2 (ebook) | DDC 621.382 – dc23/eng/20240925
LC record available at https://lccn.loc.gov/2024037419
LC ebook record available at https://lccn.loc.gov/2024037420

Cover Design: Wiley
Cover Image: © K illustrator Photo/Shutterstock

Set in 9.5/12.5pt STIXTwoText by Straive, Chennai, India

Contents

About the Authors

Dr. Nishith D. Tripathi is a Research Associate Professor at Virginia Tech. Dr. Tripathi has 24 years of hands-on industry experience and is an expert on various aspects of commercial 3G, 4G, and 5G wireless networks including design, operations, testing, and optimization. At Virginia Tech, he has led several sponsored research projects on 5G, 5G-Advanced, and 6G in the areas such as 5G O-RAN testbeds, SpaceNet testbed, O-RAN xApps, O-RAN testing, enhanced security for 5G and 6G, NTN, V2X communications, geofencing, positioning, UAV/UAS, and smart warehouses. He has coauthored five books including the world's first multimedia book on 5G, a comprehensive textbook on cellular communications, a pioneering monograph on the RRM using AI, and two books on NTN and O-RAN. He has made more than 30 contributions toward the development of the 3GPP 5G specifications. As a wireless industry expert, Dr. Tripathi has contributed to organizations such as FCC, CTIA, GSMA, NGA, NSC, Scientific American, FTC, EE Times University, and CNN. He is the founder of Aum Research and Consulting (ARC) that provides research and consulting services.

Dr. Vijay K. Shah is an Assistant Professor in the Electrical and Computer Engineering Department at North Carolina State University. He heads the NextG Wireless Lab@NC State, a leading research group specializing in cutting-edge exploration of next-generation wireless communications and networking. His research spans various critical areas including 5G/NextG cellular networks, O-RAN, dynamic spectrum sharing, and AI/ML applications in wireless technologies. He also serves as the co-founder of WiSights Lab, a startup focused on developing secure and efficient GenAI/LLM solutions for telecom and enterprise networks. He obtained his PhD in Computer Science from University of Kentucky and B. Tech in Computer Science and Engineering from National Institute of Technology, Durgapur, in 2019 and 2013, respectively. After PhD, he served as a Research Assistant Professor in the Bradley Department of Electrical and Computer Engineering at Virginia Tech until 2021, and as a tenure-track Assistant Professor at George Mason University until 2024. Dr. Shah's research on next-generation wireless communication networks has been generously supported by federal agencies, such as NSF, NIST, and NTIA.

Preface

Open Radio Access Network (O-RAN), often referred to as Open RAN, is a revolutionary technology transforming the cellular communications landscape. O-RAN is an open, programmable, interoperable, and virtualized RAN architecture. O-RAN provides significant flexibility to cellular service providers to mix and match hardware and software components from different vendors. This book provides a comprehensive view of O-RAN by addressing various aspects of O-RAN including the driving forces behind O-RAN architecture, applications such as xApps and rApps, O-RAN security considerations, internal and external ecosystems, testing and integration, R&D initiatives, O-RAN challenges, and evolution of O-RAN toward the sixth-generation (6G) cellular technology.

Chapter 1 provides a concise introduction to O-RAN. This chapter first builds a foundation for O-RAN by illustrating the evolution of cellular technologies, reviewing components of a cellular system, and describing the evolution of RAN architectures. Then, a brief introduction to the O-RAN ALLIANCE (which is in charge of defining O-RAN specifications) is given, followed by illustration of a simplified O-RAN architecture defined by the O-RAN ALLIANCE. Deployments of O-RAN use standards and frameworks developed by the O-RAN ALLIANCE and several other organizations. Roles of such non-O-RAN ALLIANCE organizations in the O-RAN ecosystem are discussed. Driving forces behind O-RAN's rising importance are explained. Numerous uses cases of O-RAN that help optimize various RAN operations are described. Finally, the overall accomplishments of O-RAN are summarized along with O-RAN's technical priorities and potential future work.

Chapter 2 provides a detailed look at the O-RAN architecture. The O-RAN ALLIANCE has defined a logical O-RAN architecture that utilizes the 3GPP-defined 5G and 4G Radio Access Network (RAN) as a baseline and introduces new nodes and interfaces in support of openness and intelligence. Key components of O-RAN are Service Management and Orchestration (which includes Non-Real-Time RAN Intelligent Controller [Non-RT RIC]), Near-Real-Time RAN Intelligent Controller (Near-RT RIC), O-RAN Central Unit – Control Plane (O-CU-CP), O-RAN Central Unit – User Plane (O-CU-UP), O-RAN Distributed Unit (O-DU), O-RAN Radio Unit (O-RU), O-RAN eNodeB (O-eNB), and O-RAN Cloud (O-Cloud). These O-RAN components are extensively discussed. The interfaces that connect O-RAN components with each other and external non-O-RAN systems are also discussed. These interfaces include O1, O2, A1, E2, Open Fronthaul (Open FH), O-Cloud Notification, and Cooperative Transport Interface (CTI). While the 3GPP has defined disaggregated Base Stations in the radio network, the O-RAN architecture further disaggregates the Base Station using the functional split 7.2x, which is described in detail. Functional distribution between O-RAN and other Standards Developing Organizations (SDOs) is explained. For O-RAN deployments, an overview of O-RAN cloudification, hierarchical cloud architecture, and deployment scenarios is also given.

Chapter 3 describes three key control loops within the O-RAN architecture. These control loops play a crucial role in dynamically optimizing network performance at different time scales. Details of each O-RAN control loop, its essential components, and associated microservices such as xApps and rApps are discussed. These microservices enable the decomposition of complex RAN functionalities into modular and independently deployable units. The control loop and microservices framework empowers network operators to adapt the network swiftly to evolving demands and promotes a cost-effective and adaptive approach to building and maintaining telecommunications infrastructure. O-RAN-defined RICs such as the Non-RT RIC and the Near-RT RIC are also elaborated. Potential enhancements to the O-RAN architecture such as the Real-Time RIC (RT RIC) and its microservices, commonly referred to as μApps or dApps, are also summarized.

Chapter 4 focuses on O-RAN security. In contrast to conventional cellular networks, O-RAN represents a paradigm shift with its open, programmable, interoperable, and virtualized RAN architecture. This innovative approach empowers network operators to seamlessly integrate hardware and software components from diverse vendors, fostering enhanced flexibility, scalability, and cost efficiency. Furthermore, this approach fosters competition among vendors. However, these O-RAN characteristics result in critical security implications. This chapter comprehensively examines various security dimensions inherent in the O-RAN architecture, shedding light on the intricate interplay between openness and safeguarding network integrity.

Chapter 5 discusses the overall ecosystem of O-RAN from two perspectives—an internal ecosystem and an external ecosystem. Design, development, integration, testing, deployment, maintenance, optimization, and promotion of O-RAN require a comprehensive internal and external ecosystem. The internal ecosystem of O-RAN carries out work within the scope of the O-RAN ALLIANCE focus streams. The external ecosystem includes SDOs, industry alliances, and government entities. Some constituents of the external ecosystem such as SDOs are of existential importance to O-RAN; without these components, O-RAN simply cannot exist. In contrast, some constituents of the external ecosystem such as government entities help make the overall ecosystem vibrant to promote the deployment of O-RAN, realizing the benefits of O-RAN for consumers, enterprises, and government entities.

Chapter 6 narrates testing and integration aspects of O-RAN. O-RAN is characterized by openness and intelligence. Openness of O-RAN is evident by the support for multiple open interfaces. Open interfaces enable a service provider to make use of products and O-RAN components from different vendors. Openness provides significant flexibility to a service provider to select vendors to meet its business objectives. However, integration of different O-RAN components is a nontrivial matter. Testing of integrated products is especially important to ensure proper operations of O-RAN in commercial deployments. A brief overview of Open Testing and Integration Centers is given. The O-RAN certification and badging program is also described. Global PlugFests that bring companies and institutions from multiple continents together are discussed. An overview of testing-related O-RAN specifications is given. Finally, examples of best practices and recommendations for O-RAN testing and integration are given.

Chapter 7 provides a glimpse of O-RAN R&D initiatives. In the last few years, O-RAN has received significant attention from governments, industry, and academia. This chapter outlines key O-RAN R&D initiatives led by industry, academia, as well as governments in the United States and across the world to propel the development and adoption of O-RAN technology for 5G and upcoming 6G networks.

Chapter 8 provides an overview of O-RAN challenges and implications of 6G on O-RAN. O-RAN is expected to be one of the prominent ways of designing and deploying a 5G Radio Access Network. However, with any new technologies, there are challenges to overcome. This chapter provides a glimpse of O-RAN challenges. Examples of potential O-RAN enhancements to address these challenges are also given. While O-RAN currently supports 5G and 4G radio interfaces at the time of this writing, it is expected to support 6G in future. A brief overview of 6G is given, followed by areas of focus in O-RAN to support 6G.

In our wireless communications industry, we cannot escape acronyms! To help you navigate through the maze of acronyms, we have included a list of acronyms in the front matter of the book. If you like to go into more details, the list of references mentioned at the end of each chapter is a good starting place. Online exercises to help you reinforce key O-RAN concepts are available at https://www.wiley.com/go/tripathioran.

We hope that this book will significantly increase your O-RAN knowledge and facilitate your professional growth. Please consider this book as the beginning of our teacher–student relationship; feel free to connect to us via social media.

November 2024

Happy O-RAN learning!

Dr. Nishith D. Tripathi and Dr. Vijay K. Shah

Acknowledgments

We are thankful to our families for their patience and sacrifice of time while we worked on the book in the evenings and weekends and during holidays. We appreciate the support and encouragement from Prof. Jeffrey H. Reed (Virginia Tech) for this endeavor.

We thank the O-RAN ALLIANCE for defining O-RAN specifications, without which this book would not exist!

We thank our employers, Virginia Tech, George Mason University, and North Carolina State University, where we got opportunities to perform Open Radio Access Network (O-RAN) research. Our research facilitated the development of this book.

We appreciate the support from National Telecommunications and Information Administration (NTIA). This book is partially based upon work supported by the NTIA's Public Wireless Supply Chain Innovation Fund (PWSCIF) under the award number (51-60-IF007). We thank National Science Foundation (NSF) for partially supporting this book through the "Open AI Cellular" project (NSF Award # 2120411).

We thank our students who contributed directly or indirectly to this book. We thank Pratheek Upadhyaya (Virginia Tech) for his early work on building 5G O-RAN testbeds at Virginia Tech and providing relevant documents. We thank Kumar Sai Bondada (Virginia Tech) for helping build a hands-on O-RAN workshop and developing some of the diagrams that are part of this book. We also thank Zac Martin (Virginia Tech) for developing nice diagrams for this book.

We extend our gratitude to our PhD students, postdocs, and colleagues at NextG Wireless Lab (NextG Lab) at North Carolina State University. Specifically, we acknowledge Abiodun Ganiyu (NextG Lab) for his contributions to the O-RAN security chapter and for his insightful feedback on a few other chapters of this book. We also appreciate Nathan Stephenson (George Mason University) for developing the xApp design and development workflow, which contributed to Chapter 3 on xApps and rApps. We thank Pranshav Gajjar for his contributions with indices and chapter exercises. Lastly, we thank Dr. Sayanta Seth (NextG Lab) for providing essential documents on O-RAN energy efficiency. We thank the reviewers of this book whose feedback helped enhance the book.

Finally, we deeply appreciate the flexibility and guidance of our publisher, IEEE Press/Wiley.

November 2024 *Dr. Nishith D. Tripathi and Dr. Vijay K. Shah*

Acronyms

1PPS	1 Pulse Per Second
3GPP	Third Generation Partnership Project
4G	Fourth Generation
5G	Fifth Generation
6G	Sixth Generation
6G-IA	6G Smart Networks and Services Industry Association
5GC	5G Core (Network)
5GCN	5G Core Network
AAL	Acceleration Abstraction Layer
AERPAW	Aerial Experimentation and Research Platform for Advanced Wireless
AF	Application Function
AH	Authentication Header
AI	Artificial Intelligence
AI/ML	Artificial Intelligence/Machine Learning
ALM	Application Life Cycle Management
AMF	Access and Mobility Management Function
AMPS	Advanced Mobile Phone System
AN	Access Network
AnLF	Analytics Logical Function
ANR	Automatic Neighbor Relation
API	Application Programming Interface
ARA	Agriculture and Rural Communities
AS	Access Stratum
AS	Application Server
ASN	Abstract Syntax Notation
ATIS	Alliance for Telecommunications Industry Solutions
AWS	Amazon Web Services
BBU	Baseband Unit
BLER	Block Error Rate
BS	Base Station
BSC	Base Station Controller
BSS	Business Support Systems
CA	Carrier Aggregation
CapEx	Capital Expenditure
CAPTCHA	Completely Automated Public Turing Test to Tell Computers and Humans Apart

CAS	Cloud Access Switch
CCC	Cell Configuration and Control
CCI	Commonwealth Cyber Initiative
CCO	Coverage and Capacity Optimization
CD	Continuous Development
CDMA	Code Division Multiple Access
CHIPS	Creating Helpful Incentives to Produce Semiconductors for America
CI	Continuous Integration
CIA	Confidentiality, Integrity, and Availability
CIA	Confidentiality, Integrity, and Authenticity
CMP	Certificate Management Protocol
CMTS	Cable Modem Termination System
CNCF	Cloud Native Computing Foundation
CNF	Cloud-Native Network Function
CNF	Containerized Network Function
CNN	Convolutional Neural Network
COTS	Commercial-Off-The-Shelf
CP	Cyclic Prefix
CP	Control Plane
C-Plane	Control Plane
CPM	Congestion Prediction and Management
CPRI	Common Public Radio Interface
CPU	Central Processing Unit
CQI	Channel Quality Indicator
CR	Cognitive Radio
CTI	Cooperative Transport Interface
C-RAN	Centralized Radio Access Network
C-RAN	Cloud RAN
CUS-plane	Control User Synchronization Plane
CUSM	Control User Synchronization Management
CU	Central Unit
CVE	Common Vulnerabilities and Exposures
DC	Dual Connectivity
DCCF	Data Collection Coordination Function
DDoS	Distributed Denial of Service
DEC	Digital Equipment Corporation
DHCP	Dynamic Host Configuration Protocol
DIX	DEC-Intel-Xerox
DL	Downlink
DMS	Deployment Management Service
DNS	Domain Name System
DOCSIS	Data Over Cable Service Interface Specification
DoD	Department of Defense
DRB	Data Radio Bearer
DRL	Deep Reinforcement Learning
DRX	Discontinuous Reception
DSS	Dynamic Spectrum Sharing

DT	Digital Twin
DTLS	Datagram Transport Layer Security
DU	Distributed Unit
DUT	Device Under Test
E1AP	E1 Application Protocol
E2AP	E2 Application Protocol
E2E	End-to-End
E2SM	E2 Service Model
EAP	Extensible Authentication Protocol
EC	Event Consumer
eCPRI	enhanced Common Public Radio Interface
EDGE	Enhanced Data Rates for GSM Evolution
EE	Energy Efficiency
EHC	EtHernet Header Compression
EI	Enrichment Information
eMBB	enhanced Mobile Broadband
EMS	Element Management System
eNB	evolved Node B
EN-DC	E-UTRA-NR Dual Connectivity
en-gNB	EN-DC gNB
EP	Event Producer
EPC	Evolved Packet Core
EPS AKA	Evolved Packet System Authentication and Key Agreement
ES	Energy Savings
ESP	Encapsulating Security Payload
ETACS	European Total Access Communication System
ETSI	European Telecommunication Standards Institute
E-UTRA	Evolved Universal Terrestrial Radio Access
E-UTRAN	Evolved Universal Terrestrial Radio Access Network
F1AP	F1 Application Protocol
FCAPS	Fault, Configuration, Accounting, Performance, and Security
FDD	Frequency Division Duplex
FFT	Fast Fourier Transform
FH	Fronthaul
FHM	Fronthaul Multiplexer
FOCOM	Federated O-Cloud Orchestration and Management
FOCOM	Fault and Configuration Management
FPGA	Field Programmable Gate Array
FTP	File Transfer Protocol
FTPES	File Transfer Protocol Explicit-mode Secure
FTPS	File Transfer Protocol Secure
GBR	Guaranteed Bit Rate
GCP	Google Cloud Platform
GDPR	General Data Protection Regulation
gNB	next-generation Node B
gNB-CU	next-generation Node B-Central Unit
gNB-CU-CP	gNB-CU-Control Plane

gNB-CU-UP	gNB-CU-User Plane
gNB-DU	next-generation Node B-Distributed Unit
GNSS	Global Navigation Satellite System
GPRS	General Packet Radio Service
GRPC	Google Remote Procedure Call
GSM	Global System for Mobile Communication
GSMA	GSM Association
GTP	GPRS Tunneling Protocol
GUTI	Globally Unique Temporary Identifier
HARQ	Hybrid Automatic Repeat reQuest
HRLLC	Hyper Reliable and Low-Latency Communication
HTC	Holographic-Type Communications
HTTP	Hyper Text Transfer Protocol
HTTPS	Hyper Text Transfer Protocol Secure
HW	Hardware
IAB	Integrated Access and Backhaul
IANA	Internet Assigned Numbers Authority
ICMP	Internet Control Message Protocol
IEFG	Industry Engagement Focus Group
IETF	Internet Engineering Task Force
IFFT	Inverse Fast Fourier Transform
IIoT	Industrial Internet of Thing
IKE	Internet Key Exchange
IMI	Internal Messaging Infrastructure
IMS	Infrastructure Management Services
IMT	International Mobile Telecommunication
ION	Intelligent Operation Network
IoT	Internet of Things
IOT	Interoperability Testing
IP	Internet Protocol
IPR	Intellectual Property Right
IPSec	Internet Protocol Security
IPT	Internet Para Todos
IQ	In-phase Quadrature
ITU	International Telecommunication Union
i-ZTA	intelligent Zero Trust Architecture
JSON	JavaScript Object Notation
JTAG	Joint Test Action Group
JWT	JSON Web Token
K8s	Kubernetes
KORA	Keysight Open RAN Architect
KPI	Key Performance Indicator
KPIMON	KPI Monitoring
KPM	Key Performance Measurement
KPM	Key Performance Metric
LAA	Licensed Assisted Access
LBT	Listen Before Talk

LCM	Life Cycle Management
LEO	Low-Earth-Orbit
LLS	Lower Layer Split
LMF	Location Management Function
LTE	Long-Term Evolution
MAC	Medium Access Control
MANO	Management and Orchestration
MCHEM	Massive Digital Channel Emulator
MCOT	Maximum Channel Occupancy Time
MCS	Modulation and Coding Scheme
ME	Managed Element
MEC	Multi-access Edge Computing
MeNB	Master evolved Node B
MFAF	Messaging Framework Adaptor Function
MIB	Master Information Block
MIMO	Multiple Input Multiple Output
MITM	Man-In-The-Middle
mIoT	massive Internet of Things
ML	Machine Learning
MLB	Mobility Load Balancing
MME	Mobility Management Entity
mMTC	massive Machine-Type Communications
MNO	Mobile Network Operator
MnS	Management Service
MORAN	Multi-Operator Radio Access Network
MPLS	Multiprotocol Label Switching
MR	Measurement Report
MRO	Mobility Robustness Optimization
mMIMO	massive MIMO
M-MIMO	Massive MIMO
MU-MIMO	Multi-User MIMO
M-plane	Management-plane
MS	Microsoft
mTLS	mutual TLS
MTI	Microelectronics Technology
MWC	Mobile World Congress
NACM	Network Configuration Access Control Model
NAS	Non-Access Stratum
NCC	Network and Computing Convergence
NDAA	National Defense Authorization Act
Near-RT RIC	Near-Real-Time RAN Intelligent Controller
NESAS	Network Equipment Security Assurance Scheme
NETCONF	Network Configuration Protocol
NF	Network Function
NFO	Network Function Orchestrator
NFV	Network Functions Virtualization
NG	Next Generation

NGA	Next G Alliance
NGAP	Next-Generation Application Protocol
NGC	Next-Generation Core
NGMN	Next-Generation Mobile Network
NG-RAN	Next-Generation RAN
NI	National Instruments
NI	Network Interface
NIB	Network Information Base
NMEA	National Marine Electronics Association
NMS	Network Management System
Non-RT RIC	Non-Real-Time RAN Intelligent Controller
NPN	Non-Public Network
NR	New Radio
NRF	Network Repository Function
NS	Network Service
NSA	Non-Stand-Alone
NSC	National Spectrum Consortium
NSF	National Science Foundation
NSSI	Network Slice Subnet Instance
NTIA	National Telecommunications and Information Administration
NTN	Non-Terrestrial Network
NVD	National Vulnerability Database
NWDAF	Network Data Analytics Function
O-CU	O-RAN Central Unit
O-CU-CP	O-CU Control Plane
O-CU-UP	O-CU User Plane
O-DU	O-RAN Distributed Unit
O-eNB	O-RAN eNB
OAI	Open Air Interface
OAIC	Open AI Cellular
OAIC-C	OAIC-Controllers
OAIC-T	OAIC-Testing
OAM	Operation, Administration, and Maintenance
OFDMA	Orthogonal Frequency Division Multiple Access
OLT	Optical Line Termination
ONAP	Open Network Automation Protocol
ONF	Open Networking Foundation
ONOS	Open Network Operating System
ONU	Optical Networking Unit
OpEx	Operating Expenditure
O-RAN	Open Radio Access Network
O-RU	O-RAN Radio Unit
OS	Operating System
OSC	O-RAN Software Community
OSS	Operations Support System
OTA	Over-The-Air
OTIC	O-RAN Testing and Integration Center

OUSD	Office of the Under Secretary of Defense
PAWR	Platform for Advanced Wireless Research
PBCH	Physical Broadcast Channel
PCI	Physical Cell Identifier
PCP	Priority Code Point
PDCP	Packet Data Convergence Protocol
PDSCH	Physical Downlink Shared Channel
PDU	Protocol Data Unit
PHY	Physical (layer)
PKI	Public Key Infrastructure
PLMN	Public Land Mobile Network
PNF	Physical Network Function
PON	Passive Optical Network
POWDER	Platform for Open Wireless Data-driven Experimental Research
PRACH	Physical Random Access Channel
PRB	Physical Resource Block
PRTC	Primary Reference Time Clock
PSK	Pre-Shared Key
PTP	Precision Time Protocol
PUCCH	Physical Uplink Control Channel
PUSCH	Physical Uplink Shared Channel
QCI	Quality of Service Class Identifier
QoS	Quality of Service
QP	QoS Prediction
QUIC	Quick UDP Internet Connection
R-NIB	Radio-Network Information Base
RACH	Random Access Channel
RADIUS	Remote Authentication Dial-In User Service
RAN	Radio Access Network
rApp	remote Application
RAT	Radio Access Technology
RC	Radio Control
REST	REpresentational State Transfer
RF	Radio Frequency
RFC	Request For Comments
RIA	RAN Intelligence and Automation
RIC	RAN Intelligent Controller
RIS	Reflective Intelligent Surface
RL	Reinforcement Learning
RLC	Radio Link Control
RMR	RIC Message Router
RNC	Radio Network Controller
ROI	Return On Investment
ROMA	RAN Orchestration and Management Automation
RPC	Remote Procedure Call
RRC	Radio Resource Control
RRM	Radio Resource Management

RS	Reference Signal
RS	Rohde & Schwarz
RSRP	Reference Signal Received Power
RSRQ	Reference Signal Received Quality
RU	Radio Unit
RX	Receiver or Receive
S-GW	Serving Gateway
S1AP	S1 Application Protocol
SA	Standalone
SBA	Service-Based Architecture
SBOM	Software Bill of Materials
SCAS	Security Assurance Specifications
SCTP	Stream Control Transmission Protocol
SDAP	Service Data Adaptation Protocol
SDFG	Standard Development Focus Group
SDK	Software Development Kit
SDL	Shared Data Layer
SDN	Software-Defined Networking
SDO	Standard Development Organization
SDR	Software-Defined Radio
SD-RAN	Software-Defined Radio Access Network
sFTP	secure File Transfer Protocol
SI	System Integrator
SIB	System Information Block
SIM	Subscriber Identity Module
SINR	Signal-to-Interference-plus-Noise Ratio
SM	Service Model
SLA	Service-Level Agreement
SLS	Service-Level Specifications
SMO	Service Management and Orchestration
SON	Self-Organizing Network
SP	Security Principlex
SP-ACC	Security Principle-Access Control
SP-ASSU	Security Principle-Assurance
SP-AUTH	Security Principle-Authentication
SP-CLD	Security Principle-Cloud
SP-CRYPTO	Security Principle-Cryptographic
SP-OPNS	Security Principle-Open Source
SP-RECO	Security Principle-Recoverability
SP-SS	Security Principle-Secure Storage
SP-SB	Security Principle-Secure Boot
SP-TCOMM	Security Principle-Trusted Communication
SP-UPDT	Security Principle-Secure Update
SP-ISO	Security Principle-Isolation
SP-PRV	Security Principle-Privacy
SP-ROB	Security Principle-Robustness
SP-SLC	Security Principle-Security Development, Testing, Logging

SPDX	Software Package Data Exchange
SQL	Structured Query Language
SRN	Standard Radio Node
SRS	Sounding Reference Signal
SSB	Synchronization Signal Block or Synchronization Signal Physical Broadcast Channel Block
SSH	Secure Shell
STC	Software Testing Certification
STIN	Space Terrestrial Integrated Network
SuFG	Sustainability Focus Group
SU-MIMO	Single-User MIMO
SUPI	Subscription Permanent Identifier
SUT	System Under Test
SW	Software
SWID	Software Identification Tag
T-GM	Telecom-Grandmaster
TCO	Total Cost of Ownership
TCP	Transmission Control Protocol
TDD	Time Division Duplex
TDMA	Time Division Multiple Access
TED	Test Equipment O-DU
TEM	Test Equipment Fronthaul Multiplexer
TER	Test Equipment O-RU
TIFG	Testing and Integration Focus Group
TIP	Telecom Infra Project
TIRO	Tactile Internet for Remote Operations
TLS	Transport Layer Security
TN	Transport Network
TNL	Transport Network Layer
ToD	Time of Day
TSC	Technical Steering Committee
TTI	Transmission Time Interval
TX	Transmitter or Transmit
UAS	Unmanned Aerial System
UAS	Unmanned Aircraft System
UAV	Uncrewed/Unmanned Aerial Vehicle
UC	User-Control
UDP	User Datagram Protocol
UE	User Equipment
UHF	Ultra High Frequency
UL	Uplink
UMTS	Universal Mobile Telecommunication System
UP	User Plane
UPF	User Plane Function
URL	Uniform Resource Locator
URLLC	Ultra-Reliable and Low-Latency Communication
US	United States

USRP	Universal Software Radio Peripheral
UTM	UAS Traffic Management
V2X	Vehicle-to-Everything
vCU	virtual CU or virtualized CU
vDU	virtual DU or virtualized DU
VLAN	Virtual Local Area Network
VM	Virtual Machine
VNF	Virtualized Network Function
VNFC	Virtualized Network Function Controller
VoIP	Voice over IP
VoLTE	Voice over LTE
VR	Virtual Reality
vRAN	virtualized RAN
WG	Work Group
X2AP	X2 Application Protocol
xApp	extended Application
XML	eXtensible Markup Language
XnAP	Xn Application Protocol
XR	eXtended Reality
XSS	Cross-Site Scripting
YAML	Yet Another Markup Language
YANG	Yet Another Next Generation
ZTA	Zero Trust Architecture

About the Companion Website

This book is accompanied by a companion website:

www.wiley.com/go/tripathioran

This website includes:

- Online Excercises

1

Introduction to O-RAN

Cellular technologies are becoming increasingly complex. For example, fifth generation (5G) is significantly more complex than fourth-generation (4G) cellular technology Long-Term Evolution (LTE). Open Radio Access Network (O-RAN), also referred to as Open RAN, is a revolutionary technology that is transforming the cellular landscape. Unlike traditional cellular networks, O-RAN is an open, programmable, interoperable, and virtualized Radio Access Network (RAN) architecture that allows network operators to mix and match hardware and software components from different vendors. O-RAN provides increased flexibility, scalability, and cost savings for network operators. Furthermore, O-RAN leads to enhanced competition among vendors, which ultimately benefits consumers and enterprises.

This chapter first creates a foundation for O-RAN by illustrating evolution of cellular technologies, reviewing components of a cellular system, and describing evolution of RAN architectures. Then, a brief introduction to the O-RAN ALLIANCE is given, followed by illustration of a simplified O-RAN architecture defined by the O-RAN ALLIANCE. Deployments of O-RAN require the use of standards and software frameworks developed by organizations other than the O-RAN ALLIANCE in addition to the specifications developed by the O-RAN ALLIANCE. Roles of such non-O-RAN ALLIANCE organizations in the O-RAN ecosystem are discussed. Driving forces behind O-RAN's rising importance are explained. Numerous use cases of O-RAN that help optimize various RAN operations are described. Finally, the overall accomplishments of O-RAN are summarized along with O-RAN's technical priorities and potential future work.

1.1 Evolution of Cellular Technologies

O-RAN is a comprehensive framework for implementing the radio access network portion of a cellular network. The O-RAN specifications are developed by the O-RAN ALLIANCE and support 4G and 5G cellular communications. As the cellular communication technologies evolve, O-RAN specifications can be augmented to support evolved cellular technologies.

Figure 1.1 shows the evolutionary path of cellular technologies from the first generation (1G) to the sixth generation (6G).

The concept of cellular communications was developed at Bell Labs in the 1970s. The 1G cellular systems were analog. Diverse types of 1G cellular systems were deployed around the world. Examples of such 1G systems included Advanced Mobile Phone System (AMPS) in the United States and NMT 500 and European Total Access Communication System (ETACS) in Europe. The 1G systems provided voice services and typically used frequency-division multiple access on the radio interface between the user device and the RAN. The analog 1G systems were replaced by

Fundamentals of O-RAN, First Edition. Nishith D. Tripathi and Vijay K. Shah.

Companion website: www.wiley.com/go/tripathioran

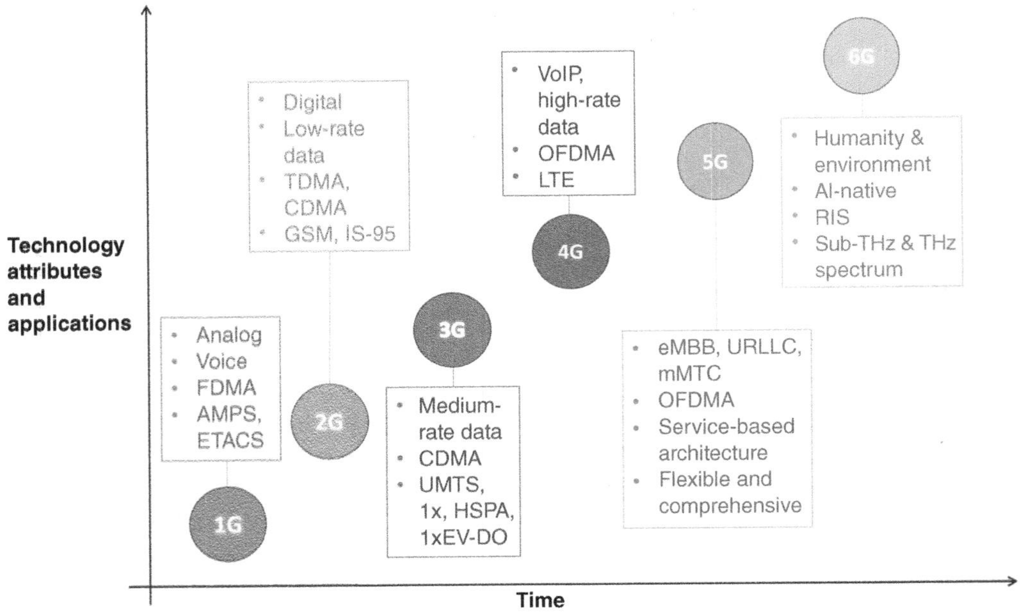

Figure 1.1 Evolutionary Path of Cellular Technologies.

digital second-generation (2G) systems. Examples of 2G systems include Global System for Mobile Communication (GSM) and Interim Standard-95 (IS-95). GSM utilized time-division multiple access (TDMA) and IS-95 used code-division multiple access (CDMA) on the radio interface. The 2G systems provided voice and low-rate data services. The 2G GSM systems typically transitioned to 2.5G General Packet Radio Service (GPRS) and Enhanced Data rates for GSM Evolution (EDGE) and then to third-generation (3G) Universal Mobile Telecommunication System (UMTS) and UMTS enhancements such as High-Speed Downlink Packet Access and High-Speed Uplink Packet Access. 2G IS-95 systems evolved to 3G CDMA200 and 1xEV-DO systems. While 2.5G systems provided higher data rates such as peak data rates of few hundred kilobits per second, 3G systems provided peak data rates of few megabits per seconds. The 3G cellular systems use two types of core networks, a circuit-switched core network to support circuit-switched voice services and a packet-switched core network to support Internet Protocol (IP)-based services such as web browsing, video streaming, and e-mail. The competing 3G technologies of CDMA2000 and UMTS evolved to the common 4G LTE.

4G LTE utilizes orthogonal frequency-division multiple access (OFDMA) on the radio interface. Furthermore, LTE utilizes a single type of core network, a packet-switched core network called Evolved Packet Core (EPC). Since there is no circuit-switched core network in LTE, voice calls are supported using Voice over IP (VoIP), widely known as Voice over LTE (VoLTE). The use of OFDMA and wide radio channel bandwidths enable LTE to support much higher data rates such as tens of megabits per seconds.

5G is a technology with the potential to transform many industries and segments of the economy. While 5G continues to use OFDMA on the radio interface, it significantly increases the flexibility such as large channel bandwidths, variable subcarrier spacing, support for millimeter wave spectrum, and cartier bandwidth parts. A high-performance and flexible radio interface enables 5G to support a much broader set of use cases. These use cases may belong to one or more ITU usage categories of enhanced mobile broadband (eMBB), ultra-reliable and low-latency communication (URLLC), and massive Internet of Things (mIoT). 5G introduces network slicing, where different logical networks are created to meet diverse service and customer requirements using the same physical infrastructure. 5G also introduces a Service-Based Architecture (SBA) that makes use of virtualization technologies for efficiency, scalability, and flexibility.

6G is the first cellular technology that emphasizes humanity and environment right from the requirements phase instead of focusing primarily on user experience and network performance like previous generations of cellular technologies (NGA 2022). 6G is expected to be even more transformational than 5G. 6G is likely to introduce new technologies such as native artificial intelligence (AI), Reflective Intelligent Surface (RIS), and higher-frequency spectrum, such as sub-THz spectrum and THz spectrum.

1.2 Components of Cellular Systems

While wireless communication is one of the most prominent features of a cellular system, the overall cellular system includes both wireless and wired links. Figure 1.2 illustrates a high-level architecture of a 5G system (Tripathi and Reed 2019).

A 5G system consists of the user equipment (UE), the 5G access network (AN), and the 5G core network (5GC or 5GCN) (3GPP TS23.501 2024). Different types of 5G UEs are supported including smartphones, Internet of Things (IoT) devices, smart watches, and augmented reality/virtual reality headsets. The 5G access network may be a Third Generation Partnership Project (3GPP)-defined RAN such as next-generation RAN (NG-RAN) or a non-3GPP access network. The NG-RAN may have next-generation-Node Bs (gNBs) or next-generation-evolved Node Bs (ng-eNBs). The gNB communicates with 5G UEs using a New Radio (NR) air interface, while the ng-eNB communicates with UEs using the 4G LTE air interface. Both the gNB and the ng-eNB connect to the 5GC, which is also referred to as the next-generation core (NGC). *A 5G RAN can be defined as the RAN that utilizes the 5G NR air interface for communications with UEs.* Hence, 5G RAN is a subset of NG-RAN, because NG-RAN can encompass both gNBs and ng-eNBs. An example of a non-2GPP access network is a Wi-Fi access network. This book primarily focuses on the 5G NR air interface between the 5G UE and the gNB while discussing the O-RAN.

A 5G UE exchanges Access Stratum (AS) signaling such as Radio Resource Control (RRC) signaling with the gNB in support of various RAN operations such as RRC connection setup, bearer setup for user traffic exchange, and handover. A 5G UE exchanges Non-Access Stratum (NAS) signaling with the NGC in support of operations such as registration, mutual authentication,

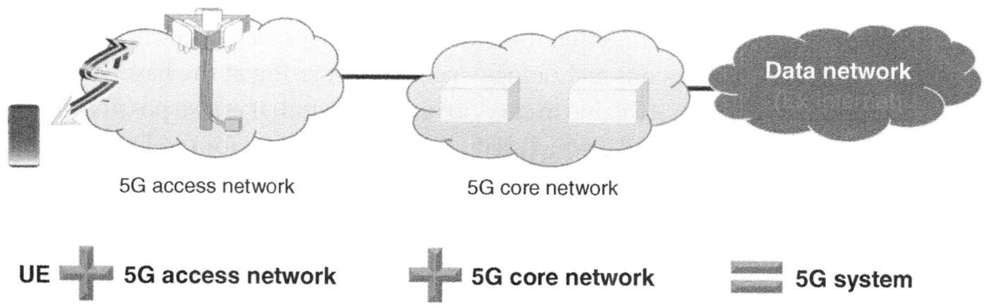

Figure 1.2 5G System: A High-Level Overview.

security activation, and protocol data unit (PDU) session setup. The gNB transparently transports the NAS signaling messages between the UE and the NGC. A UE can connect to and exchange user traffic with data networks (DNs) such as the internet through the 5G AN and the NGC.

The 5G O-RAN represents one possible way of designing and deploying a 5G AN. An operator or a service provider may deploy a 5G RAN with or without O-RAN, although O-RAN is expected to be a dominant way of deploying 5G RAN. The 5G O-RAN can be viewed as the realization of the 5G RAN, where gNBs or gNB entities such as gNB-Central Units (gNB-CUs) and gNB-Distributed Units (gNB-DUs) comply with O-RAN specifications. The O-RAN specifications utilize concepts such as disaggregation, openness, AI, and virtualization and cloud technologies. The 5G O-RAN does not alter the 3GPP-defined NR air interface between the UE and the 5G RAN in any way. A 5G UE does not need to be aware of the O-RAN and does not need to do anything differently to be able to work with the 5G O-RAN. A 5G O-RAN appears to a 5G UE as a 3GPP-defined 5G RAN. Similarly, the NGC does not need to be aware of the O-RAN and does not need to do anything differently to be able to work with the 5G O-RAN.

1.3 Evolution of the RAN

The 3GPP has defined the architectures for 3G, 4G, and 5G. Such architectures are referred to as logical architectures, where logical links between the network nodes, network elements, or network functions (NFs) are defined. For example, two gNBs can communicate with each other using the Xn interface. Specifically, two gNBs exchange Xn Application Protocol (XnAP) signaling with each other. The user traffic can also traverse on the Xn interface between two gNBs via a GTP[1] tunnel in support of packet forwarding during the inter-gNB handover. The 3GPP does not dictate or mandate any specific physical implementation or realization of the architecture. Hence, the operator has flexibility in deciding how to design, provision, and deploy the network architecture to meet target objectives.

Figure 1.3 illustrates how RAN evolution has occurred through generations of cellular technologies.

The older generations of cellular technologies from 1G to 3G utilize the RAN architecture with centralized RAN controllers called Base Station Controllers (BSCs) or Radio Network Controllers (RNCs). A BSC or RNC typically controls hundreds of base stations (BSs) or Node Bs. Certain radio resource management (RRM) tasks, such as the handover algorithm, are implemented at the BSC/RNC. The RAN thus consists of one or more BSCs and numerous BSs or Node Bs. The BSs are distributed across the cellular service area.

In commercial deployments, it is common to divide a BS into two parts, a baseband unit (BBU) and a radio unit (RU). An RU includes components such as radio frequency (RF) amplifiers and filters and is often situated next to the antenna that transmits and receives RF signals in support of wireless communications. A BBU carries out technology-specific processing at the baseband. The BBU and the RU can be connected via an RF cable to carry an analog signal. It is also possible to use an optical fiber between the BBUs and the RU, which reduces the overall loss for the link between the mobile device and the BS. Common Public Radio Interface (CPRI) is a widely used protocol between the BBU and the RU.

In a distributed RAN, BSs are distributed across the service area and a centralized RAN controller is absent. In such flat or distributed architecture, the BS manages its own radio resources and can

1 GTP stands for GPRS Tunneling Protocol. GPRS is General Packet Radio Service.

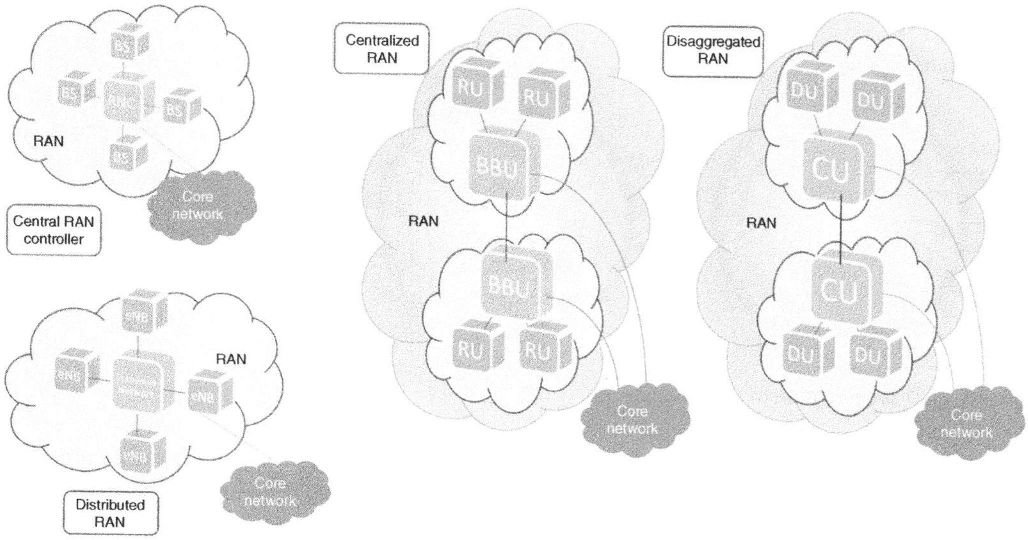

Figure 1.3 Evolution of the RAN Architecture.

communicate with other BSs through an interface. For example, in 4G LTE, the BSs called evolved Node Bs (eNBs or eNodeBs) communicate with each other using the Xn interface. An internal IP network of the service provider is used to implement the Xn interface between two eNBs. Such distributed or flat RAN is quite scalable, and new eNBs can be added to meet capacity and coverage needs in a given service area. Each eNB needs to connect to the core network. In commercial deployments, the eNB can be deployed using the BBU and the RU with relatively short BBU-RU distance such as few or tens of meters. Several 4G LTE and 5G deployments have used such distributed RAN.

In a centralized RAN (C-RAN) approach, a set of BBUs is placed in a relatively central data center and remote radio units (RRUs) or remote radio heads (RRHs) are placed at cell sites. Since the distance between the BBU and the RU is quite long (e.g., hundreds of meters or few kilometers), such RUs are called remote radio units. The C-RAN architecture is commercially deployed in LTE networks. The C-RAN provides benefits such as energy cost savings due to colocation of multiple BBUs, real estate cost savings due to a smaller footprint at the cell site, and performance enhancement due to enhanced collaboration of radio resources across multiple cells.

1.3.1 Horizontal Disaggregation of RAN: A Protocol Perspective

A disaggregated RAN divides the BS into a central unit (CU) and a distributed unit (DU). For example, while 5G gNBs have been deployed using the distributed RAN architecture, the 3GPP allows supports disaggregated gNBs, where a gNB is divided into a gNB-CU and a gNB-DU. The NR radio interface protocol stack is implemented partially at the gNB-CU and partially at the gNB-DU. Time-sensitive processing (e.g., acknowledgments for received data and any needed retransmissions) is carried out at the gNB-DU. Relatively less time-sensitive processing such as AS signaling is carried out at the gNB-CU. The gNB-DUs are placed at cell sites. While gNB-CUs can be colocated with gNB-DUs, gNB-CUs are often located in data centers. A gNB-CU connects to other gNB-CUs and the NGC. The 3GPP also allows decomposition of the gNB-CU into the gNB-CU-control plane (gNB-CU-CP) and the gNB-CU-user plane (gNB-CU-UP).

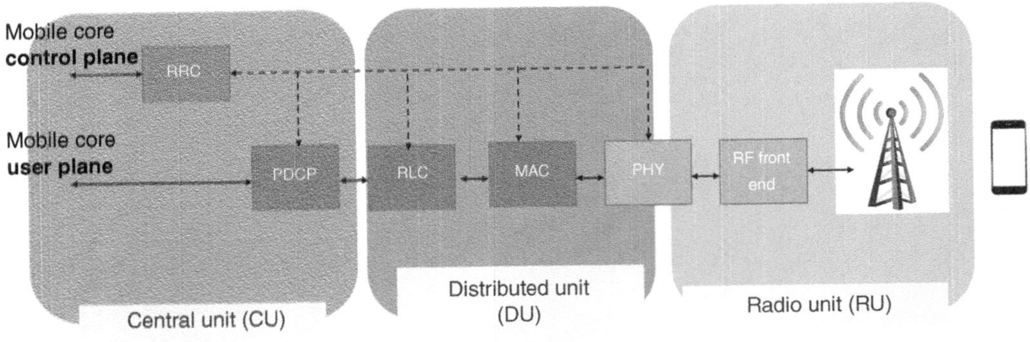

Figure 1.4 Disaggregation of the Base Station.

Figure 1.4 illustrates disaggregation of the BS from the perspectives of the radio interface protocol stack and control plane (CP) and user plane (UP) (Peterson et al. 2023). The UP, responsible for carrying user traffic, utilizes the protocols such as Packet Data Convergence Protocol (PDCP), Radio Link Control (RLC), Medium Access Control (MAC), and Physical (PHY) layer with the RF front end. In Figure 1.4, the CU implements RRC and PDCP protocols; the DU implements RLC, MAC, and selected PHY layer functions; and the RU implements selected PHY layer functions and the RF front end.

Software-Defined RAN: Figure 1.5 illustrates the implementation of the RAN using software-defined networking (SDN) principles, which leads to the development of a software-defined RAN. This approach facilitates the incorporation of a programmatic API that supports software-based control over the RAN UP pipeline. In the O-RAN architecture, this programmatic API is commonly known as the RAN Intelligent Controller (RIC). Refer to Chapter 2 for more details on RICs.

The 5G O-RAN makes use of the 3GPP-defined disaggregated 5G RAN architecture, carries out further disaggregation of the gNB-DU, and facilitates management of the RAN using intelligent controllers as described in Section 1.4 and Chapters 2 and 3.

When the RAN components are implemented using suitable RAN component software (e.g., the gNB-CU software) on generic commercial-off-the-shelf (COTS) hardware, such RAN implementation is termed as virtualized RAN (vRAN). The vRAN implementation no longer requires the RAN software to be coupled with specific or proprietary hardware. Hence, a service provider can purchase COTS hardware such as servers and instantiate suitable RAN software of its network vendors on such COTS servers to create a vRAN. While a vRAN can be deployed on a small scale using in-house network resources, cloud resources can also be exploited to build a vRAN. When cloud

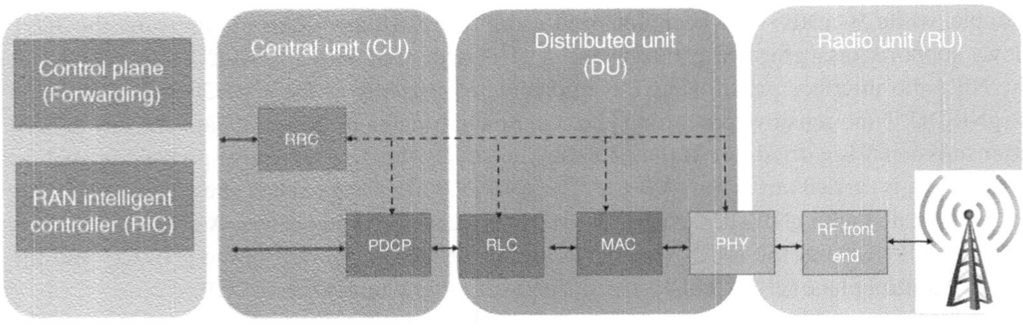

Figure 1.5 Disaggregated Base Station with SDN Principles.

resources are used to build a vRAN, such vRAN is also known as Cloud-RAN. For example, cloud resources of traditional cloud service providers or cloud vendors, such as Amazon Web Services (AWS), Microsoft Azure, and Google Cloud Platform (GCP), can be used to create Cloud-RAN.

1.4 Introduction to the O-RAN ALLIANCE and the O-RAN Architecture

The O-RAN ALLIANCE is an operator-driven industry alliance that has defined the O-RAN architecture. In this book as well as general public literature, O-RAN usually refers to the O-RAN architecture defined the O-RAN ALLIANCE. Section 1.4.1 provides a brief overview of the O-RAN ALLIANCE. Section 1.4.2 describes a high-level O-RAN architecture. Section 1.4.3 summarizes key focus areas of various groups operating in the O-RAN ALLIANCE. Further details on the O-RAN ALLIANCE and its groups can be found in Chapter 5. A comprehensive view of the O-RAN architecture is given in Chapter 2.

1.4.1 O-RAN ALLIANCE: A Brief Overview

The O-RAN ALLIANCE was founded in February 2018 by AT&T, China Mobile, Deutsche Telekom, NTT DOCOMO, and Orange. It was established as a German entity in August 2018. Today's O-RAN ALLIANCE is a global community of mobile network operators (MNOs), vendors, and research and academic institutions. The mission of the O-RAN ALLIANCE is to "*re-shape the RAN industry towards more intelligent, open, virtualized and fully interoperable mobile networks*" (O-RAN ALLIANCE 2024a).

The O-RAN ALLIANCE emphasizes *openness* and *intelligence* for the next-generation wireless networks and beyond (O-RAN ALLIANCE 2018). Open interfaces enable operators to build cost-effective and agile RAN through multivendor deployments in a more competitive RAN supplier ecosystem. The use of open-source software and reference hardware designs can accelerate the pace of RAN innovations. Intelligence is important to manage increasing complex RAN by minimizing human involvement, harnessing the power of AI/machine learning (ML), and automating network operations. Embedded intelligence at the component and the network level facilitates dynamic and efficient RRM, optimizing networkwide efficiency. The development of O-RAN specifications by the O-RAN ALLIANCE enables a more competitive and diverse RAN supplier ecosystem and can accelerate the pace of RAN innovations to increase efficiency, reduce costs, and enhance user experience.

Figure 1.6 shows the three focus streams of the O-RAN ALLIANCE. The specification efforts involve extending RAN specifications such as the 3GPP 5G specifications to achieve openness and intelligence. Several work groups (WGs)[2] have been formed to address different aspects of O-RAN and define the overall O-RAN specifications. The O-RAN Software Community (OSC) of the O-RAN ALLIANCE, in cooperation with the Linux Foundation, is in charge of the release and the development of open-source software for O-RAN. The testing and integration efforts facilitate O-RAN members in testing and integration of O-RAN implementations to ensure interoperability of products and compliance with O-RAN specifications.

There are several governing bodies in the O-RAN ALLIANCE. The O-RAN ALLIANCE board consists of up to 15 members with 5 founding members and up to 10 elected members. The elections for the (elected) board members take place every two years. The executive committee

2 Section 1.4.3 provides a summary of O-RAN work groups. Chapter 5 provide more details about the work groups.

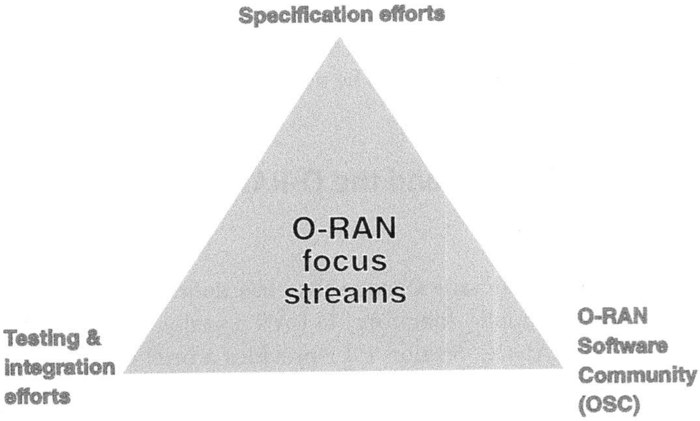

Figure 1.6 The Three Focus Streams of the O-RAN ALLIANCE.

(EC) supports the board by proposing agendas, priorities, projects, and releases for the board to consider and approve. The EC consists of representatives of the five O-RAN founding members and two elected representatives from the board members. The Technical Steering Committee (TSC) decides or gives guidance on O-RAN technical topics and approves O-RAN specifications prior to the Board approval and publication. The TSC consists of member representatives and the WG and focus group cochairs.

1.4.2 The O-RAN Architecture in a Nutshell

Figure 1.7 illustrates a simplified O-RAN architecture (O-RAN ALLIANCE 2024c).

The 3GPP-defined gNB-CU-CP and gNB-CU-UP are transformed into O-RAN-CU-CP (O-CU-CP) and O-RAN-CU-UP (O-CU-UP) by the O-RAN ALLIANCE, respectively. Furthermore,

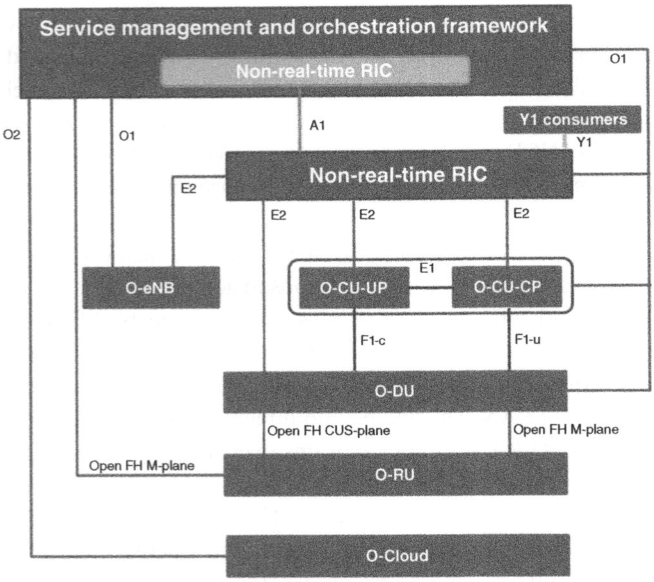

Figure 1.7 Simplified Logical O-RAN Architecture.

the 3GPP-defined gNB-DU is disaggregated and transformed into O-RAN-Distributed Unit (O-DU) and O-RAN-Radio Unit (O-RU) by the O-RAN ALLIANCE. A new interface called the open fronthaul between the O-DU and the O-RU is defined by the O-RAN ALLIANCE. Such disaggregation of the gNB-DU is often referred to as functional split 7.2x. The 3GPP had studied numerous functional splits of the NR radio protocol stack and finally standardized functional split Option 2 corresponding to the gNB-CU and the gNB-DU. The O-RAN ALLIANCE, in addition to supporting and augmenting such functional split, utilizes an additional split 7.2x. The O-DU and the O-RU interact with each other using two interfaces, open fronthaul control–user plane–synchronization (CUS)-plane and open fronthaul management (M)-plane. Details on this functional split 7.2x are given in Chapter 5.

The O-RAN ALLIANCE introduces two intelligent controllers in the spirit of the intelligence aspect of the O-RAN architecture, Non-Real-Time RAN Intelligent Controller (Non-RT RIC) and Near-Real-Time RAN Intelligent Controller (Near-RT RIC). The Non-RT RIC is part of the Service Management and Orchestration (SMO) framework and operates with a time scale exceeding 1 second. The Near-RT RIC operates with a time scale between 10 ms and 1 second. The O-RAN ALLIANCE-defined E2 interface enables the O-CU-CP, the O-CU-UP, and the O-DU to provide measurements to the Near-RT RIC. The Near-RT RIC executes suitable software applications called extended applications (in short, xApps) to manage the O-CU-CP, the O-CU-UP, and the O-DU. For example, an xApp in the Near-RT RIC can influence the scheduling algorithm implemented in the O-DU. Similarly, the Non-RT RIC executes suitable software applications called remote applications (in short, rApps) to manage the O-CU-CP, the O-CU-UP, and the O-DU. For example, an rApp may adapt handover parameters to optimize handover performance. In addition to supporting 5G RAN, the O-RAN also supports 4G LTE as seen from the existence of the O-RAN-eNB (O-eNB) in the O-RAN architecture.

The SMO framework interacts with the Near-RT RIC, the O-CU-CP, the O-CU-UP, O-DU, and the O-eNB through the O1 interface. The O-RAN-Cloud (O-Cloud) can be viewed as the cloud infrastructure on which suitable O-RAN functions, such as the O-CU-CP and the O-CU-UP, are implemented.

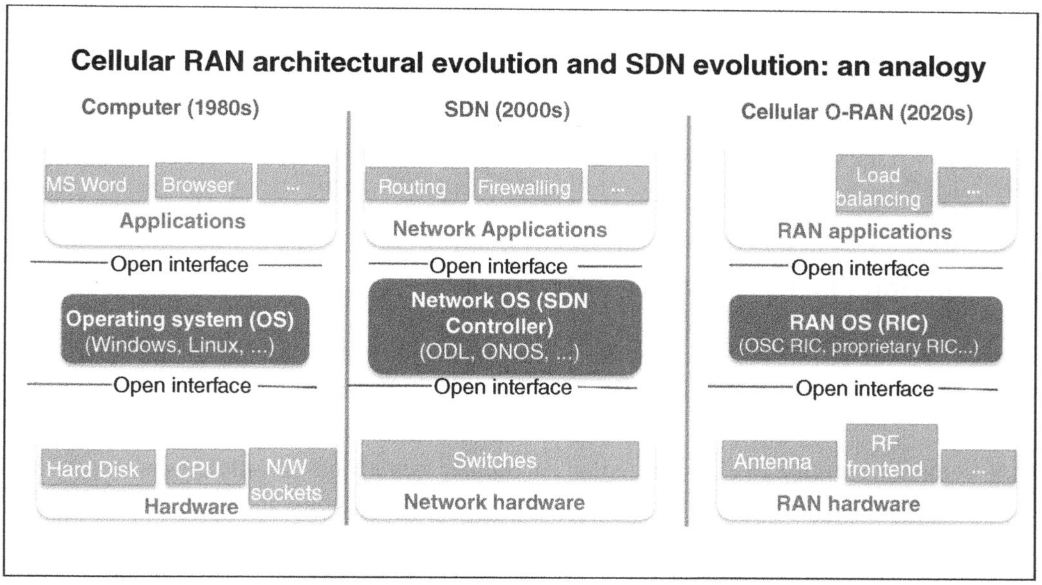

(Continued)

(Continued)

Let us draw parallels between the cellular RAN architectural evolution and computer and software-defined networking (SDN) evolution.

Consider what happened with computers back in the 1980s, when the proprietary, black box "Mainframes" were disaggregated into (i) computer hardware (e.g., hard disk and CPU) and (ii) computer applications (e.g., MS Word and a browser) with operating system (OS). The OS acts as a glue between the applications and computer hardware. This changed the computer world for better forever and set the foundation of today's computer world where every device, whether it is smartphone, smart watch, computer, or tablet, adopts the disaggregated computer architecture with OS as the middleware, which ensures that the applications (developed by independent application developers) run on the system without them worrying about how to work the computer hardware.

The same thing happened to the internet world in the early 2000s when SDN was introduced. Internet switches/routers were disaggregated into hardware switches (for data forwarding only), and all the intelligence in form of network applications such as routing and firewalling were moved to servers. And the middleware called SDN controller (informally referred to as network operating system) serves as the glue between network applications and hardware switches.

Similar evolution is happening with wireless networks with the introduction of the O-RAN paradigm. With O-RAN, the cellular RAN is disaggregated into RAN hardware (e.g., antennas and RF frontend) and RAN applications called xApps and rApps. The RAN Intelligent Controller (RIC) serves as the RAN Operation System (RAN OS) and functions as the glue between RAN applications and RAN hardware.

1.4.3 Groups in the O-RAN ALLIANCE

The technical O-RAN work is performed by O-RAN WGs and O-RAN focus groups and research groups. Table 1.1 lists these groups and summarizes their key focus areas.

As seen from Table 1.1, each of the WGs focuses on a part of the O-RAN architecture. In contrast, the focus and research groups deal with the topics that are overarching the technical WGs or relevant for the entire organization. See Chapter 5 for more details about the groups listed in Table 1.1.

1.5 Driving Forces Behind O-RAN

Several forces are driving the emergence of O-RAN as an approach to design and deploy 5G (and 4G) RAN as listed in Figure 1.8. O-RAN driving forces include RAN vendor diversity, complexity of RAN, accelerated innovations, and geopolitical environment.

RAN Vendor Diversity. In early generations of cellular technologies, a cellular service provides numerous RAN vendors to choose from. Examples of these vendors included Alcatel, Lucent Technologies, Motorola, Siemens, Nokia, Nortel Networks. Huawei, Samsung, and ZTE. Due to factors such as rising competition and economic environments, consolidation among RAN vendors occurred over time and the number of 4G and 5G RAN vendors has reduced significantly.

Table 1.1 Groups in the O-RAN ALLIANCE.

Group	Key Area of Focus
WG1: Use Cases and Overall Architecture Work Group	Define overall O-RAN architecture and use cases
WG2: The Non-Real-Time RAN Intelligent Controller and A1 Interface Work Group	Support Non-RT intelligent radio resource management, higher layer procedure optimization, policy optimization in RAN, and transfer of AI/ML models to Near-RT RIC
WG3: The Near-Real-Time RIC and E2 Interface Work Group	Define the Near-RT RIC architecture and support for data collection and actions over E2 interface
WG4: The Open Fronthaul Interfaces Work Group	Define open fronthaul interfaces to enable multivendor DU-RU interoperability
WG5: The Open F1/W1/E1/X2/Xn Interface Work Group	Define 3GPP-compliant multi-vendor profile specifications for F1/W1/E1/X2/Xn interfaces and propose 3GPP specification enhancements (if any)
WG6: The Cloudification and Orchestration Work Group	Enable decoupling of RAN software from the underlying hardware platforms to produce technology and reference designs that leverage commodity hardware platforms
WG7: The White-box Hardware Work Group	Specify and release a complete reference design to foster a decoupled software and hardware platform
WG8: Stack Reference Design Work Group	Develop the software architecture, design, and release plan for the O-CU and O-DU based on O-RAN and 3GPP specifications for the NR protocol stack
WG9: Open X-haul Transport Work Group	Specify transport equipment, physical media, and control/management protocols associated with the transport network
WG10: OAM Work Group	Specify the OAM requirements, OAM architecture, and the O1 interface
WG11: Security Work Group	Address security aspects of the O-RAN ecosystem
SDFG: Standard Development Focus Group	Determine the standardization strategies and interface with other Standard Development Organizations (SDOs)
OSFG: Open-Source Focus Group	Successfully launched the O-RAN Software Community (OSC) (The group is now dormant group because the open-source software development activities are being carried out by the OSC)
TIFG: Testing and Integration Focus Group	Define the overall approach for testing and integration including coordination of test specifications across various WGs
nGRG: Next-Generation Research Group	Carry out research of open and intelligent RAN principles in 6G and future network standards
IEFG: Industry Engagement Focus Group	Promote and accelerate O-RAN based technology adoption, proliferation and continuous innovation by engaging with industry/ecosystem
SuFG: Sustainability Focus Group	Optimize energy consumption, reduce environmental impact, and create more energy-efficient and environmentally friendly mobile networks

O-RAN provides opportunities for new players to be an integral part of the O-RAN ecosystem due to open interfaces, a high degree of disaggregation, and support for independent development of AI-based applications RAN controllers.

Complexity of the RAN. 5G is significantly more complex than 4G LTE. Remember that flexibility is a euphemism for complexity! There are several newer RAN operations and degrees of

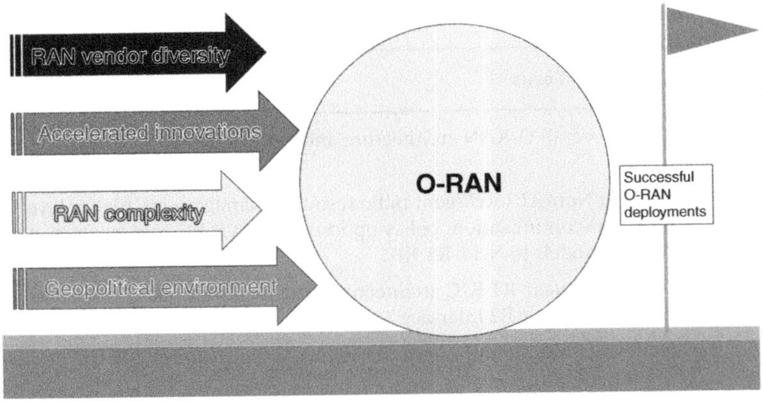

Figure 1.8 Major Driving Forces Behind O-RAN.

freedom in 5G relative to LTE such as flexible OFDM numerologies, carrier bandwidth parts, highly granular time-frequency radio resources for resource allocation, and support for network slicing for diverse applications and customer requirements. The definitions of open interfaces and support for intelligent RAN controllers enable the service providers to automate RAN operations and optimization to a large degree, reducing the probability of human errors and enhancing the overall RAN performance.

Accelerated Innovations. According to an anonymous Facebook philosopher,[3] when two people with one idea each work separately, each has one idea. But when they work together, they both have two ideas! That is the power of collaboration. When more companies in the ecosystem, the pace of innovations can accelerate significantly compared to the case of limited RAN vendors.

Geopolitical Environment. Security concerns of the US government about Chinese RAN vendors such as Huawei led to the banning of the Huawei equipment in the United States. Indeed, the US Congress passed a bill to help operators replace the Huawei equipment. Additionally, supply chain management has become an important policy consideration for the US government. Governments of several other countries such as Germany and the United Kingdom also share the security concerns of the US government. There is a significant push by the US government to support O-RAN to increase supplier diversity, ensure security and trustworthiness of RANs, and alleviate supply chain management issues, with the US congress passing a bill to allocate funding to support adoption of O-RAN. In summary, a geopolitical environment is a major driving force behind O-RAN.

1.6 O-RAN Ecosystem

While the O-RAN ALLIANCE defines a complete architecture for the RAN, design and implementation of such architecture in practice require support from other organizations. The overall O-RAN ecosystem includes (i) entities directly working on the three focus streams of O-RAN, (ii) organizations that develop specifications, and (iii) organizations that develop, maintain, and

3 People share humorous and often philosophical quotes on social media such as Facebook. Some of these quotes are extremely valuable but quite difficult to fully implement in practice. All that matters is we move in the right direction!

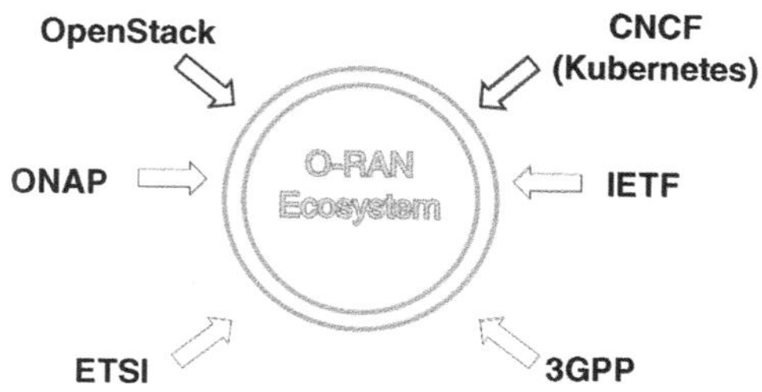

Figure 1.9 Building Blocks of the O-RAN Ecosystem.

publish software packages that are utilized by O-RAN. Figure 1.9 mentions examples of key entities that are building blocks for designing, deploying, and managing the O-RAN architecture. These entities include 3GPP, European Telecommunication Standards Institute (ETSI), Internet Engineering Task Force (IETF), Cloud Native Computing Foundation (CNCF), OpenStack, and Open Network Automation Protocol (ONAP).

3GPP. The O-RAN ALLIANCE utilizes the logical RAN architecture defined by the 3GPP as the baseline. For example, the O-CU performs all the functions of the gNB-CU defined by the 3GPP. Similarly, the O-DU and O-RU together carry out all the tasks of the gNB-DU defined by the 3GPP. The O-CU, the O-DU, and the O-RU together implement the 5G NR radio interface protocol.

ETSI. O-RAN functions or nodes such as the O-CU, the O-DU, the Non-RT RIC, and the Near-RT RIC can be implemented using the cloud infrastructure O-Cloud. O-Cloud makes use of technologies such as network functions virtualization (NFV) can be used in O-Cloud to implement O-RAN nodes. The NFV architecture is defined by ETSI. See O-RAN ALLIANCE (2024g) for a list of the relevant ETSI NFV standards documents that are useful for realizing O-RAN. These documents are related to virtualized network function (VNF) packaging, Life Cycle Management (LCM) operations, and security for managing virtual machines (VMs) and containers. Examples of ETSI NFV standards relevant to the O-RAN include (i) SOL003, SOL005, and SOL006 standards for VNFs and (ii) ETSI GS NFV-SOL 013 for O2 interface security. Hence, ETSI plays an important role in the realization of O-RAN.

IETF. O-RAN nodes can interact with each other using IP. To ensure security of communications between the O-RAN nodes, IP-level security mechanisms can be applied. Examples of such mechanisms applicable to O-RAN include IPsec and Transport Layer Security (TLS) (O-RAN ALLIANCE 2024f). The 3GPP also supports the use of these mechanisms in a 5G network to provide security at the IP layer. The IETF is responsible for specifying these IP-based mechanisms.

CNCF. Kubernetes, abbreviated as K8s, is a portable, extensible, and open-source platform for managing containerized workloads and services (Kubernetes 2024). K8s is a container orchestration engine for automating deployment, scaling, and management of containerized applications. Google originally developed K8s and open sourced it in 2014. The K8s project is hosted by the CNCF. K8s can be used in the design and operation of an O-RAN platform such as the Near-RT RIC platform.

OpenStack. OpenStack is a widely deployed open-source cloud software. OpenStack controls large pools of compute, storage, and networking resources through APIs or a dashboard. OpenStack provides functionalities such as infrastructure-as-a-service, orchestration, fault management, and service management (OpenStack 2024). Since O-RAN is highly virtualized, OpenStack can be used to manage O-Cloud compute, storage, and networking resources.

ONAP. ONAP is an open-source platform that can be used to manage functions and/or components of a virtualized network. Since virtually all (!) O-RAN nodes can be virtualized, ONAP can be viewed as a candidate platform for automation and orchestration in an O-RAN implementation. ONAP can be used for a specific purpose such as facilitating SON and network slicing use cases. For example, the sixth release of ONAP, called Frankfurt, focuses on support for 5G network slicing (Rethink 2024). It also features increased support for the O-RAN O1 interface for fault, performance, and configuration management. New functions to support A1 policies have been added to ONAP (Bonneau and Keeney 2023). A1 policies enable the Non-RT RIC to provide policy guidance to the Near-RT RIC. Specifically, ONAP supports O-RAN's A1 interface, between the orchestration layer (which could be ONAP or another MANO framework) and the RIC.

1.7 O-RAN Use Cases

The O-RAN architecture is intended to facilitate RAN operations and optimization. The O-RAN ALLIANCE has published white papers and specifications related to O-RAN use cases. Initial O-RAN use cases are discussed in a 2020 White Paper (O-RAN ALLIANCE 2020), and more use cases are described in a 2021 White Paper (O-RAN ALLIANCE 2021). Furthermore, multiple use cases are briefly described at a high level by O-RAN WG1 as part of Release R003 (O-RAN ALLIANCE 2024d). Details of the O-RAN use cases can be found in O-RAN ALLIANCE (2024e).

Table 1.2 lists the O-RAN use cases mentioned in O-RAN ALLIANCE (2024d). *These use cases provide a glimpse of the types of RAN management that the O-RAN architecture can facilitate. As new 5G features and capabilities are defined, O-RAN use cases are expected to grow.*

These use cases are briefly summarized below. Note that openness and intelligence are two central themes of O-RAN as mentioned in Section 1.4.1. The use cases briefly described below should be viewed as examples of potential ways in which intelligence aspects of O-RAN could be utilized. An operator may use RICs or other means to exploit intelligence in the RAN. Furthermore, operators may or may not utilize these use cases in their networks. Additionally, RAN functions other than the ones mentioned below could be exploited in O-RAN. Since RAN functions are typically implementation specific and proprietary and not standardized, there is a significant degree of flexibility and innovations in realizing RAN functions. However, motivation and typical processing are briefly explained for each use case below.

1.7.1 Context-Based Dynamic HO Management for V2X

The use of vehicle-to-everything (V2X) communication helps realize benefits such as increased safety, reduced emissions, and shorter commute times. However, the V2X environment is quite dynamic due to the dynamic nature of vehicular traffic and heterogenous environment, resulting in a challenging situation for handover. This use case aims to prevent or minimize performance degradation of V2X applications due to ineffective handover using two main functionalities. The *first functionality* makes use of long-term analytics, where the Non-RT RIC utilizes the UE-based handover events and mobility data maintained by the V2X application server. The Non-RT RIC identifies causes of handover anomalies and determines optimal handover sequences. A database records resolutions to anomalies. The *second functionality* involves a trained ML model in the Near-RT RIC that performs real-time (RT) optimization. The Near-RT RIC monitors UE-specific mobility context, detects or predicts unexpected handover events, and generates a target handover sequence to address the handover anomaly. The Near-RT RIC creates and optimizes the UE-specific Neighbor Relation Table (NRT) that contains the list of neighbor cells.

Table 1.2 Examples of O-RAN Use Cases.

Use Case ID	Brief Use Case Description	Use Case ID	Brief Use Case Description
1	Context-Based Dynamic HO Management for V2X	14	Massive MIMO SU/MU-MIMO Grouping Optimization
2	Flight Path-Based Dynamic UAV Radio Resource Allocation	15	O-RAN Signaling Storm Protection
3	Radio Resource Allocation for UAV Application Scenario	16	Congestion Prediction and Management
4	QoE Optimization	17	Industrial IoT Optimization
5	Traffic Steering	18	BBU Pooling to Achieve RAN Elasticity
6	Massive MIMO Beamforming Optimization	19	Integrated SON Function
7	RAN Sharing	20	Shared O-RU
8	QoS-Based Resource Optimization	21	Energy Saving
9	RAN Slice SLA Assurance	22	MU-MIMO Optimization
10	Multivendor Slices	23	Sharing Non-RT RIC Data with the Core
11	3GPP 4G/5G Dynamic Spectrum Sharing (DSS)	24	Industrial Vision SLA Assurance
12	NSSI Resource Allocation Optimization	25	Non-Public Network (NPN) RAN-Sharing via Midhaul for Multi-Operator Coverage
13	Local Indoor Positioning in RAN	26	Interference Detection and Optimization

1.7.2 Flight Path-Based Dynamic UAV Radio Resource Allocation

An uncrewed/unmanned aerial vehicle (UAV) experiences a unique radio environment compared to a traditional device on the ground. A UAV's radio propagation environment is characterized by line-of-sight propagation, increased amount of interference, signal variations due to the sidelobes of antenna radiation patterns, and more sudden changes in signal strengths. This use case aims to address the radio environment challenges for UAVs by exploiting flight path-based radio resource allocation. The Non-RT RIC obtains UAV-related metrics from suitable measurement reports from UAV UEs. The Non-RT RIC also obtains supporting information from a suitable application (e.g., unmanned traffic management or UTM), including the flight path information of the UAV, climate information, forbidden or restricted area information, and space load information, and builds (and later updates) an AI/ML model. The AI/ML model may consider uplink (UL)/downlink (DL) interference from/to UAVs, detection of UAVs, and prediction of available radio resources. The Near-RT RIC deploys and executes the AI/ML model developed by the Non-RT RIC and allocates radio resources allocation to UAV UEs.

1.7.3 Radio Resource Allocation for UAV Application Scenario

This use case envisions a UAV control vehicle that houses 5G O-RAN to control operations of various UAV UEs to provide reliable communications services. A large amount of data needs to be processed locally in a UAV control vehicle so that the overall UAV system can meet its operational objectives. Examples of data being communicated between the UAV UE and the UAV control vehicle include both the application data and the control data. The application data such as 4k high-definition videos require asymmetric DL/UL data rates and high UL data rates. In contrast, the control data such as navigation commands, configuration changes, and flight status data reporting require low latency, but low data rates are adequate. The Non-RT RIC obtains UE-level radio resource requirements from a suitable AS to determine UE-specific radio resource allocation policies. The Near-RT RIC executes these policies and creates configuration parameters for the E2 nodes. The E2 nodes adjusts radio resources and provide information about the UE status to the Near-RT RIC.

1.7.4 QoE Optimization

Diverse quality of experience (QoE) requirements of applications such as Cloud VR, industrial automation, online multiplayer gaming, and connected vehicles cannot be met by a semi-static quality of service (QoS) framework. QoE requirements can often vary during an application's lifetime. QoE estimation/prediction from the application layer and exposure of RAN performance to the application layer can be used to address such uncertainty, increase the efficiency of radio resource utilization, and enhance user experience. This use case aims to develop an automated closed-loop approach to optimize QoE. Specifically, the use case envisions a "User RAN Policy" that resides at the Non-RT RIC or Near-RT RIC and applies the operator's desired QoE configuration for a specific user, slice or 5QI flow in response to (i) requests from external systems or (ii) UE mobility. This policy supports reporting of the service-level agreement (SLA) information to external systems hosting the user applications. The use case exposes per-user or per-cell radio performance analytics information to external applications so that these applications can optimize user QoE.

1.7.5 Traffic Steering

Multiple access technologies such as 5G NR, LTE, and Wi-Fi with different amounts of spectrum may be available for use in a given geographic area. 5G NR may use licensed spectrum, unlicensed spectrum, or shared spectrum. Variations in the traffic load across cells of different access technologies or variances in available bandwidths and QoS attributes may lead to suboptimal spectrum utilization, limiting the achievable user experience. This use case aims to configure the desired optimization policies and utilize suitable performance criteria to proactively manage user traffic across different access technologies. The use case also enables the use of the enrichment information such as radio fingerprint information that is based on the data analytics of the historical RAN data. The Non-RT RIC observes the user experience by the UE-level performance measurements and the cell-level resource utilization and compares the observed performance against the expected service-level requirements. If the expected requirements are not met, the Non-RT RIC determines the related cell and evaluates the radio resource conditions in suitable cells. The Non-RT RIC may decide to relocate one or more users to other cells or even split user traffic across technologies. The Non-RT RIC creates traffic management policies to relocate UEs along with the prioritized list of cells. These policies from the Non-RT RIC are used by the Near-RT RIC to manage radio resources of different technologies.

1.7.6 Massive MIMO Beamforming Optimization

This use case aims to adjust coverage of massive MIMO beams proactively to enhance capacity/throughput and mobility performance by adapting beam configurations in a non-RT or near-RT situations. Example parameters that may be adjusted include the number of beams, beam boresights, vertical beamwidths, horizontal beamwidths, beam black lists or white lists, and beam mobility thresholds. Non-RT situations involve factors such as 3D construction, 3D terrain topology, network, weather seasons, intraday cell splitting/merging/shaping, traffic distribution, beam conflicts. Near-RT situations include moving users or hotspots, changing traffic distribution, and crowd source data. The O-RAN framework is highly attractive in this use case due to the large number of configuration parameters per antenna array, the amount of available measurements, and the overall complexity of MIMO. This use case envisions three optimization loops. An outer Non-RT loop implements massive MIMO grid of beams (GoB) beamforming optimization. An inner Near-RT loop implements massive MIMO beam-based mobility robustness optimization. Another inner Near-RT loop handles massive MIMO beam selection optimization.

1.7.7 RAN Sharing

Network sharing is often considered to be an efficient and sustainable way to accelerate the deployment of 5G due to a common pool of physical infrastructures shared by multiple operators. O-RAN can enable efficient RAN sharing including whole or part of O-CU and O-DU due to open interfaces and RICs. In one implementation approach, two operators share the same RAN but have their own independent core networks. Operators can share the state of their radio resources to facilitate optimization of radio resource utilization.

In an example RAN sharing approach, multiple service providers with business relationship with their customers can share the same RAN. Such a RAN can advertise public land mobile network (PLMN) identities of the supported service providers. Such a RAN sharing arrangement can be quite attractive to service providers economically, because they do not need to deploy their dedicated RANs to serve their customers. We observe that a Multi-Operator Radio Access Network (MORAN) is a widely used term to refer to RAN sharing that allows multiple operators to share the same RAN infrastructure. In an example implementation of MORAN, the infrastructure resources such as the antennas, tower, cell site, and power are shared among multiple operators, but operators have their dedicated radio channels or carrier frequencies. Operators can still independently implement desired features on a given MORAN.

1.7.8 QoS-Based Resource Optimization

This use case aims to optimize resource allocation for a special category of users such as public safety personnel (e.g., firefighters). In a statis implementation, a certain amount of radio resources may be typically used or reserved as part of a networks slice for such a special category of users. However, statis provisioning may be inadequate to fulfill the service-level specifications (SLS). In this use case, the Non-RT RIC monitors the performance of a RAN slice by obtaining QoS metrics from the E2 nodes (e.g., O-CU and O-DU) and the SMO functions. The Non-RT RIC may also receive information about user priorities from an external server. When needed, the Non-RT RIC modifies policies to fulfill the SLS. The Near-RT RIC receives the policies and provide suitable guidance to the E2 nodes so that the E2 nodes can execute the QoS enforcement decisions issued by the Near-RT RIC.

1.7.9 RAN Slice SLA Assurance

5G supports end-to-end network slicing, with the RAN being an important component of a network slice. This use case clarifies mechanisms and parameters for the SLA assurance for the RAN slice. In an example implementation, the SMO obtains RAN slice SLA. The Non-RT RIC observes long-term trends of the performance of RAN slices, evaluates this performance against the SLAs, and applies AI/ML techniques to adjust policies. The SMO can influence suitable policies, or the Non-RT RIC can provide policies to the Near-RT RIC. It is possible for the Non-RT RIC to develop AI/ML models that are executed by the Near-RT RIC. The policies may include information such as the network slice ID and key performance indicator (KPI) targets. The Near-RT RIC can optimize actions of E2 nodes by considering policies and slice-specific measurements and executing the deployed AI/ML models.

1.7.10 Multivendor Slices

This use case enables deployments of multivendor slices, where one slice is provided by a combination of O-CU(s) and O-DU(s) from one vendor and another slice is provided by a combination of O-CU(s) and O-DU(s) from another vendor. Different vendors may have unique features or advantages for different network slices. Hence, such multivendor slices provide the operator with significant flexibility in choosing a vendor for a given network slice. Supply chain risks are also mitigated because of the availability of products from multiple vendors.

1.7.11 3GPP 4G/5G Dynamic Spectrum Sharing (DSS)

The 3GPP has defined 4G/5G dynamic spectrum sharing (DSS) to enable spectrum sharing between 5G NR and LTE. This use case aims to exploit O-RAN to realize 4G/5G DSS. Specifically, an RIC can coordinate between an LTE scheduler and a 5G NR scheduler to share the same spectrum efficiently. The 4G/5G DSS can be realized using multiple applications. A resource management application, the 4G/5G DSS-App in the Non-RT RIC, manages the shared spectrum resources by adapting to dynamic 4G and 5G workload requirements. The 4G/5G DSS-App in the Non-RT RIC creates a long-term policy or intent as a scheduling guidance to 4G and 5G schedulers to reflect business, user, spatial, and temporal workload factors. The 4G/5G DSS-App in the Non-RT RIC converts the global 4G/5G DSS objectives such as workload requirements for a region and time-of-day into 4G/5G DSS policies (e.g., the maximum or minimum bandwidth threshold). Another application, RAT-App in the Non-RT RIC, maps the global 4G/5G DSS policies from the Non-RT 4G/5G DSS-App to RAT-specific policies. The 4G/5G DSS-App in the Near-RT RIC is involved in a closed-loop decision-making process by using the KPIs from the RAN and adapting to the needs of the 4G and 5G cells. The RAT-App in the Near-RT RIC performs RAT-specific configuration, control, and data subscription with the E2 nodes.

1.7.12 NSSI Resource Allocation Optimization

5G supports diverse use case scenarios, such as eMBB, URLLC, and mMTC, and utilizes Network Slice Subnet Instances (NSSIs) to offer these services. This use case trains an AI/ML model based on the huge volumes of performance data collected at O-RAN nodes over a long period (e.g., days, weeks, or even months) to predict the traffic demand patterns of 5G networks for each network slice subnet at target times and geographic locations. The AI/ML model then automatically reallocates the network resources to optimize network performance and prevent occurrence of any network performance issues. In an example implementation, the Non-RT RIC monitors the network performance by collecting measurements from the radio nodes such as O-CU and O-DU and utilizes an AI/ML model to process the measurements and predict the future traffic demand. The AI/ML model at the Non-RT RIC also determines the actions needed to adjust radio resources for the RAN NFs. The Non-RT RIC reconfigures the NSSI attributes (e.g., those related to the RRM policy) via the Operation, Administration, and Maintenance (OAM) functions in the SMO to configure the E2 nodes.

1.7.13 Local Indoor Positioning in RAN

The 3GPP supports a comprehensive framework to support positioning. For example, the 3GPP has defined the location management function (LMF) in the core network to support 5G NR-based positioning. However, in certain scenarios such as industrial manufacturing, it is more effective to carry out positioning locally in the RAN itself to reduce latency and expose such location to local applications. This use case utilizes the O-RAN architecture to deploy a positioning application in the Near-RT RIC. This positioning application determines the UE location and velocity using the measurements obtained through the E2 interface. This application can potentially make use of an AI/ML model that is trained based on historical signal measurements and known user locations. The positioning application can then expose the UE location to an edge application securely and/or to the SMO.

1.7.14 Massive MIMO SU/MU-MIMO Grouping Optimization

5G supports both single-user MIMO (SU-MIMO) and multi-user MIMO (MU-MIMO). In a given cell, different users are allocated different Physical Resource Blocks (PRBs) when SU-MIMO is used for such users. In contrast, MU-MIMO allocates the same PRB(s) to multiple users in the cell. The SU-MIMO and MU-MIMO are suitable for different scenarios. The performance of MU-MIMO is highly susceptible to the UE speed compared to the SU-MIMO performance. Furthermore, the MU-MIMO performance may be relatively lower in a low-traffic volume scenario compared to the SU-MIMO performance. Additionally, SU-MIMO and MU-MIMO have different resource requirements (e.g., reference signals). This use case adapts the transmission method (i.e., SU-MIMO versus MU-MIMO) for a given user based on factors such as the radio environment and traffic.

In an example implementation, the Non-RT RIC obtains necessary configurations, performance indicators, measurement reports, and auxiliary information such as user GPS information and traffic information to create suitable AI/ML models. The Non-RT RIC uses such trained AI/ML model to determine a list of UEs for an SU-MIMO group and an MU-MIMO group. The Non-RT RIC can also determine RRC configurations (e.g., sounding reference signal [SRS] and connected mode discontinuous reception [DRX] configurations) for the SU-MIMO and the MU-MIMO groups. The Near-RT RIC can send the configurations to E2 nodes by policy. The RAN nodes convey RRC configurations to UEs in both SU-MIMO and MU-MIMO groups and carry out scheduling for SU-MIMO and MU-MIMO groups. The RAN nodes also collect and report performance measurements to SMO related to SU-MIMO and MU-MIMO spectral efficiency (e.g., average number of spatial multiplexing layers, rank, and throughput for SU-MIMO and MU-MIMO groups).

1.7.15 O-RAN Signaling Storm Protection

Some devices may aggressively attempt to register or attach to the network at a rate of a few thousand times per hour. An attacker that can manipulate a large number of such vulnerable devices remotely can cause an attach or registration-related signaling storm, leading to potentially long outages. Intelligent and fine-grained security control at the RAN can stop such attacks, insulating the core network from in the core.

In an example implementation, an xApp can be built to address the signaling storm with two main functionalities, a distributed denial-of-service (DDoS) detection capability and a DDoS mitigation capability. A near-RT detection portion of the DDoS detection capability occurs in an xApp, and a non-RT detection portion (which relies on information available in the 5GC) occurs at the SMO.

The detection algorithms detect an abnormally high volume of CP messages originating from a group of devices. The detection algorithm may identify (i) a list of devices responsible for the signaling storm and/or (ii) geographic areas or gNBs expiring the signaling storm. Normal behavior may be learned by an AI/ML algorithm over time.

1.7.16 Congestion Prediction and Management

This use case proactively addresses congestion in a gNB by analyzing the radio resource utilization and taking a timely corrective action to mitigate any potential congestion. Congestion in a commercial cellular network leads to problems such as Radio Link Failures (RLFs), handover failures, and low data rates. Operators currently do not have a well-defined mechanism to predict congestion. This use case aims to utilize the embedded intelligence of O-RAN to predict congestion to initiate a cell congestion mitigation solution.

This use case proposes congestion prediction and management (CPM) architecture to proactively detect and mitigate congestion. In the CPM architecture, E2 node statistics such as counters are collected and preprocessed (e.g., addition of cell IDs and conversion of counters into KPIs) by the SMO. The Non-RT RIC invokes a suitable training model in an AI server inside SMO for data cleaning and training. Predicted KPIs are obtained by the CPM rApp. Suitable AI/ML models can be used to learn and predict the future traffic for the next hour/day/month. The CPM rApp in the Non-RT RIC forms the inference about cell congestion.

Two options to mitigate cell congestion are envisioned by this use case. In Option A, the CPM rApp in the Non-RT RIC transfers the inference to the CPM xApp in the Near-RT RIC through A1 interface, and the Near-RT RIC decides the inference-based mitigation solutions such as switching to dual connectivity mode, debarring of user access, and load sharing. These solutions can be controlled using the E2 interface. In Option B, the Non-RT RIC directly helps to mitigate congestion using the O1 interface using mitigation solutions such as cell splitting, addition of more carrier frequencies, switching to higher-order MIMO, and transferring some users to Wi-Fi.

1.7.17 Industrial IoT Optimization

In support of high reliability requirements for the industrial Internet of Thing (IIoT) scenarios such as factory automation and transportation industry, 5G makes use of features including data duplication (e.g., PDCP duplication), Time-Sensitive Networking (e.g., Ethernet header compression [EHC]), and different prioritized transmissions (e.g., preemption of lower priority transmissions by higher priority transmissions).

This IIoT optimization use case aims to guarantee reliability and resource utilization efficiency by optimizing mechanisms of PDCP duplication, EHC, and different prioritized transmissions. The enhancement mechanisms for IIoT are semi-static or dynamically configured for certain users and certain services based on the AI/ML-based service prediction and performance prediction. The solutions envisioned by this use case include Non-RT RIC optimization solution and Near-RT RIC optimization solution.

In the Non-RT RIC-based solution, the Non-RT RIC retrieves network data (e.g., network load, performance metrics, and 5QIs) and external service information from the SMO. The SMO also obtains external enrichment information such as service type, traffic period, duration, packet size, and KPI requirements from sources such as a multi-access edge computing (MEC) server, APP server, or industrial control platform. The Non-RT RIC retrieves configured policy information such as cell edge user reliability KPI target, Ethernet header compression efficiency from the SMO. The Non-RT RIC deploys and updates an AI/ML model received from the SMO. The Non-RT RIC uses the AI/ML model to infer and decide suitable IIoT configurations and actions (e.g., mapping between the QoS flow and the DRB that supports duplication and EHC and preemption decision). The Near-RT RIC retrieves IIoT-related configurations from Non-RT RIC and sends the configurations to E2 nodes by policy. In support of this use case, E2 nodes such as the O-CU and the O-DU provide UE measurement reports and performance metrics to the SMO. E2 nodes enforce the E2 control policy or O1 configuration.

In the Near-RT RIC solution of IIoT optimization, the SMO obtains external enrichment information such as service type, traffic period, duration, packet size, and KPI requirements from an MEC server, APP server, or industrial control platform. The SMO then provides to the Non-RT RIC the information including external service information and network data such as network load, performance metrics, and 5QIs The Non-RT RIC trains an AI/ML model for the Near-RT RIC. Furthermore, the Non-RT RIC also evaluates the collected data and A1 policy feedback and generates or updates suitable optimizations (e.g., reliability targets) and sends it to the Near-RT RIC.

The Near-RT RIC deploys or updates the AI/ML model. The Near-RT RIC also receives an A1 policy from the Non-RT RIC and initiates relevant optimization procedure(s). The Near-RT RIC evaluates the performance data from the E2 nodes to determine compliance to the A1 policy-dictated KPI targets. For example, cell edge user performance cannot meet the A1 policy. The Near-RT RIC considers a variety of factors such as service characteristics, channel quality estimation, and A1 policy to determine actions such as PDCP duplication on/off, RLC entity selection, and high priority traffic preemption. The Near-RT RIC may generate or update E2 policies. The Near-RT RIC sends reports to the Non-RT RIC for evaluation and optimization as needed. The E2 nodes such as the O-CU and the O-DU provide network metrics to the SMO. The E2 nodes also provide UE measurement reports, performance metrics, and cell resource utility to the Near-RT RIC through E2 interface. The E2 nodes enforce the received E2 control policy.

1.7.18 BBU Pooling to Achieve RAN Elasticity

This use case focuses on cloudified baseband units (BBUs) that are deployed on a common hardware pool, centralized at a given cloud location. Such BBU pooling enables flexible mapping between O-RUs and BBUs, achieving RAN elasticity and providing benefits such as reduced costs, increased resiliency, increased agility of the RAN to adapt to traffic demands, and increased RAN utilization efficiency. Both the O-CUs and the O-DUs can be pooled.

In an example implementation of O-DU pooling, O-DU resources for multiple cell sites are pooled at a given centralized location to create an O-DU pool. The Cloud Access Switch (CAS) at the O-DU pool aggregates traffic from the multiple fronthaul gateways and routes the traffic to a suitable O-DU blade to realize load balancing. For the O-DU pooling use case, it is assumed that an open fronthaul exists with O-RAN-defined 7.2x split. Based on the association between O-RUs and O-DUs, and the granularity at which the traffic is assigned from the O-RUs to the O-DUs, we define the following.

Three classes of O-DU pooling are defined based on the association between O-RUs and O-DUs and the granularity at which the traffic is assigned from the O-RUs to the O-DUs. In Class 0 pooling, an O-RU is assigned to a single-specific O-DU statically. Reassigning an O-RU to a different O-DU requires significant manual operations and is performed infrequently and during specific maintenance windows. In Class 1 pooling, a set of O-RUs is initially assigned to a single O-DU for a specific period, but these O-RUs can be reassigned to a different O-DU at any time using SMO-assisted orchestration. Finally, in case of Class 2 pooling, a set of O-RUs is assigned to a set of O-DUs, and a cluster aware scheduler dynamically distributes subsets of traffic from one O-RU to the O-DU resources within the O-DU pool.

1.7.19 Integrated SON Function

A self-organizing network (SON) includes self-configuration (e.g., configuration of parameters and automated interface setup), self-optimization (e.g., optimization of parameters), and self-healing (i.e., recovery from a failure). When a SON algorithm is executed in the 3GPP management system, it is called centralized SON. When a SON algorithm is executed in the NF layer, it is called distributed SON. Finally, when part of the SON algorithm is executed at the NF layer and part of the SON algorithm is executed at the management layer, it is called hybrid SON.

This SON use case focuses on following SON functions: (i) initial PCI allocation, conflict resolution, and Automatic Neighbor Relations (ANR); (ii) mobility load balancing (MLB) and mobility robustness optimization (MRO); (iii) RACH optimization; (iv) energy savings (ES); and (v) Coverage and Capacity Optimization (CCO).

In centralized SON deployments, one challenge is that the operator may need to choose the same vendor for both the SON solution and the RAN nodes in case of centralized SON. In the case of distributed SON, the interface between the SON and RAN nodes could be proprietary and internal to the vendor's product. The performance of such vendor product-specific SON solutions affects overall KPIs, because typical 5G deployments utilize products from multiple vendors.

This use case proposes the realization of SON algorithms are as rApps or as xApps or as a combination of rApps and xApps. For example, the A1 and E2 and optionally O1 interfaces are utilized. SON functions operating with long time scales such as initial PCI, initial ANR, and ES can be implemented at the Non-RT RIC using an rApp or in the management entities like EMS, NMS, or at the SMO. In contrast, SON functions that need to operate with short time scales are realized at the Near-RT RIC using one or more xApps of RT-RIC.

1.7.20 Shared O-RU

The current O-RAN architecture associates a given O-RU node with a single O-DU node. Such constraint limits the ability of fronthaul systems to be used effectively. For example, use cases such as RAN sharing and multivendor network slices involve a common O-RU that works with multiple O-DUs. A shared O-RU provides enhanced scalability and/or enhanced node availability in fronthaul systems, where multiple O-DU nodes deployed by a single operator are connected to a shared O-RU. When multiple O-DU nodes are deployed by different operators, the benefit of a shared O-RU is enhanced sharing capabilities across operators.

The initial phase of the proposed O-RU sharing is to enhance the current open fronthaul architecture to enable a common O-RU to work with multiple O-DUs with static allocation of resources to distinct O-DUs. This initial phase supports sharing of the O-RU by O-DUs of (i) a single-operator case or (ii) multiple operators.

1.7.21 Energy Saving

Energy consumption of the RAN is important to network operators to manage costs. RAN ES depends on suitable planning and configuration. Due to the varying nature of traffic load and user mobility, optimization of energy consumption of the RAN is quite complex, and it is entirely possible that the RAN equipment would consume a significant amount of energy even when traffic is light.

The motivation behind this use case is to leverage O-RAN's AI/ML capabilities and open interfaces to introduce optimized ES solutions that switch off/on different network components at different time scales when appropriate. The ES use case is divided into *three sub-use cases* based on the time scale of the control and the controlled system.

The first sub-use case of "carrier and cell switch off/on ES" operates in non-RT. Its goal is to reduce O-CU/DU/RU power consumption by switching off/on one or more carriers or a cell of a given technology. Implementation-specific AI/ML-assisted solutions in the Non-RT RIC can be used to control the traffic load of the carriers and the cell and determine when to switch off/on one or more carriers or a cell using O1 and/or open fronthaul management plane (M-plane) parameter configurations. Suitable traffic steering is also used in conjunction with off/on switching for service continuity and QoS. The carrier switch off/on can be carried out using (i) a hibernate mode where the radio's power amplifiers remain on but with minimum electric current draw or (ii) a complete switch off mode.

The second sub-use case of "RF channel switch off/on ES" operates in non-RT or near-RT. This sub-use case aims to reduce power consumption of an O-RU with massive MIMO by switching off/on certain RF channels. An rApp or xApp, with assistance from AI/ML algorithms triggers switching off/on certain RF channels, based on information such as traffic load, user location, and mobility. The number of RF channels, the maximum number of spatial multiplexing layers, and number of MU-MIMO users can be reduced. The O-RU can be reconfigured using the open fronthaul M-plane from an E2 node or SMO.

The last sub-use case of "Advanced Sleep Mode ES" operates in near-RT and aims to reduce power consumption by partially switching off O-RU components. The Near-RT RIC configures suitable cell parameters (e.g., the SSB periodicity) based on traffic load, user service type, and energy efficiency measurements.

1.7.22 MU-MIMO Optimization

MU-MIMO beamforming can be optimized using the Near-RT RIC for both DL and UL and TDD and FDD, enhancing user performance and cell performance for a wide range of vehicular speeds. The Near-RT RIC collects information from the E2 nodes; selects users; determines suitable time and frequency resources, precoding coefficients and modulation and coding schemes for selected users; and sends control messages to the E2 nodes using the E2 interface. Specifically, the Near-RT RIC collects UE states from the O-CU and configures the O-CU/O-DU to report RRC parameters and UL/DL traffic and radio channel information and provides users and transmission parameters for selected users to the O-DU over E2 interface. The O-DU uses the information received from the Near-RT RIC to schedule MU-MIMO transmissions.

1.7.23 Sharing Non-RT RIC Data with the Core

Cellular networks are becoming more complex with 5G being significantly more complex than 4G LTE. Increased network complexity necessitates complex analytics to be available in the RAN as well as the core network. For example, the 3GPP-defined network data analytics function (NWDAF) provides analytics to other 5GC NFs so that they can optimize their processing.

The 3GPP has defined an analytics logical function (AnLF) in support of the NWDAF, which can use AI/ML algorithms to generate analytics information useful to 5GC NFs.

rApps can produce analytics for multiple data consumers, such as other rApps, Near-RT RIC/xApps, general external entities (e.g., an application function [AF]). This use case seeks to add to this list the 5G Core NFs. In such a multiconsumer environment, it is a good architectural practice to avoid coupling of the data producer with the data consumer, in fact even hiding from the producer the identity of its data consumers or even whether the data consumers are single or multiple. In keeping with this principle, rApps can structure the information content of the data published over R1 such that it is "generally useful" and not customized for a specific data consumer.

This use case envisions a scenario where the Non-RT RIC-produced analytics are useful to the 5GC. Hence, this use case proposes a mechanism whereby such analytics can be shared with the 5GC. Two options (called Option 1 and Option 2) for the SMO/Non-RT RIC to perform the role of a 5GC NWDAF without a separate intervening NWDAF are suggested by this use case. Furthermore, two options (called Option 3 and Option 4) for the SMO/Non-RT RIC to interact with a separate intervening NWDAF are also suggested.

Option 1. SMO/Non-RT RIC Direct to 5GC Consumer with façade of an NWDAF. In this approach, the SMO/Non-RT RIC exposes the analytic directly to the consuming 5GC NF, acting as an NWDAF relative to such 5GC NF. The AnLF functionality exists only in the Non-RT RIC. Option 1 requires an rApp that makes its data available to 5G core NFs and registers its R1 data type as an analytic ID. The 5GC NF discovers the appropriate "NWDAF" (associated with SMO/Non-RT RIC) through the UDM for UE-related analytics and the NRF for non-UE-related analytics.

Option 2. SMO/Non-RT RIC to DCCF/MFAF with façade of an NWDAF. In this approach, the SMO/Non-RT RIC shares its analytics with a DCCF/MFAF that makes those analytics available to 5GC NFs. Like Option 1, the AnLF functionality is not duplicated but exists only in the Non-RT RIC. The SMO/Non-RT RIC acts as an NWDAF toward the DCCF/MFAF to leverage the NWDAF services for data subscription and delivery. The DCCF/MFAF receives analytics requests from 5GC NFs, forwards them to the SMO/Non-RT RIC, receives analytics from the SMO/Non-RT RIC, and provides these analytics to the 5GC NFs. In this approach, the 5GC NFs only need to discover via the NRF the DCCF/MFAF instances. The DCCF/MFAF contacts suitable NWDAF façade instances.

Option 3. SMO/Non-RT RIC to Separate NWDAF with façade of OA&M. In this approach, the SMO/Non-RT RIC shares its analytics with a separate NWDAF that provides these analytics to 5GC NFs. The AnLF functionality is duplicated in the Non-RT RIC and the NWDAF. In this approach, the SMO/Non-RT RIC acts as the OAM system toward the NWDAF. The separate NWDAF receives analytics requests from 5GC NFs, determines the source data to be collected, constructs analytics, and forwards the analytics to 5GC NFs.

Option 4. SMO/Non-RT RIC to Separate NWDAF with façade of RAN NF. In this approach, as in Option 3, the SMO/Non-RT RIC shares its analytics with a separate NWDAF that provides these analytics to 5GC NFs. Also, like Option 3, the AnLF functionality is duplicated in the Non-RT RIC and the NWDAF. However, the SMO/Non-RT RIC acts as a RAN NF instead of the OAM system in Option 3. The SMO/Non-RT RIC provides performance management OAM relative to the NWDAF. As in Option 3, the separate NWDAF receives analytics requests from 5GC NFs, determines the source data to be collected, constructs analytics, and forwards the analytics to 5GC NFs.

1.7.24 Industrial Vision SLA Assurance

Industrial vision is an image recognition technology that uses machines instead of human eyes for target measurements and evaluations to facilitate decision-making. It is suitable for automatic inspection, work piece processing and assembly automation, and monitoring and control of production process. The industrial vision system obtains images of the products being tested or processed and transmits these images to the vision server for detection and providing the feedback with the results of a suitable analysis.

O-RAN enables adaptive prescheduling parameters configuration to efficiently meet the QoS requirements of this use case. Key components for the industrial vision use case include the Non-RT RIC, Near-RT RIC, E2 nodes, industrial cameras, and the industrial vision server. Non-RAN data such as production line and camera information and data transmission information are transmitted to Non-RT RIC as enrichment information. With enrichment information from the vision server, the Near-RT RIC dynamically calculates prescheduling parameters (e.g., prescheduling data size, prescheduling period, and prescheduling start time) to meet the requirements of the industrial vision use case so as to configure the E2 node via E2 control/policy service with these prescheduling parameters as scheduling recommendations. The E2 node can synchronize the camera data UL resource allocations with work piece arrival times through dynamic prescheduling.

1.7.25 Non-Public Network (NPN) RAN-Sharing via Midhaul for Multioperator Coverage

This use case enables multiple operators to securely and cost-effectively share a RAN in the case of Non-Public Network (NPN) small cell (e.g., microcell or picocell) deployments in private spaces such as buildings, airports, hospitals, and hotels. Both the NPN owner and the MNO derive the benefits. The NPN owner obtains better coverage and better return on investment (ROI), while the MNO can extend its coverage without spending CapEx. O-RAN can facilitate the implementation of this use case by creating multivendor interoperability profiles for the 3GPP-defined F1 interface.

In an example architecture, a hierarchical cell structure can be used, where the 3GPP-defined F1 interface (i.e., the midhaul) and the 3GPP feature of the Integrated Access and Backhaul (IAB) are utilized to integrate the NPN-owned small cells with the macrocells of MNOs. A new NPN-RAN sharing manager function is defined in the O-RAN SMO and RAN-orchestrator to divide the NPN-RAN into multiple RAN partitions through the distribution of NPN-RAN PRBs among partner MNOs. Specifically, the shared NPN-RAN O-DU performs PRB distribution for NPN O-RUs and connects to the partner MNO O-CUs. The NPN-RAN spectrum is partitioned in a suitable manner to support partner MNOs.

1.7.26 Interference Detection and Optimization

In scenarios such as spectrum sharing between 4G and 5G, heterogeneous networks, and ultra-dense networks, effective interference management can increase the spectrum utilization efficiency and enhance performance. This use case exploits UE and network level data collection and implements interference detection, interference relationships construction, and interference optimization to optimize overall O-RAN performance.

The Near-RT RIC enables UE-level interference detection through proper configuration of the reference signal for the serving cell and relevant neighboring cells for the UE. The Near-RT RIC determines the E2 control or policy containing resource allocation for the reference signal for the configuration of the UE based on the interference detection-related A1 policy specified by the Non-RT RIC.

The Near-RT RIC utilizes QoS metrics from the SMO and network-level measurement data to construct interference relationships (e.g., interference graphs) that describe the interference relationships among UEs, groups of UEs, or RAN slices.

In one interference optimization approach, the Near-RT RIC optimally allocates radio resources to UEs, groups of UEs, RAN slices, or cells based on the interference relationships. The Non-RT RIC provides the interference optimization-related A1 policy to the Near-RT RIC. The Near-RT RIC obtains suitable measurement data through E2 interface and creates an interference optimization policy for the serving cell and relevant intra-frequency neighboring cells by allocating suitable radio resources. Note that interference optimization can be realized via slow loop optimization by the Non-RT RIC or fast loop optimization by the Near-RT RIC depending on traffic trends or use cases.

1.8 O-RAN: Accomplishments, Technical Priorities, and Potential Future Work

The current state of O-RAN is summarized in the form of accomplishments and technical priorities in Sections 1.8.1 and 1.8.2, respectively. Section 1.8.3 comments on the potential areas on which the O-RAN ALLIANCE may need to work on as cellular technologies advance from 5G to 5G-Advanced and 6G.

1.8.1 O-RAN Accomplishments

O-RAN has carried out significant work in its three workstreams—specifications, testing and integration, and OSC. As mentioned earlier, the O-RAN specifications work is carried out by O-RAN WGs with each WG focusing on a specific aspect of O-RAN. Release 3 versions of specifications are available from July 2023. As mentioned earlier, the testing and integration WG focuses its efforts on the overall approach for testing and integration and coordinates test specifications across various WGs. Several PlugFests have been held by the O-RAN ALLIANCE to validate compliance of products to the O-RAN specifications and interoperability between RAN vendors. The OSC typically releases new software every six months. The OSC software is aligned with the architecture and specifications released by the O-RAN ALLIANCE WGs. The OSC software is available publicly on GitHub. The first OSC software release, Amber, occurred in December 2019. As of early 2024, the first release occurred in December 2023.

1.8.2 O-RAN Technical Priorities

The O-RAN ALLIANCE has published three releases for *technical priorities* as of early 2024 (O-RAN ALLIANCE 2024b). The overall goal of these priorities is to accelerate the development of competitive O-RAN solutions in Europe and other regions to accelerate the global adoption of O-RAN. The technical priorities are the priorities for O-RAN solutions determined by the signatories (e.g., Deutsche Telekom, Orange, Telefónica, TIM, and Vodafone). The technical priorities provide guidance to the RAN supplier industry to accelerate market deployments.

These priorities serve as an input into TIP's O-RAN Release Framework to develop requirements as the basis for certification to promote an efficient supply chain.

Release 1 technical priorities were published in June 2021, and Release 2 technical priorities were published in March 2022. Release 3 updated Release 1 and Release 2 documents of technical priorities in April 2023. Release 1 focused on the main scenarios and technical requirements for the building blocks of a multivendor RAN. Release 2 focused on intelligence, orchestration, transport and cloud infrastructure, and energy efficiency goals to support a sustainable O-RAN. The focus of Release 3 is on developing additional requirements on the SMO and the RIC. Furthermore, significant enhancements are targeted for the cloud infrastructure, the O-CU, the O-DU, and the O-RU. Release 3 also focuses on details of security topics and the challenges of disaggregation in O-RAN.

1.8.3 Role of O-RAN Beyond 5G

The 3GPP is enhancing 5G specifications to introduce new features and capabilities and enhance existing features and capabilities compared to Release 15 Phase 1 5G and Release 16 Phase 2 5G. For example, the 3GPP introduced the nonterrestrial network (NTN) with a transparent payload in Release 17 and may consider the NTN with a regenerative payload in Release 19. The 3GPP has envisioned mobile IAB. Such new architectures can impact O-RAN design and deployments. Implementation of network slicing for diverse deployment scenarios can also affect the O-RAN design, configurations, and operational guidelines for optimal performance.

Organizations around the globe such as ATIS in North America are defining a vision and requirements for 6G (NGA 2022). 6G is anticipated to be deployed in 2030s. 6G is expected to be AI native and sharing native. In other words, 6G is expected to have built-in AI right from the design stage. Similarly, 6G would support sharing natively by supporting features such as spectrum sharing. 6G is also expected to have an even more advanced radio interface than 5G NR. For example, 6G may use RISs, which would alter the overall RAN design. See Section 8.2 in Chapter 8 for a brief overview of 6G.

In summary, O-RAN would need to keep up with evolving cellular technologies including 5G-Advanced and 6G. As mentioned earlier in the chapter, the O-RAN ALLIANCE has created a focus group called nGRG to monitor developments in 6G and future standards and carry out research of open and intelligent RAN principles so that suitable actions can be taken in the O-RAN ALLIANCE to keep O-RAN relevant to future technologies beyond 5G.

1.9 Major Takeaways of the Chapter

O-RAN is an emerging trend to deploy 5G RANs. O-RAN can also be used to deploy 4G LTE. Here are the major takeaways of this chapter.

- Cellular technologies have evolved from analog 1G technology to high-performance and flexible 5G technology. 5G is a transformational cellular technology that is expected to transform multiple industries. O-RAN is expected to play a significant role in 5G RAN deployments.
- Key components of a cellular system are the UE, the RAN, and the core network.
- RAN architectures have evolved from highly centralized architectures in 1G, 2G, and 3G to distributed architectures in 4G and 5G. Furthermore, 5G introduced disaggregation of the gNB such as the gNB-CU and the gNB-DU to distribute the radio protocol stack protocols. The gNB-CU

can be further disaggregated into the gNB-CU-CP and the gNB-CU-UP to separate CP and UP functions and provide additional deployment flexibility.

- The O-RAN ALLIANCE was founded in February 2018 by AT&T, China Mobile, Deutsche Telekom, NTT DOCOMO, and Orange. It was established as a German entity in August 2018. Today's O-RAN ALLIANCE is a global community of MNOs, vendors, and research and academic institutions.
- The O-RAN ALLIANCE has defined the O-RAN architecture that emphasizes openness and intelligence to cost-effectively deploy RANs and optimize RANs using AI/ML mechanisms inherently supported in the O-RAN architecture. For example, Non-RT RIC and Near-RT RIC provide intelligence to O-RAN by making use of AI/ML model to optimize various RAN operations.
- Deployments of O-RAN require the use of standards and software frameworks developed by organizations such as 3GPP, ETSI, IETF, CNCF/Kubernetes, OpenStack, and ONAP.
- Major forces driving O-RAN include RAN vendor diversity, complexity of RAN, accelerated innovations, and geopolitical environment.
- O-RAN helps optimize various RAN operations. Examples of O-RAN use cases are context-based dynamic handover management for V2X, flight path-based dynamic UAV radio resource allocation, traffic steering, massive MIMO optimization, RAN sharing, RAN slice management, DDS, local indoor positioning in RAN, signaling storm protection, CPM, industrial IoT optimization, BBU pooling, integrated SON functions, shared O-RU, ES, sharing Non-RT RIC data with the core network, and industrial vision SLA assurance.
- O-RAN has conducted significant work in its three workstreams—specifications, testing and integration, and OSC. The O-RAN specifications work is conducted by O-RAN WGs with each WG focusing on a specific aspect of O-RAN.
- The testing and integration WG focuses its efforts on the overall approach for testing and integration and coordinates test specifications across various WGs. Several PlugFests have been held by the O-RAN ALLIANCE to validate compliance of products to the O-RAN specifications and interoperability between RAN vendors.
- The OSC typically releases new software every six months. The OSC software is aligned with the architecture and specifications released by the O-RAN ALLIANCE WGs. The OSC software is available publicly on GitHub.
- The overall goal of O-RAN's technical priorities is to accelerate the development of competitive O-RAN solutions in Europe and other regions to accelerate the global adoption of O-RAN. The technical priorities provide guidance to the RAN supplier industry to accelerate market deployments.
- nGRG is an O-RAN focus group that monitors developments in 6G and future standards and conducts research of open and intelligent RAN principles to keep O-RAN relevant to future technologies beyond 5G.

References

3GPP TS23.501 (2024). *System Architecture for the 5G System (5GS).* The Third Generation Partnership Project. https://portal.3gpp.org/desktopmodules/Specifications/SpecificationDetails.aspx?specificationId=3144 (accessed 18 July 2024).

Bonneau, M. and Keeney, J. (2023). *O-RAN A1 Policies in ONAP.* April 2023, https://wiki.onap.org/display/DW/O-RAN+A1+Policies+in+ONAP (accessed 18 July 2024).

Kubernetes (2024). *Overview*. https://kubernetes.io/docs/concepts/overview/ (accessed 18 July 2024).

(NGA (2022). *Roadmap to 6G. Next G ALLIANCE Report*. February 2022, https://www.nextgalliance.org/wp-content/uploads/2022/02/NextGA-Roadmap.pdf (accessed 18 July 2024).

O-RAN ALLIANCE (2018). O-RAN: Towards an Open and Smart RAN. White Paper, October 2018, https://mediastorage.o-ran.org/white-papers/O-RAN.White-Paper-2018-10.pdf (accessed 18 July 2024).

O-RAN ALLIANCE (2020). O-RAN Use Cases and Deployment Scenarios: Towards Open and Smart RAN. White Paper, February 2020, https://mediastorage.o-ran.org/white-papers/O-RAN.WG1.Use-Cases-and-Deployment-Scenarios-White-Paper-2020-02.pdf (accessed 18 July 2024).

O-RAN ALLIANCE (2021). O-RAN Minimum Viable Plan and Acceleration towards Commercialization. White Paper, June 2021, https://mediastorage.o-ran.org/white-papers/O-RAN.Minimum-Viable-Plan-and-Acceleration-towards-Commercialization-white-paper-2021-06.pdf (accessed 18 July 2024).

O-RAN ALLIANCE (2024a). *About Us*. https://www.o-ran.org/about (accessed 18 July 2024).

O-RAN ALLIANCE (2024b). *O-RAN Ecosystem Resources*. https://www.o-ran.org/ecosystem-resources (accessed 18 July 2024).

O-RAN ALLIANCE (2024c). *O-RAN Architecture Description*. O-RAN.WG1.OAD-R003-v09.00. The O-RAN ALLIANCE. https://specifications.o-ran.org (accessed 18 July 2024).

O-RAN ALLIANCE (2024d). *Use Cases Analysis Report*. O-RAN.WG1. Use-Cases-Analysis-Report-v14.00. The O-RAN ALLIANCE, March 2024, https://specifications.o-ran.org (accessed 18 July 2024).

O-RAN ALLIANCE (2024e). *Use Cases Detailed Specification*. O-RAN.WG1.Use-Cases-Detailed-Specification-R003-v14.00. The O-RAN ALLIANCE, March 2024, https://specifications.o-ran.org (accessed 18 July 2024).

O-RAN ALLIANCE (2024f). *Security Protocols Specifications*. O-RAN.WG11.Security-Protocols-Specification.O-R003-v05.00. The O-RAN ALLIANCE, https://specifications.o-ran.org (accessed 18 July 2024).

O-RAN ALLIANCE (2024g). *O2 Interface General Aspects and Principles*. O-RAN.WG6.O2-GA&P-R003-v03.00. The O-RAN ALLIANCE. https://specifications.o-ran.org (accessed 18 July 2024).

OpenStack (2024). *Software*. https://www.openstack.org/software/ (accessed 18 July 2024).

Peterson, L., Oguz Sunay, O. and Bruce Davie, B. (2023). *Private 5G: A Systems Approach*. Systems Approach LLC, *2023*, https://5g.systemsapproach.org/ran.html (accessed 18 July 2024).

Rethink (2024). *ONAP's New Release Focuses on 5G Slicing and O-RAN Integration*. June 2020, https://rethinkresearch.biz/articles/onaps-new-release-focuses-on-5g-slicing-and-o-ran-integration/ (accessed 18 July 2024).

Tripathi, N. and Reed, J. (2019). *5G Cellular Communications—Journey and Destination*. 1st edn. The Wireless University. https://thewirelessuniversity.com/ (accessed 18 July 2024).

2

O-RAN Architecture

The Open Radio Access Network (O-RAN) ALLIANCE has defined a logical O-RAN architecture that utilizes the Third Generation Partnership Project (3GPP)-defined 5G and 4G RAN and introduces new nodes and interfaces in support of openness and intelligence. This chapter provides a comprehensive foundation of the O-RAN architecture. Key nodes, network elements, functions, or components of O-RAN are Service Management and Orchestration (SMO) (including Non-Real-Time RAN Intelligent Controller [Non-RT RIC]), Near-Real-Time RAN Intelligent Controller (Near-RT RIC), O-RAN Central Unit-Control Plane (O-CU-CP), O-RAN Central Unit-User Plane (O-CU-UP), O-RAN Distributed Unit (O-DU), O-RAN Radio Unit (O-RU), O-RAN evolved Node B (O-eNB), and O-RAN Cloud (O-Cloud). These O-RAN components are extensively discussed. The interfaces that connect O-RAN components with each other and external non-O-RAN systems are also discussed. These interfaces include O1, O2, A1, E2, open fronthaul (FH), O-Cloud Notification, and Cooperative Transport. While the 3GPP has defined disaggregated Base Stations in the radio network, the O-RAN architecture further disaggregates the Base Station using the functional split 7.2x. Functional distribution between O-RAN and other Standards Development/Developing Organizations (SDOs) is explained. In the context of O-RAN deployments, an overview of O-RAN cloudification, hierarchical cloud architecture, and deployment scenarios is also given.

2.1 O-RAN Logical Architecture

Figure 2.1 illustrates a logical O-RAN architecture (O-RAN ALLIANCE 2024b). Key nodes, network elements, functions, or components of O-RAN are SMO including Non-RT RIC, Near-RT RIC, O-CU-CP, O-RAN-Central Unit-User Plane (O-CU-UP), O-DU, O-RU, O-eNB, and O-Cloud. These O-RAN components are briefly summarized below (O-RAN ALLIANCE 2024b). *Details of these O-RAN components and associated interfaces are discussed in Sections* 2.2 *and* 2.3.

O-RAN considers Near-RT-RIC, O-CU-CP, O-CU-UP, O-DU, and O-RU to be O-RAN Network Functions as enclosed by dashed lines in Figure 2.1. Furthermore, the 5G UE communicates using the 5G New Radio (NR) Uu interface with O-RAN as if it were communicating with the 3GPP-defined next-generation-Node Bs (gNB). The O-RAN components included in the shaded box (i.e., the O-RU, the O-DU, the O-CU-CP, and the O-CU-UP) provide all the 3GPP-defined NR Uu protocol stack functions.

Fundamentals of O-RAN, First Edition. Nishith D. Tripathi and Vijay K. Shah.

Companion website: www.wiley.com/go/tripathioran

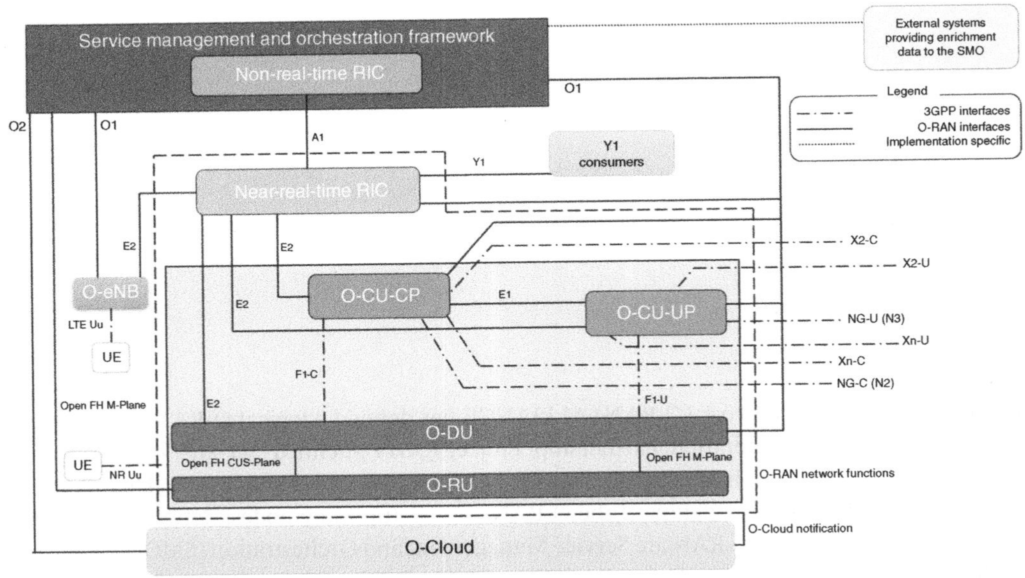

Figure 2.1 O-RAN Logical Architecture.

What is the dependency relationship between the 3GPP-defined RAN and O-RAN as illustrated in Figure 2.1? Well, the 3GPP RAN can exist without O-RAN, but O-RAN cannot exist without the 3GPP RAN! O-RAN starts with the 3GPP-defined RAN as the baseline network architecture, further disaggregates the 3GPP-defined RAN using a specific radio protocol functional split,[1] and adds intelligence aspects using RAN Intelligent Controllers (RICs). O-RAN transforms the 3GPP-defined next-generation-Node B-central unit-control plane (gNB-CU-CP) and gNB-CU-UP into O-RAN-CU-CP (O-CU-CP) and O-RAN-CU-UP (O-CU-UP), respectively. Such transformation involves the support for O-RAN-defined interfaces and functions. Furthermore, O-RAN disaggregates the 3GPP-defined gNB-DU and transforms it into O-DU and O-RU using Functional Split 7.2x.

The **SMO** provides Fault, Configuration, Accounting, Performance, and Security (FCAPS) interface to O-RAN Network Functions, houses the Non-RT RIC to facilitate RAN optimization, and carries out management, orchestration, and workflow management of the O-Cloud. The SMO supports O-RAN-defined (i) O1 interface toward the Near-RT RIC, the O-CU-CP, the O-CU-UP, the O-DU, and the O-eNB; (ii) O2 interface toward the O-Cloud; and (iii) the open fronthaul management-plane (open FH M-plane) toward the O-RU.

The **Non-RT RIC**, residing in the SMO, influences the content transported on the A1 interface toward the Near-RT RIC. It consists of the **Non-RT RIC framework** and the Non-RT RIC Applications called *remote Apps* **(rApps)**.

The **Non-RT RIC framework** terminates the A1 interface and exposes a set of internal SMO services to rApps via the R1 interface. The Non-RT RIC framework provides rApps with AI/ML workflow that includes model training, inference, and relevant updates.

rApps are Non-RT RIC applications that leverage the functionality exposed through the R1 interface to provide value-added services for RAN operations. Examples of these services include

1 This functional split is 7.2x, which is discussed in Section 2.4. See O-RAN ALLIANCE (2024k) for details of the functional split 7.2x.

determining the content to be transported on the A1 interface, recommending values and actions that may be applied over the O1/O2 interface, and generating Enrichment Information (EI) for the use of other rApps. rApps enable non-real-time control and optimization of RAN and policy-based guidance to the Near-RT RIC.

The **Near-RT RIC** is a logical function that enables near-real-time control and optimization of RAN through suitable data collection and actions over the E2 interface. The Near-RT RIC, in an implementation-specific manner, may exploit AI/ML-based model training, inference, and updates. The Near-RT RIC supports O-RAN-defined (i) A1 interface toward the Non-RT RIC; (ii) O1 interface toward the SMO; (iii) E2 interface toward the O-CU-CP, the O-CU-UP, the O-DU, and the O-eNB; and (iv) Y1 interface toward Y1 consumers. The Near-RT RIC houses extended Applications (xApps) at the core of its intelligence. xApps typically consist of one or more microservices and identify which data it consumes and which data it provides at the time of onboarding. An xApp is independent of the Near-RT RIC and may be provided by a third party. The E2 interface enables a direct association between the xApp and the RAN functionality being influenced by the xApp.

As mentioned earlier, the **O-CU-CP** hosts the 3GPP-defined gNB-CU-CP-containing Radio Resource Control (RRC) and Packet Data Convergence Protocol-Control Plane (PDCP-C or PDCP-CP). The O-CU-CP supports (i) 3GPP-defined interfaces such as the E1 interface toward the O-CU-UP; the F1-C interface toward the O-DU; the NG-C or N2 interface toward the Access and Mobility Management Function (AMF); the Xn-C interface toward an O-CU-CP, gNB-CU-CP, gNB-CU, or gNB; the X2-C interface toward an eNodeB (eNB), O-eNB, or O-eNB-CU and (ii) O-RAN-defined interfaces such as the O1 interface toward the SMO and the E2 interface toward the Near-RT RIC.

The **O-CU-UP** hosts the 3GPP-defined gNB-CU-UP-containing Service Data Adaptation Protocol (SDAP) and Packet Data Convergence Protocol-User Plane (PDCP-U or PDCP-UP). The O-CU-UP supports (i) 3GPP-defined interfaces such as the E1 interface toward the O-CU-CP; the F1-U interface toward the O-DU; the NG-U or N3 interface toward the User Plane Function (UPF); the Xn-U interface toward an O-CU-UP, gNB-CU-UP, gNB-CU, or gNB; the X2-U interface toward an eNB, O-eNB, or O-eNB-CU; and (ii) O-RAN-defined interfaces such as the O1 interface toward the SMO and the E2 interface toward the Near-RT RIC.

The **O-DU** hosts a sizable portion of the 3GPP-defined gNB-DU functions and contains radio link control (RLC), medium access control (MAC), and high-PHY layers based on the functional split 7.2x. The O-DU supports (i) 3GPP-defined interfaces such as the F1-C interface toward the O-CU, gNB-CU, or gNB and the F1-U interface toward the O-CU-UP, gNB-CU-UP, gNB-CU, or gNB and (ii) the E2 interface toward the Near-RT RIC and the open FH control user synchronization (CUS) plane and the open FH M-plane toward the O-RU.

The **O-RU** hosts lower PHY layer of the 3GPP-defined gNB-DU based on the functional split 7.2x and includes suitable Radio Frequency (RF) processing. The O-RU supports O-RAN interfaces such as the open FH CUS-plane and the open FH M-plane toward the O-DU and the open FH M-plane toward the SMO.

The **O-eNB** is a transformed version of the 3GPP-defined eNB or next-generation-eNB (ng-eNB) that supports O-RAN-defined interfaces such as the E2 interface toward the Near-RT RIC and the O1 interface toward the SMO. The O-eNB communicates with UEs using the Long-Term Evolution (LTE)-based radio interface. While Figure 2.1 shows an integrated O-eNB, the O-eNB also supports O-DU and O-RU functions with an open FH interface between them.

The **O-Cloud** is a cloud computing platform that hosts (i) O-RAN functions such as the Near-RT RIC, O-CU-CP, O-CU-UP, and O-DU; (ii) the supporting software components such as Operating

System, Virtual Machine (VM) Monitor, and Container Runtime; and (iii) suitable management and orchestration functions. It consists of physical infrastructure nodes. The O-Cloud includes the O-Cloud Notification interface that can be used by relevant O-RAN network functions such as the Near-RT RIC, O-CU-CP, O-CU-UP, and O-DU to receive O-Cloud-related notifications.

Y1 consumers are entities within or outside of the PLMN trust domain that consumes the Y1 services produced by the Near-RT RIC. Y1 consumers are not considered a logical O-RAN function. Y1 consumers obtain RAN analytics information services from the Near-RT RIC via the Y1 service interface. These services can be consumed by Y1 consumers after mutual authentication and authorization by subscribing to or requesting the RAN analytics information via the Y1 service interface. Y1 consumers outside the PLMN trusted domain may use Y1 services in a secure manner via an exposure function. The framework for mutual authentication and authorization between Y1 consumers and the Near-RT RIC is developed in an implementation-specific manner.

The O-RAN network functions can be implemented as Virtualized Network Functions (VNFs) in the form of VMs or Containers sitting above the O-Cloud and/or Physical Network Functions (PNFs) utilizing customized hardware. The O-RAN network functions except the O-RU support the O1 interface toward the SMO framework.

One of the important terms in O-RAN is an *E2 node*. An E2 node is a logical node terminating the E2 interface toward the Near-RT RIC. E2 nodes include the O-CU-CP, the O-CU-UP, and the O-DU or their combination for the NR access and the O-eNB for Evolved Universal Terrestrial Radio Access.

2.2 O-RAN Components: A Closer Look

As mentioned in Section 2.1, O-RAN components include SMO with Non-RT RIC, Near-RT RIC, O-CU-CP, O-CU-UP, O-DU, O-RU, O-eNB, and O-Cloud. These nodes are discussed in detail below.

2.2.1 SMO and Non-RT RIC: A Closer Look

Figure 2.2 illustrates the architecture of the SMO framework (O-RAN ALLIANCE 2024b).

The SMO performs RAN domain management. The key SMO capabilities or services are the FCAPS interface to O-RAN Network Functions, the Non-RT RIC for RAN optimization, and O-Cloud Management, Orchestration and Workflow Management (O-RAN ALLIANCE 2024b). The A1 interface between the Non-RT RIC in the SMO and the Near-RT RIC enables the SMO to perform RAN optimization. The SMO provides the FCAPS support using (i) the O1 interface between the SMO and the O-RAN Network Functions and (ii) the open FH M-Plane interface between the SMO and the O-RU. The SMO utilizes the O2 interface between the SMO and the O-Cloud to provide platform resources and workload management. As shown in Figure 2.2, the R1 services being exposed may be from the Non-RT RIC framework or the portion of the SMO framework outside the Non-RT RIC.

2.2.1.1 SMO Support for FCAPS to O-RAN Network Functions

The SMO provides support for O-RAN network function FCAPS via the O1 interface. The SMO carries out the operations and management of multi-vendor Open RAN Network Elements using the O1 interface including startup installation, software management, configuration management, performance management, fault management, and file management. O-RAN aims to align the O1 interface with the 3GPP specifications for RAN element management. Examples of

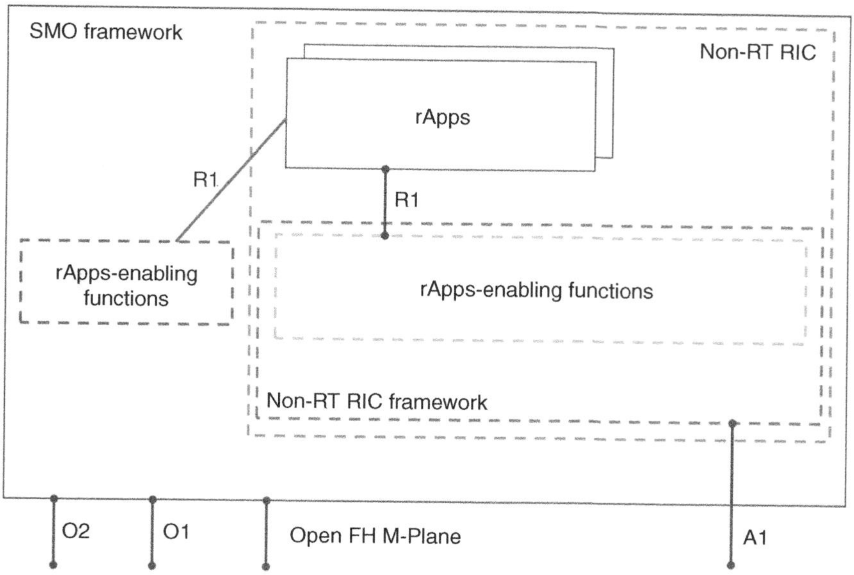

Figure 2.2 SMO Framework: A Closer Look.

the FCAPS functions defined in the O1 specifications include performance management, configuration management, fault management, file management, communications surveillance (e.g., heartbeat), trace, PNF discovery, and PNF software management (O-RAN ALLIANCE 2024c).

The SMO utilizes the open FH M-Plane interface to support FCAPS to the O-RU. Examples of the O-RU-related FCAPS functions include start-up installation, software management, configuration management, performance management, fault management, and file management (O-RAN ALLIANCE 2024l).

2.2.1.2 SMO: O-Cloud Management, Orchestration, and Workflow Management

The SMO uses the O2 interface to support cloud infrastructure management and deployment operations. The SMO, through the O2 interface, supports (i) the orchestration of O-Cloud infrastructure resource management such as inventory, monitoring, provisioning, software management, and lifecycle management and (ii) deployment of the Open RAN network functions using cloud resources including the lifecycle management of deployments. Examples of O-Cloud-related SMO functionalities include discovery and administration of O-Cloud resources, scale-in and scale-out for the O-Cloud, FCAPS of the O-Cloud, software management of the cloud platform, creation and deletion of deployments and associated O-Cloud resources, scale-in and scale-out of deployments and allocated O-Cloud resources, FCAPS of deployments and allocated O-Cloud resources, and software management of deployments (O-RAN ALLIANCE 2024b).

2.2.1.3 Non-RT RIC

Figure 2.3 illustrates the reference architecture of the Non-RT RIC along with the overall SMO framework (O-RAN ALLIANCE 2024f). Recall that the SMO framework includes the Non-RT RIC as well as the functionality outside the Non-RT RIC.

Note that *R1 services are a set of services produced by logical functions in the Non-RT RIC framework or the SMO framework functionality outside the Non-RT RIC framework.* These logical functions are mentioned in Figure 2.3. A service is a set of related capabilities (with a common

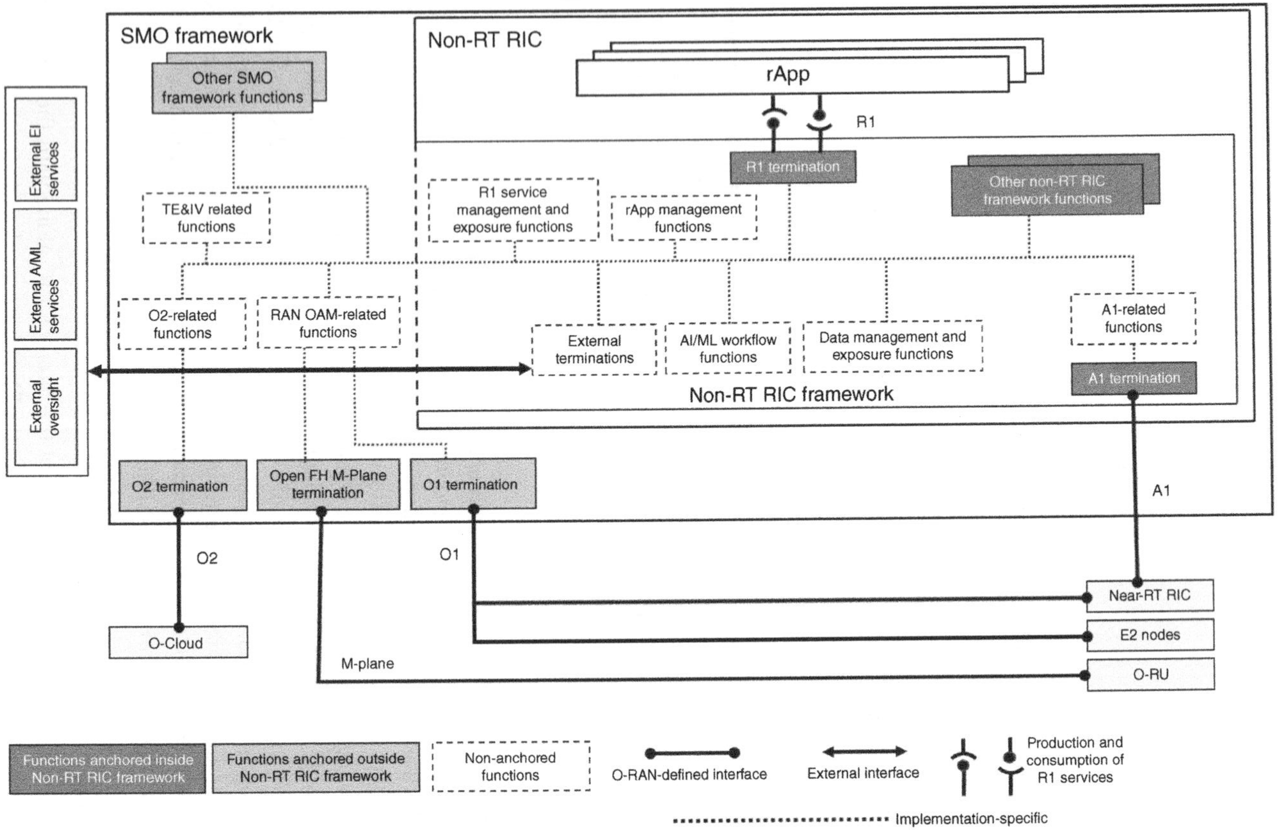

Figure 2.3 Reference Architecture for the SMO Framework and the Non-RT RIC. Source: O-RAN ALLIANCE (2024f). Reproduced with permission.

purpose) offered by a service producer to one or more service consumers via service endpoints such as application programming interfaces (APIs). Examples of R1 services include service management and exposure services, data management and exposure services, A1-related services, AI/ML workflow services, RAN Operations and Maintenance (OAM)-related services, O2-related services, and rApp management services. Service management and exposure services include service registration and service discovery services, authentication services, authorization services, and communication support services. Data management and exposure services deliver data created or collected by Data Producers to Data Consumers. A1-related services provide access to the functionality related to A1. AI/ML workflow services provide access to AI and ML workflow. RAN OAM-related services provide access to the RAN OAM functionality related to the O1 and open FH M-Plane interfaces and the OAM functionality related to RAN-specific network slicing in the SMO framework. O2-related services provide access to functionality related to O2. rApp management services provide functionality for managing rApps. Examples of logical functions that provide these R1 services include R1 service management and exposure functions, R1 termination, Data management and exposure functions, AI/ML workflow functions, A1-related functions, RAN OAM-related functions, O2-related functions, A1 termination, O1 termination, open FH M-Plane termination, O2 termination, Other SMO functions, and External terminations. These logical functions are briefly discussed later in this Section 2.2.1. Details of these R1 services and associated logical functions can be found in O-RAN ALLIANCE (2024f).

As shown in Figure 2.3, three categories of logical functions are defined in the Non-RT RIC framework and SMO framework (O-RAN ALLIANCE 2024f). Some functions are anchored inside the Non-RT RIC framework and shown in light gray boxes. Some functions are anchored outside the Non-RT RIC framework and are shown in dark gray boxes. Finally, some are nonanchored functions shown as boxes with dashed lines. Note that the decomposition of logical functions is up to vendor's implementation. An overview of these logical functions is given later in this section. The R1 services are defined to be a set of services produced by logical functions in the Non-RT RIC framework, the SMO framework outside the Non-RT RIC framework, or rApps. The services exposed to rApps via the R1 interface enable rApps to obtain information relevant to their specific functions and trigger actions such as policies and reconfiguration using the A1, O1, O2, and open FH M-Plane interfaces. The implementation-specific interfaces shown in Figure 2.3 enable the logical functions in the SMO or Non-RT RIC framework to communicate with each other in an implementation-specific manner.

The Non-RT RIC is defined as a logical function within the SMO framework that enables non-real-time control and optimization of RAN elements and resources, AI/ML workflow including model training and updates, and policy-based guidance of applications/features in Near-RT RIC (O-RAN ALLIANCE 2024f). One of the major functions of the Non-RT RIC is to support intelligent RAN optimization by providing policy-based guidance, ML model management, and enrichment information to the Near-RT RIC to facilitate optimization of functions (e.g., Radio Resource Management [RRM]) by the RAN. Another major function of the Non-RT RIC is to perform intelligent RRM function in non-real-time (i.e., with a time interval greater than one second). Note that the Non-RT RIC is the functionality internal to the SMO. The Non-RT RIC can use data analytics and AI/ML training and/or inference to determine the RAN optimization actions by leveraging the SMO services such as data collection and provisioning services of the O-RAN nodes and O-RAN-defined O1 and O2 interfaces.

As shown in Figure 2.3, the Non-RT RIC consists of two subfunctions, **Non-RT RIC framework** and **rApps** (O-RAN ALLIANCE 2024f). The Non-RT RIC framework is the functionality internal to the SMO framework. It terminates the A1 interface and exposes a set of R1 services to rApps using the R1 interface. Note that the Non-RT RIC framework exposes all the required functionality from the Non-RT RIC framework itself or the SMO framework to the rApps. rApps are modular applications that leverage the functionality exposed by the Non-RT RIC framework to perform RAN optimization and other functions. rApps can be viewed as Non-RT RIC applications that leverage the functionalities available in the SMO/Non-RT RIC framework to provide value-added services such as RRM, data analytics, and providing enrichment information in support of RAN operation and optimization.

2.2.1.4 Logical Functions in the Non-RT RIC Framework or the SMO Framework

Figure 2.4 summarizes the mon-RT RIC functions (O-RAN ALLIANCE 2024f).

R1 service management and exposure functions include service registration and discovery functions and the authentication and authorization functions. The service registration and discovery functions produce the service registration, service deregistration, service discovery, service notification, and communication support services and optionally bootstrap and heartbeat services. Examples of service producers are (i) service management and exposure services producer and (ii) A1-related services producer. rApps are examples of service consumers. However, rApps can also be service producers, and logical functions can also be service consumers. The authentication and authorization functions produce the authentication and authorization services.

The data management and exposure functions produce the data registration and discovery service, the data request and subscription service, the data offer service, and optionally data processing services.

A1-related functions include A1 policy functions and A1 EI functions. A1 policy functions produce the A1 policy-related services, which are exposed to rApps via the R1 interface, perform

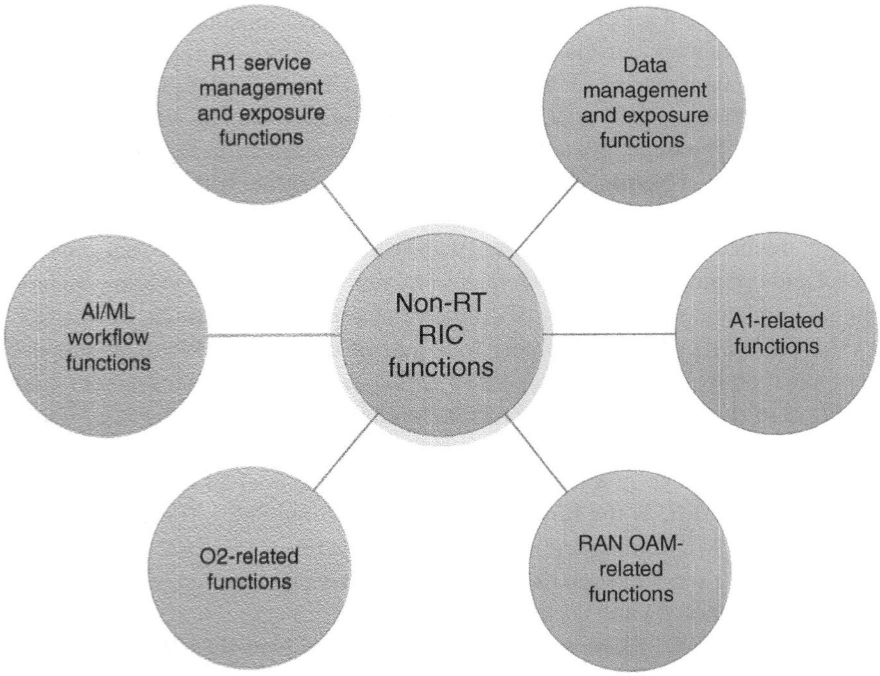

Figure 2.4 Non-RT RIC Functions.

A1 policy conflict mitigation, and interface with Near-RT RICs through A1 termination. A1 EI functions produce the A1 EI-related services, which are exposed to rApps via the R1 interface. They also produce the A1 EI service, which is exposed to Near-RT RICs via the A1 interface. They also collect data that are to be delivered as EI job results to Near-RT RICs via the data management and exposure functions.

RAN OAM-related functions produce the RAN OAM-related services, which are exposed to rApps via the R1 interface. They produce RAN OAM-related services, perform conflict mitigation, interface with Near-RT RICs and E2 nodes through O1 termination for tasks such as collection of performance data and configuration changes related to Near-RT RICs and E2 nodes, interface the O-RUs through open FH M-Plane termination for tasks such as collection of performance data and configuration changes related to the O-RUs, and interface with the RAN-specific slice management functionality in the SMO framework for tasks such as performance data and configuration changes related to RAN-specific network slicing.

O2-related functions produce the O2-related services and include Federated O-Cloud Orchestration and Management (FOCOM), Network Function Orchestration (NFO), and O2-related OAM functions.

Finally, the AI/ML workflow functions produce the AI/ML training services and AI/ML model management and exposure services to enable AI/ML model registration/deregistration, AI/ML model discovery, AI/ML model change subscription, AI/ML model storage, and (optionally) AI/ML model training capability registration.

2.2.2 Near-RT RIC

The Near-RT RIC facilitates near real-time (about 10 ms to 1 second) control and optimization of E2 nodes functions and resources through fine-grained data collection and actions over the E2 interface (O-RAN ALLIANCE 2024b). It contains one or more xApps that use the E2 interface to collect near real-time information on a UE or cell basis or a cell basis and provide value-added services. The Near-RT RIC obtains suitable policies and the enrichment data from the Non-RT RIC via the A1 interface. The Near-RT RIC also generates the RAN analytics information and exposes such information to Y1 Consumers via the Y1 interface.

In support of a given set of use cases, suitable RRM functional allocation between the Near-RT RIC and the E2 Node is needed. Such allocation depends on the capabilities of E2 nodes exposed over the E2 interface by an E2 Service Model (E2SM).[2] The E2SM describes the E2 node's functions that may be controlled by the Near-RT RIC and the related procedures. The E2SM specifies a function-specific RRM split between the E2 node and the Near-RT RIC. The Near-RT RIC may monitor, suspend, stop, override, or control the behavior of E2 node via policies for a function exposed in the E2SM. Note that the E2 node continues to provide its services even if the Near-RT RIC fails although value-added services provided the Near-RT RIC would be unavailable.

Figure 2.5 illustrates the Near-RT RIC architecture (O-RAN ALLIANCE 2024h). As shown in Figure 2.5, the Near-RT RIC platform hosts several functionalities to provide its value-added services. Details of these functions can be found in O-RAN ALLIANCE (2024h).

The *database and Shared Data Layer* (SDL) services enable reading and writing of the RAN and UE information and other auxiliary information in support of target use cases. The User Equipment-Network Information Base (UE-NIB) portion of the database maintains a list of UEs and associated data and tracks the UE identities associated with the connected E2 nodes.

2 See Chapter 3 for details of the E2SM.

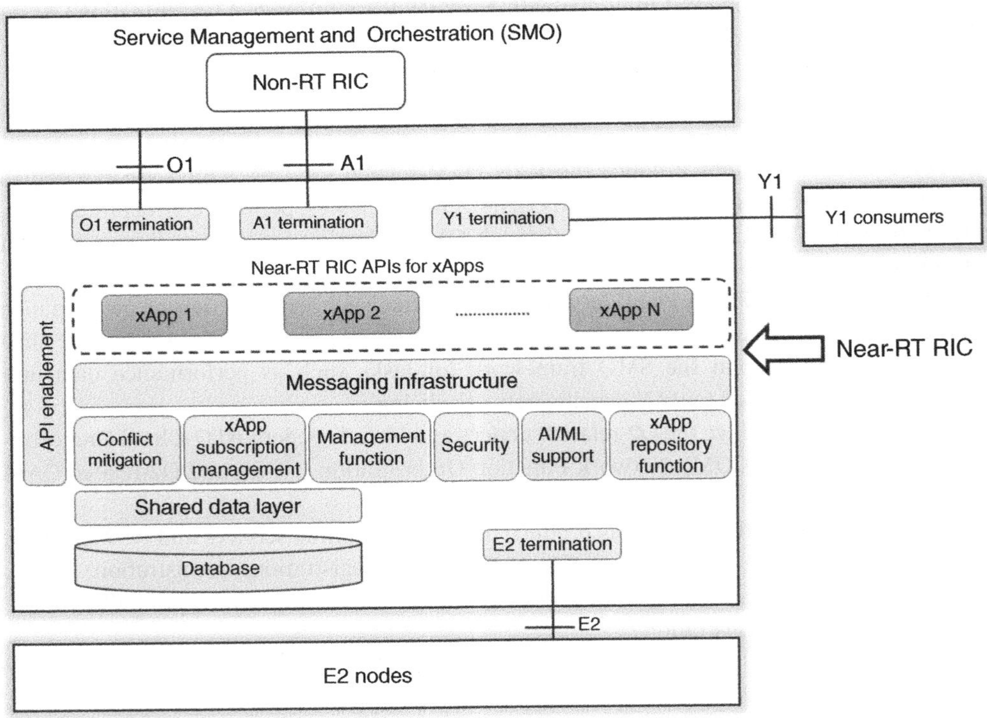

Figure 2.5 Near-RT RIC Architecture: A Closer Look.

The Radio-Network Information Base (R-NIB) portion of the database stores the configurations and near-real-time information related to connected E2 nodes and the mappings between these E2 nodes. xApps utilize the SDL to subscribe to database notification services and to read, write, and modify information stored on the database. The UE-NIB database, the R-NIB database, and other use case-specific information may be exposed using the SDL services.

The *xApp subscription management* manages subscriptions from different xApps to E2 nodes and provides a unified data distribution to xApps. It enforces authorization of policies controlling xApp access to messages and enables merging of identical subscriptions from different xApps into a single subscription toward an E2 node.

The *Conflict Mitigation* functionality resolves potentially overlapping or conflicting requests from different xApps. When applications individually change one or more parameters to optimize their own specific metrics, conflicting actions may occur. *Direct Conflicts* are the conflicts that can be observed directly by Conflict Mitigation. For example, two or more xApps may request different settings for the very same configuration of one or more parameters. In such cases, the Conflict Mitigation function processes these requests and determines a resolution. The conflicts that cannot be observed directly are *Indirect Conflicts*; it may be possible to observe some dependence among the parameters and resources targeted by xApps target. In this case, the Conflict Mitigation function may anticipate the possible conflicts and take actions to mitigate them. For example, one xApp may modify antenna tilts, and another xApp may modify measurement offsets, but these actions by different xApps influence the handover boundary. Implicit Conflicts are the conflicts that cannot be observed directly or dependence between xApps is not obvious. For example, different xApps may optimize different metrics and modify different parameters but such independent actions by xApps may have implicit, unwanted, and potentially adverse side effects on the metrics optimized by other xApps. An xApp that aims to protecting throughput metrics for Guaranteed Bit Rate (GBR) users may degrade non-GBR metrics or cell throughput.

The *Messaging Infrastructure* functionality enables low-latency and robust message delivery service between Near-RT RIC endpoints. It supports registration, discovery, and deletion of endpoints. It provides an APIs to send/receive messages to/from the messaging infrastructure. Messaging modes such as a point-to-point mode (where messages are exchanged between the endpoints) and a publish/subscribe mode (where real-time data are dispatched from E2 termination to multiple subscriber xApps) are supported.

The *Security* functionality is intended to provide the security scheme for xApps. It aims to prevent malicious xApps from abusing radio network information (e.g., exporting to unauthorized external systems) and/or control over RAN functions.

The Management functionality supports Operation, Administration, and Maintenance (OAM) management of the Near-RT RIC. OAM management comprises accounting, performance, file, security and other management plane services. In support of these OAM management services, the Near-RT platform provides the capabilities including fault management, configuration management, logging, tracing, and metrics collection.

The *interface termination* functionalities support E2 termination toward an E2 node, A1 termination toward the Non-RT RIC, the O1 termination toward the SMO, and the Y1 termination toward Y1 consumers.

The functions hosted by xApps enable execution of services in the Near-RT RIC and provide outcomes to E2 nodes. xApps collaborate with the Near-RT RIC platform functions to support target use cases. Upon registration to the Near-RT RIC platform, the xApp informs the Near-RT RIC platform about its OAM and control information to enable suitable functionalities. Note that the Life Cycle Management (LCM) of xApps is performed by the SMO and the O-Cloud. See Chapter 3 for details of xApps.

The *API Enablement* functionality supports capabilities related to the Near-RT RIC API operations such as API repository/registry, authentication, discovery, and generic event subscription. The Near-RT RIC APIs may be related to E2-related services, A1-related services, Management Function, and Database services. xApp accesses the API enablement services via one or more enablement APIs.

The *AI/ML support* functionality supports data pipelining, training, and performance monitoring for xApps. The AI/ML data pipeline in the Near-RT RIC offers data ingestion and preparation for xApps. Example inputs to the AI/ML data pipeline include E2 node data collected via the E2 interface, enrichment information via the A1 interface, information from xApps, and data retrieved from the Near-RT RIC database through the messaging infrastructure. The output of the AI/ML data pipeline may be provided to the AI/ML training capability in Near-RT RIC. The AI/ML training capability provides generic and use case-independent training capabilities to AI/ML-based xApps useful for one or more use cases.

The *xApp Repository* functionality selects xApps for A1 message routing based on A1 policy types and operator policies and performs access control of A1 EI types for xApps based on operator policies.

Well-defined Near-RT RIC APIs enable hosting of third-party xApps on different Near-RT RIC platforms. Figure 2.6 shows how these APIs are defined for specific interfaces (O-RAN ALLIANCE 2024h).

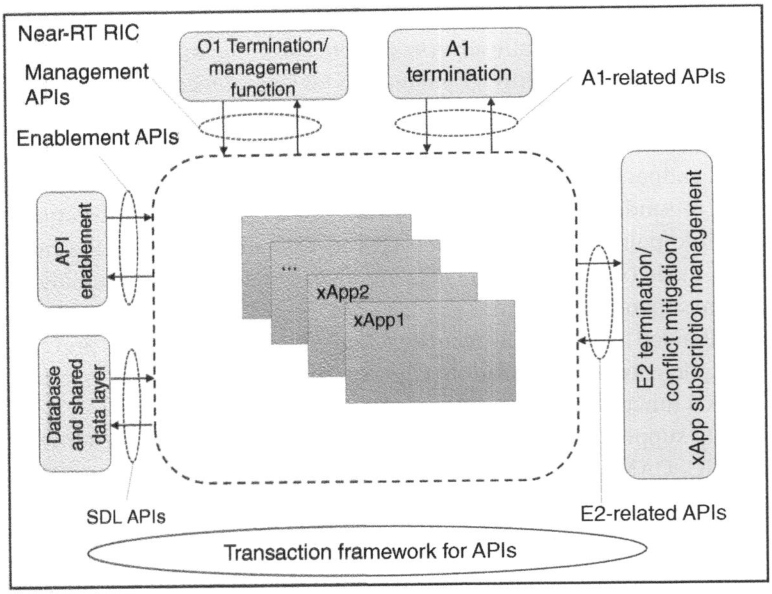

Figure 2.6 Near-RT RIC APIs: An Overview.

A1-related APIs allow access to A1-related functionality. E2-related APIs allow access to E2-related functionality and associated xApp Subscription Management and Conflict Mitigation functionality. Management APIs enable access to management-related functionality. SDL APIs enable access to the SDL-related functionality. The Enablement APIs are the APIs between xApps and API enablement functionality.

The Near-RT RIC APIs may be implemented using two approaches, Network API approach and Software Development Kit (SDK) approach (O-RAN ALLIANCE 2024h). In the Network API approach, each relevant Near-RT RIC endpoint exposes a network endpoint and specifies a particular data encoding protocol and a network transport protocol for communications. In the SDK approach, the Near-RT RIC vendor provides an SDK, which is a software library that handles management of all connections and exposes a simple API for the xApp to interact with the Near-RT RIC. The interface between the SDK library embedded in the xApp and the Near-RT RIC Platform may be vendor proprietary or aligned to a specific Network API.

2.2.3 O-CU-CP

As mentioned earlier, the O-CU-CP utilizes the 3GPP-defined gNB-CU-CP as baseline and adds support for O-RAN-defined interfaces and functionalities. Figure 2.7 illustrates the architecture of the O-CU-CP with both 3GPP-defined and O-RAN-defined interfaces (O-RAN ALLIANCE 2024b).

The O-CU-CP supports the 3GPP-defined interfaces such as the NG-C interface toward the AMF, the X2-C interface toward the eNB or the EN-DC gNB (en-gNB), the Xn-C interface toward the gNB or the ng-eNB, the F1-C interface toward the gNB-DU or the O-DU, and the E1 interface toward the gNB-CU-UP or the O-CU-UP using relevant terminations. EN-DC stands for E-UTRA-NR Dual Connectivity, where E-UTRA is Evolved Universal Terrestrial Radio Access and NR is New Radio. The O-CU-CP also supports the RRC and PDCP-Control Plane protocols of the NR Uu interface toward the UE.

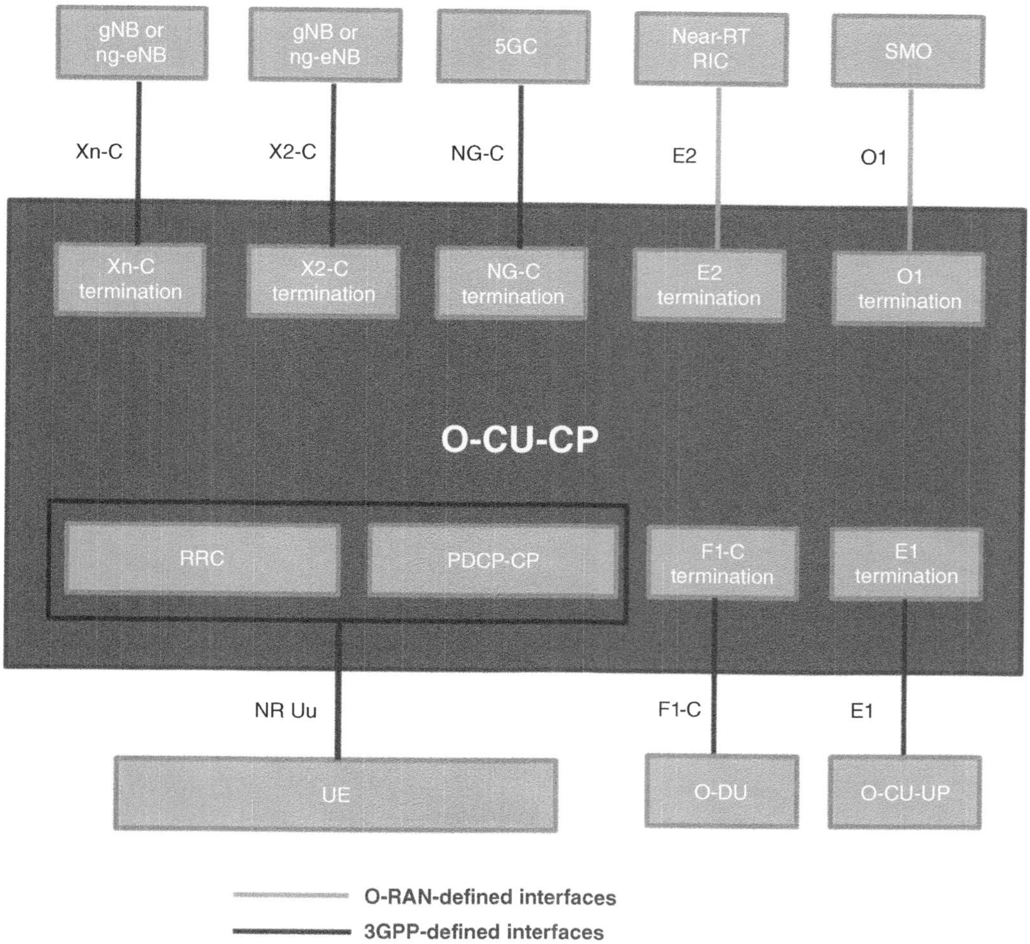

Figure 2.7 O-CU-CP: A Closer Look.

The O-CU-CP supports the O-RAN-defined interfaces such as the E2 interface toward the Near-RT RIC and the O1 interface toward the SMO using relevant terminations.

2.2.4 O-CU-UP

The O-CU-CP utilizes the 3GPP-defined gNB-CU-UP as baseline and adds support for O-RAN-defined interfaces and functionalities. Figure 2.8 illustrates the architecture of the O-CU-UP with both 3GPP-defined and O-RAN-defined interfaces (O-RAN ALLIANCE 2024b).

The O-CU-UP supports the 3GPP-defined interfaces such as the NG-U interface toward the UPF, the X2-U interface toward the eNB or the en-gNB, the Xn-U interface toward the gNB or the ng-eNB, the F1-U interface toward the gNB-DU or the O-DU, the S1-U interface toward the Evolved Packet Core (EPC) (specifically Serving gateway or S-GW), and the E1 interface toward the gNB-CU-CP or the O-CU-CP using relevant terminations. The O-CU-CP also supports the SDAP and PDCP-User Plane protocols of the NR Uu interface toward the UE.

The O-CU-UP supports the O-RAN-defined interfaces such as the E2 interface toward the Near-RT RIC and the O1 interface toward the SMO using relevant terminations.

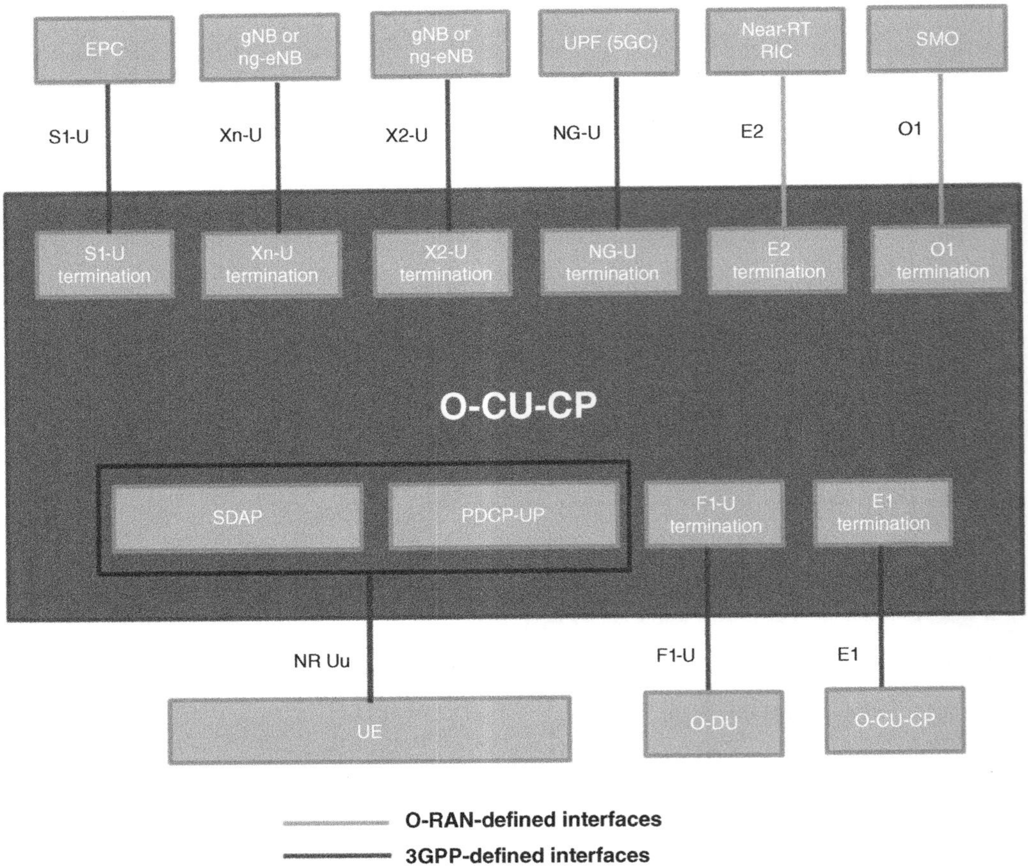

Figure 2.8 O-CU-UP: A Closer Look.

2.2.5 O-DU

The O-DU, combined with one or more O-RU(s) connected to it, performs all the functions of the 3GPP-defined gNB-DU. Furthermore, the O-DU supports the O-RAN-defined interfaces and functions. The O-DU may or may not be implemented using virtualization although a virtualized implementation is expected to be popular in O-RAN deployments. Figure 2.9 illustrates the O-DU architecture (O-RAN ALLIANCE 2024b).

The O-DU supports the 3GPP-defined interfaces such as the F1-C interface toward the gNB-CU or the O-CU-CP and the F1-U interface toward the gNB-CU or the O-CU-UP using relevant terminations. The O-DU implements the RLC and MAC protocols of the NR Uu interface toward the UE. The O-DU also implements the High-PHY portion of the 3GPP-defined PHY layer protocol of the NR Uu interface toward the UE. Note that the 3GPP does not separately define High-PHY and Low-PHY; the 3GPP-defined PHY layer is realized by dividing the PHY layer into High-PHY and Low-PHY, which is defined by O-RAN as part of the functional split 7.2x.

The O-DU supports the O-RAN-defined interfaces such as the E2 interface toward the Near-RT RIC, the O1 interface toward the SMO, and the open FH interface toward the O-RU[3] using relevant terminations. The open FH M-Plane interface between the O-DU and the O-RU supports

3 The Open Fronthaul interface is also known as the Lower Layer Split (LLS) interface.

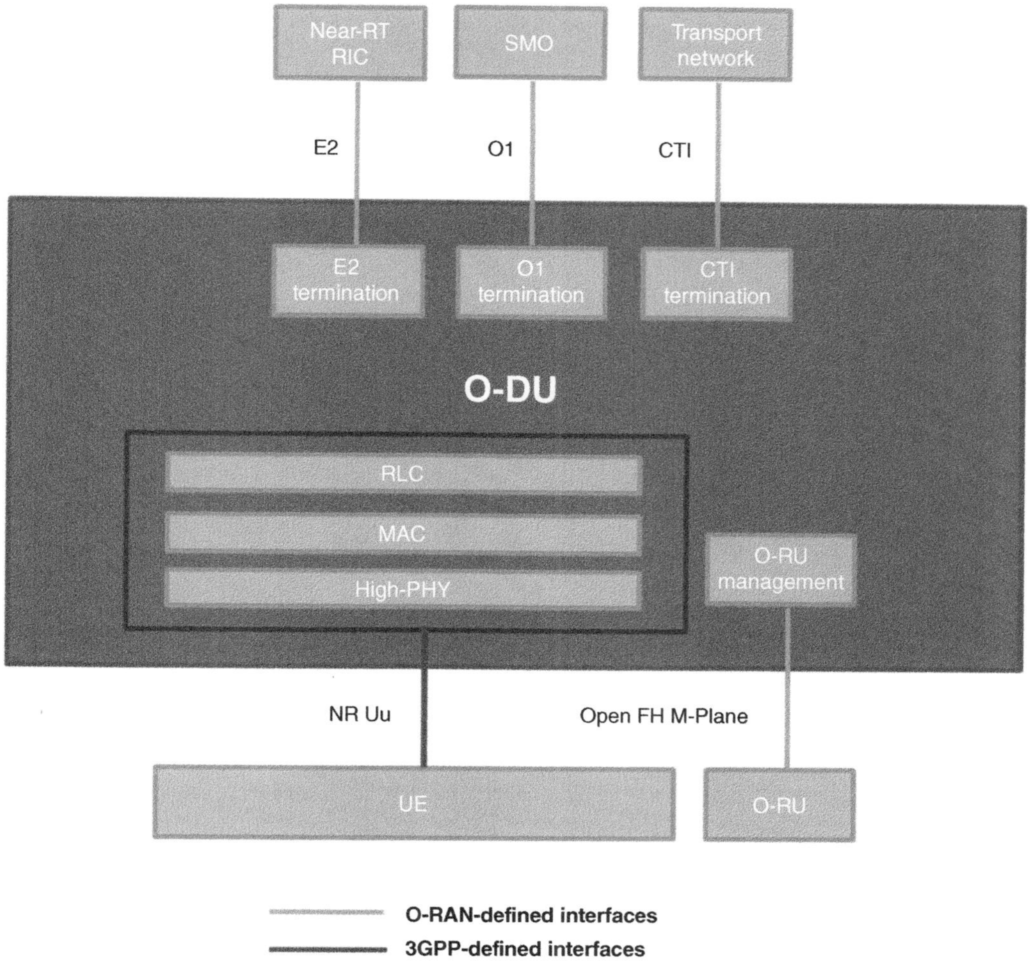

Figure 2.9 O-DU: A Closer Look.

O-RU management. The open FH CUS-Plane interface supports transfer of information related to the Control Plane, the User Plane, and synchronization. Furthermore, the O-DU may support the cooperative transport interface (CTI) to a Transport Network (TN) to control UL bandwidth allocation to Transport Units (TUs) for UL LLS traffic on shared point-to-multipoint transport network. Examples of a TN include a Passive Optical Network (PON) Optical Line Termination (OLT) and data over cable service interface specification (DOCSIS) Cable Modem Termination System (CMTS). Related TUs are a PON Optical Networking Unit (ONU) or DOCSIS cable modem.

2.2.6 O-RU

The O-RU, combined with the O-DU connected to it, performs all the functions of the 3GPP-defined gNB-DU. Furthermore, the O-RU supports the O-RAN-defined interfaces and functions. Since O-RUs involves physical entities required to process RF signals such as power amplifiers, frequency oscillators, and antennas, O-RUs cannot be implemented using virtualization. Hence, the O-RU is a physical node. Figure 2.10 illustrates the O-RU architecture (O-RAN ALLIANCE 2024b).

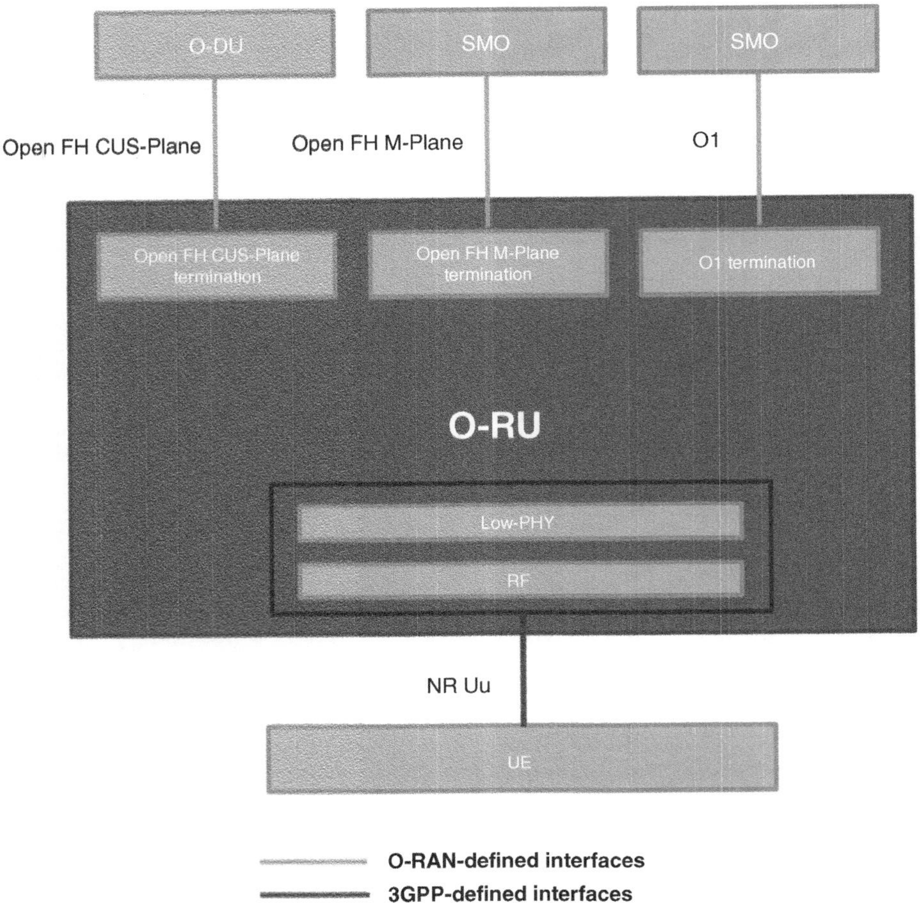

Figure 2.10 O-RU: A Closer Look.

The O-RU implements the Low-PHY portion of the 3GPP-defined PHY layer protocol of the NR Uu interface toward the UE based on the functional split 7.2x.

The O-RU supports the O-RAN-defined interfaces such as open FH M-Plane interface toward the O-DU and the SMO and the open FH CUS-Plane interface toward the O-DU.

2.2.7 O-eNB

The O-eNB utilizes the 3GPP-defined eNB or ng-eNB as baseline and adds support for O-RAN-defined interfaces and functionalities. Figure 2.11 illustrates the architecture of the O-eNB with both 3GPP-defined and O-RAN-defined interfaces (O-RAN ALLIANCE 2024b).

When the O-eNB implements the 3GPP-defined eNB, it supports the 3GPP-defined interfaces such as the S1-MME interface toward the MME of the EPC, the S1-U interface toward the S-GW of the EPC, and the X2 interface toward the eNB or the O-eNB. Such O-eNB also supports the RRC, PDCP, RLC, MAC, and PHY layers of the LTE-Uu radio interface toward the UE. Furthermore, such O-eNB supports the O-RAN-defined interfaces such as the E2 interface toward the Near-RT RIC and the O1 interface toward the SMO.

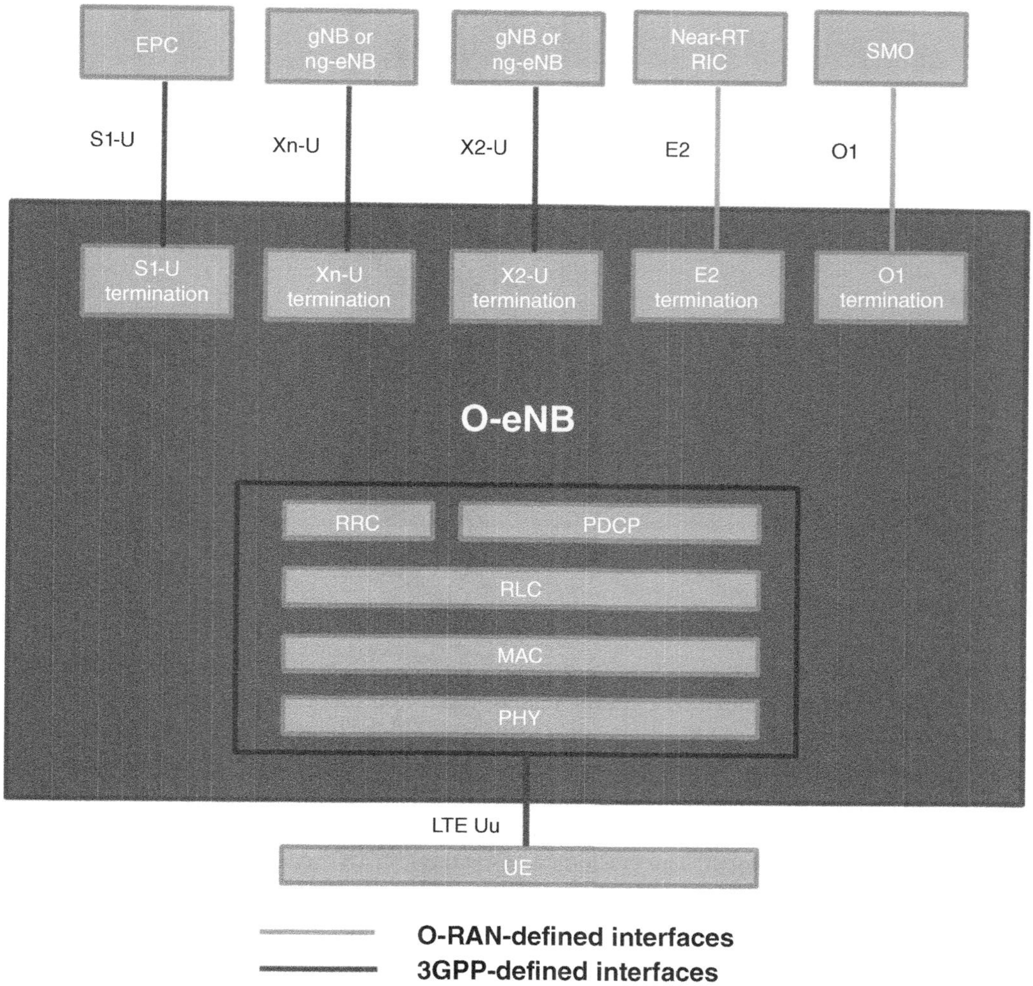

Figure 2.11 O-eNB: A Closer Look.

When the O-eNB implements the 3GPP-defined ng-eNB, it supports the 3GPP-defined interfaces such as the NG-C or N2 interface toward the AMF, the NG-U or N3 interface toward the UPF, and the Xn interface toward the gNB or the ng-eNB. Such O-eNB also supports the RRC, SDAP, NR PDCP, RLC, MAC, and PHY layers of the LTE-Uu radio interface toward the UE. Furthermore, such O-eNB supports the O-RAN-defined E2 interface toward the Near-RT RIC and the O1 interface toward the SMO.

While Figure 2.11 depicts an integrated eNB with O-RAN extensions, the O-eNB can also be realized using gNB-like disaggregation. In such case, the O-eNB supports O-DU and O-RU functions with the open FH interface between them.

2.2.8 O-Cloud

The O-Cloud is a cloud computing platform that consists of (i) a set of *physical infrastructure nodes* that host the relevant O-RAN functions (i.e., Near-RT RIC, O-CU-CP, O-CU-UP,

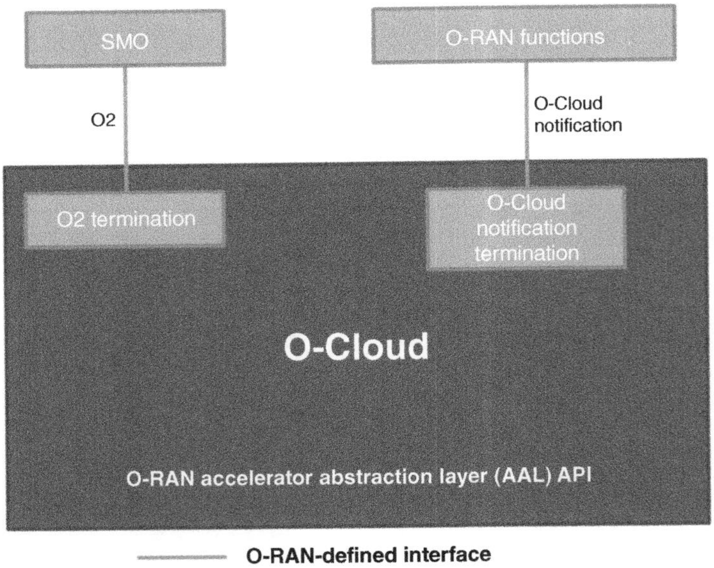

Figure 2.12 O-Cloud: A Closer Look.

and O-DU), (ii) the supporting software components (e.g., Operating System, VM Monitor, and Container Runtime), and (iii) relevant management and orchestration functions (O-RAN ALLIANCE 2024b). Figure 2.12 illustrates the O-Cloud interfaces (O-RAN ALLIANCE 2024b).

The O-Cloud supports the O-RAN-defined O2 interface toward the SMO for cloud and workload management to provide functions such as infrastructure discovery, registration, software lifecycle management, workload lifecycle management, fault management, performance management, and configuration management. The O-Cloud exports the O-RAN Accelerator Abstraction Layer API toward the hosted O-RAN workloads for hardware accelerator management. The O-Cloud also supports the O-Cloud Notification interface toward the hosted O-RAN workloads to notify the workloads of critical notifications such as Precision Time Protocol (PTP) synchronization states. A REpresentational State Transfer (REST) API or RESTful API allows Event Consumers (ECs) such as a vDU or Cloud-Native Network Function (CNF) to subscribe to events/status from the O-Cloud (O-RAN ALLIANCE 2024o). The cloud infrastructure provides Event Producers (EP) to enable cloud workloads to receive events/status known only to the O-Cloud infrastructure.

As illustrated in Figure 2.13, the O-Cloud is a distributed cloud composed of O-Cloud resource pools, where each resource pool is a collection of O-Cloud Nodes at a specific location (O-RAN ALLIANCE 2024n). An O-Cloud Node is the basic computational resource designator and can be viewed as a server. An operator may have several O-Cloud resource pools from different vendors but can manage them as a single entity. These O-Cloud resource pools are viewed by the SMO framework as a Federated O-Cloud. The O-Cloud provides a single (logical) reference point for all the O-Cloud Resource Pools within the O-Cloud boundary.

O-Cloud Nodes are provisioned into resources that are assigned for a specific use by the O-Cloud Management Software according to a blueprint. The O-Cloud can be visualized as having three planes, Management Plane, Control Plane, and Deployment Plane. The Control and Management planes may (i) be combined or (ii) use common tools and hence act as one plane. In the Management Plane, O-Cloud Nodes host the software that is responsible for managing the O-Cloud. In the Control Plane, O-Cloud Nodes host the software that manages the

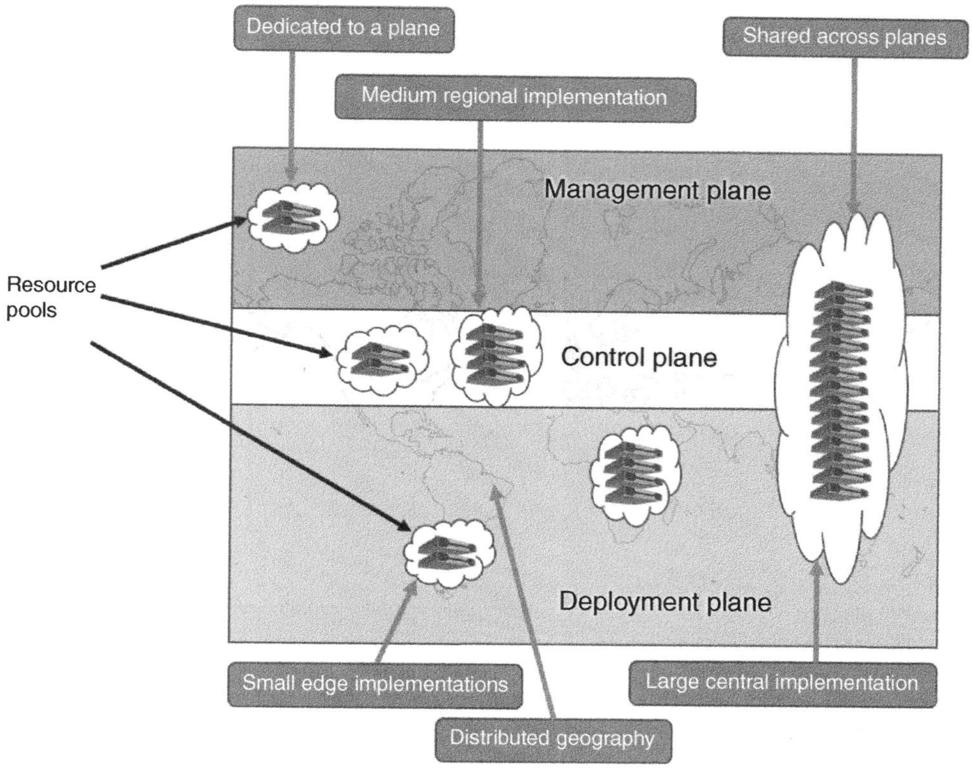

Figure 2.13 O-Cloud Planes (O-RAN ALLIANCE 2024n). Source: O-RAN ALLIANCE (2024n). Reproduced with permission.

resources assigned in the deployment plane to specific deployment instances. In the Deployment Plane, O-Cloud Nodes host service model deployments.

The O-Cloud Inventory consists of (i) the physical infrastructure used to create the O-Cloud, (ii) the logical clouds that are interfaces for deployments, and (iii) the inventory of deployments using the logical cloud. The SMO is responsible for financial accounting of physical inventory. The SMO inquires the O-Cloud about the mapping of resources to their use.

2.3 O-RAN Interfaces: A Closer Look

Now that we have discussed all the nodes or network functions of O-RAN, let us take a closer look at O-RAN interfaces to gain deeper understanding of how O-RAN works.

2.3.1 O1 Interface

The O-RAN OAM architecture identifies management services, managed functions, and managed elements of O-RAN. The O1 interface and the O2 interface support the O-RAN OAM. The O1 interface is between O-RAN Managed Element (ME) and the management entity, and the O2 interface is for management of the O-Cloud. Figure 2.14 shows the O-RAN OAM architecture (O-RAN ALLIANCE 2024c).

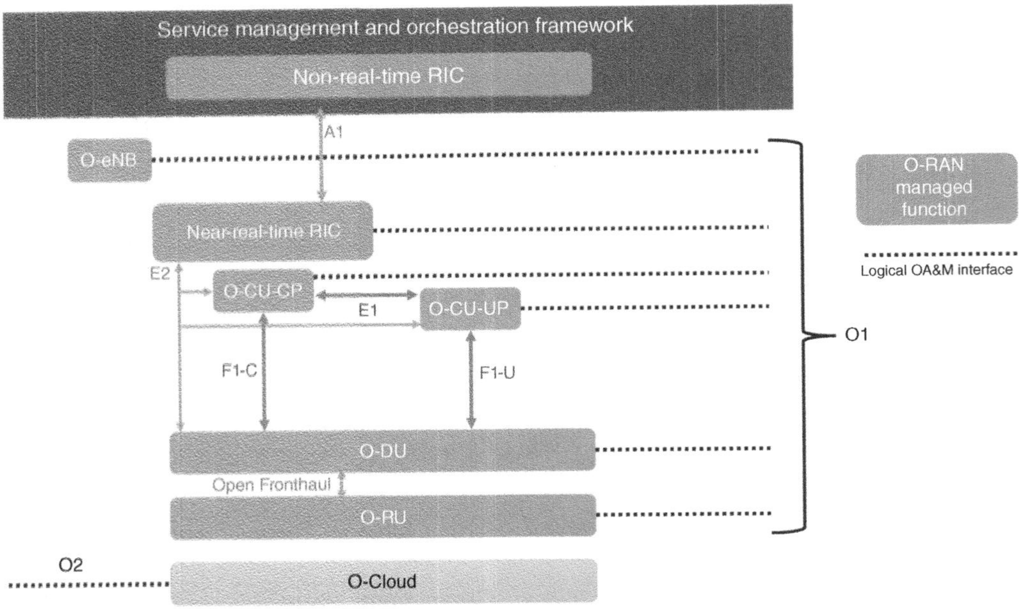

Figure 2.14 O-RAN OAM Logical Architecture.

There are three potential models for the OAM architecture, flat management model, hierarchical management model, and hybrid model. A network may deploy one or more models. In the flat management model, all the O-RAN Managed Functions are also MEs and expose an O1 interface to the SMO. For example, the O-RU is managed by the SMO through the O1 interface. NM/orchestration platforms provide a distributed deployment model of NM functions with benefits such as increased scaling and reduced latency functions. The open FH M-Plane does not support a flat management model for the O-RU. In a hierarchical management model, a higher-level ME manages a subnetwork of MEs. For example, the O-DU manages the O-RU using the open FH M-Plane interface; the SMO does not manage the O-RU. In the Hybrid management architecture, an ME is managed partially by (i) another ME and (ii) the SMO. For example, in such hybrid model, the O-RU is managed partially by the O-DU and partially by the SMO.

The O-RAN-compliant SMOs support the O1 interfaces as described in O-RAN ALLIANCE (2024a). The Management Service (MnS) providers and consumers that use Transport Layer Security (TLS) support TLS v1.2 or higher. MnS providers and consumers that use Hyper Text Transfer Protocol (HTTP) support HTTP v1.1 or higher; O-RAN prefers HTTP v2.0.

As shown in Figure 2.15, the O1 interface supports numerous management services including provisioning management services, fault supervision MnSs, Performance Assurance MnSs, Trace MnSs, File MnSs, Heartbeat MnSs, PNF Startup and Registration MnSs, PNF Software MnSs. See O-RAN ALLIANCE (2024a) for details of these services.

Provisioning management services enable (i) a Provisioning MnS Consumer to configure attributes of managed objects (on the Provisioning MnS Provider) that change the Provisioning MnS Provider's capabilities in facilitating end-to-end network services and (ii) a Provisioning MnS Provider to report configuration changes to the Provisioning MnS Consumer.

Fault supervision management services enable (i) a Fault Supervision MnS Provider to report errors and events to a Fault Supervision MnS Consumer and (ii) a Fault Supervision MnS Consumer to perform fault supervision operations on the Fault Supervision MnS Provider (e.g., to obtain an alarm list).

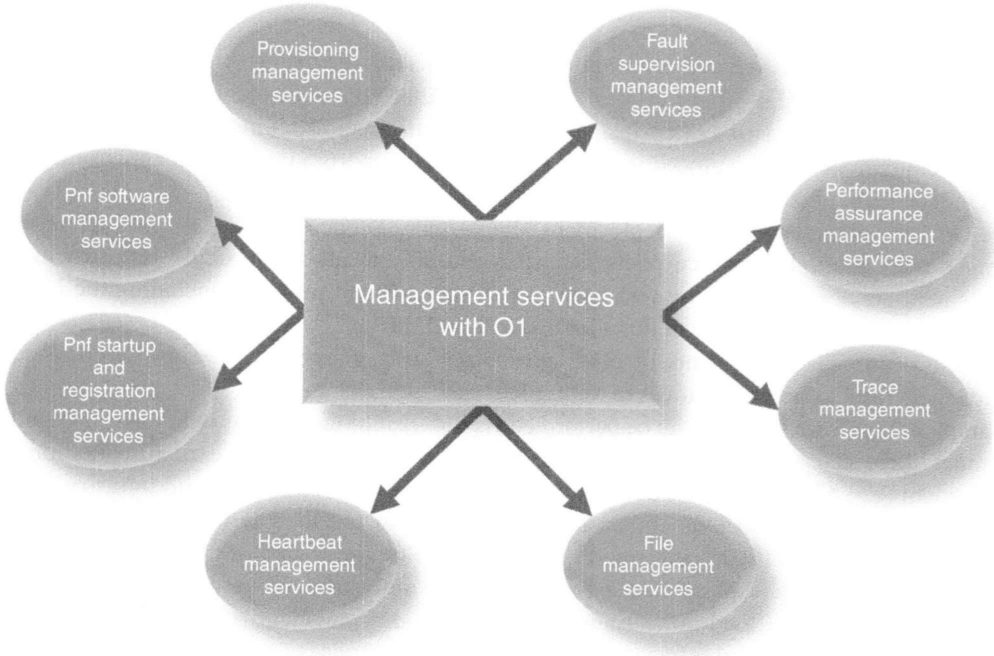

Figure 2.15 Management Services of the O1 Interface.

Performance Assurance MnSs enable (i) a Performance Assurance MnS Provider to report bulk file-based and/or real-time streaming performance data to a Performance Assurance MnS Consumer and (ii) a Performance Assurance MnS Consumer to perform performance assurance operations on the Performance Assurance MnS Provider (e.g., selection of the measurements to be reported and configuration of the frequency of reporting).

Trace management services enable a Trace MnS Provider to report file-based or streaming trace records to the Trace MnS Consumer. Trace Control enables the Trace Consumer to start a trace session by configuring a Trace Job via the Trace Control or by establishing a trace session that provides trace parameters to other trace management providers via signaling. Examples of trace MnSs include Call Trace, Minimization of Drive Testing, RRC Connection Establishment Failure, and Radio Link Failure.

File management services enable a File Management MnS Consumer to request the transfer of files between the File Management MnS Provider and the File Management MnS Consumer.

Heartbeat MnS enables (i) a Heartbeat MnS Provider to send heartbeats to the Heartbeat MnS Consumer and (ii) Heartbeat MnS Consumer to configure the heartbeat services on the Heartbeat MnS Provider.

PNF Startup and Registration management services enable (i) a physical PNF Startup and Registration MnS Provider to obtain its network layer parameters during startup. During this process, the PNF Startup and Registration MnS Provider also obtains the IP address of the PNF Startup and Registration MnS Consumer for PNF Startup and Registration MnS Provider registration. Upon registration by the PNF Startup and Registration MnS Provider, the PNF Startup and Registration MnS Consumer can make the PNF Startup and Registration MnS Provider operational.

Software management services enable (i) a PNF Software MnS Consumer to request a physical PNF Software MnS Provider to download, install, validate, and activate a new software package and (ii) a physical PNF Software MnS Provider to report its software versions.

2.3.2 O2 Interface

The O2 interface provides secured communications and enables the management of the O-Cloud infrastructure and the deployment LCM of O-RAN cloudified NFs implemented in the O-Cloud. The O2 interface is defined to be extensible to support new information or functions without changing the overall protocol or the procedures. The O2 interface supports a multi-vendor environment and is independent of specific implementations of the SMO and O-Cloud.

The O-Cloud provides multiple Deployment Management Services (DMSs). A DMS manages resources from multiple resource pools and span multiple locations. A single Infrastructure Management Services (IMS) framework manages all the O-Cloud resources including the DMS resources and the resources that are not allocated to any DMS.

The SMO can utilize blueprints or other preconfiguration templates to assign resources to resource pools and to assign resources to a specific DMS.

Examples of the functions that can be performed via the O2 interface for an operational O-Cloud fall under three categories (O-RAN ALLIANCE 2024n): (i) O-Cloud Infrastructure Resource Management (e.g., O-Cloud infrastructure Discovery and administration, O-Cloud infrastructure Scale-In and Scale-Out, O-Cloud infrastructure FCAPS, and O-Cloud infrastructure Platform Software Management), (ii) Management of Abstracted Resources and DMSes (e.g., Creation, Scale-In, and Scale-Out of abstracted assigned O-Cloud infrastructure resources, deployment FCAPS for abstracted O-Cloud infrastructure resources, and deployment DMS such as creation, deletion, and lease of O-Cloud infrastructure), and (iii) NF & Services Deployment Orchestration (e.g., [a] deployment Software Management, [b] FCAPS for NF and Services deployment resources, and [c] deployment, termination, Scaling, and Healing of NF & Services deployment resources).

Figure 2.16 Illustrates how logical functional blocks of the SMO and the O-Cloud connect via the O2 interface (O-RAN ALLIANCE 2024n).

The O-Cloud is logically composed of two functional blocks—IMS and DMS. The IMS block includes the functions that manage the cloud infrastructure, and the DMS block includes the functions that manage deployments on the cloud infrastructure. These functional blocks are supported by the O2 interface. The O2 interface consists of two service-based interfaces (SBIs). These SIBs support functional blocks such as FOCOM, NFO, OAM Functions, IMS, and DMS (O-RAN ALLIANCE 2024n).

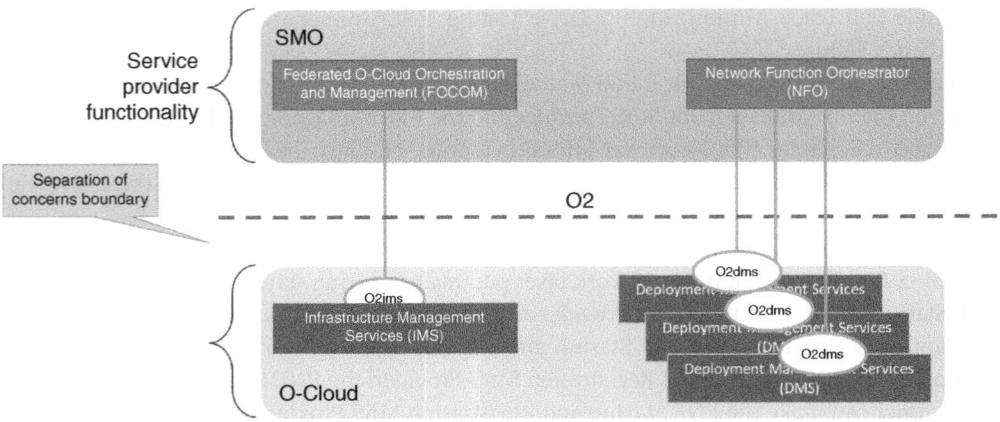

Figure 2.16 Uses of the O2 Reference Point-Based Interfaces. Source: O-RAN ALLIANCE (2024n). Reproduced with permission.

The FOCOM performs accounting and asset management of the resources in the cloud. It is a primary consumer of IMS services. The FOCOM maintains information about the O-Cloud resources management (e.g., domain of services such as the operator domain or an external domain). The NFO orchestrates the assembly of the network functions as a composition of NF deployments. The NFO may use OAM Functions to access the O1 interface toward deployed NFs. The NFO is a primarily consumer of the DMS services. The OAM Functions perform FCAPS management of O-RAN-managed entities. They receive faults and/or performance data from a subscription for O2ims or O2dms procedures. The IMS manages the O-Cloud resources and associated software. The DMS manages NF deployments into the O-Cloud. The DMS enables instantiation, monitoring, and termination of NF Deployments.

2.3.3 A1 Interface

The A1 interface is between the Non-RT-RIC residing in the SMO and the Near-RT RIC. The A1 interface supports three types of services to facilitate RAN optimization: A1 Policy MnS, A1 EI Service, and A1 ML Model MnS as shown in Figure 2.17 (O-RAN ALLIANCE 2024d).

2.3.3.1 A1 Policies and A1 Policy Management

A policy is a set of rules that are used to manage and control the changing and/or maintaining of the state of one or more managed objects (O-RAN ALLIANCE 2024d). A1 policies are declarative policies that contain statements on policy objectives and policy resources applicable to entities such as UEs and cells (O-RAN ALLIANCE 2024d). A1 policies may influence an individual UE or dynamically defined groups of UEs. They have temporary validity but take precedence over the configuration. They are not critical to traffic. Furthermore, they are nonpersistent and hence need to be redefined if the Near-RT RIC restarts.

The purpose of the A1 policies is to guide the RAN performance toward the overall goal expressed in RAN intent. The Non-RT RIC manages the A1 policies based on A1 policy feedback and the network status available via the O1 interface. The Non-RT RIC utilizes the O1 observables to evaluate the impact of the current A1 policies on the RAN intent and updates the goals in the A1 policies. To support the policy enforcement in the Near-RT RIC, the Non-RT RIC can provide EI via the A1 interface.

An A1 policy consists of a scope identifier and one or more policy statements. The scope identifier represents what the policy statements are applicable to such as UEs, Quality of Service (QoS) flows, or cells. The policy statements represent the goals specified for the Near-RT RIC and includes policy objectives and policy resources.

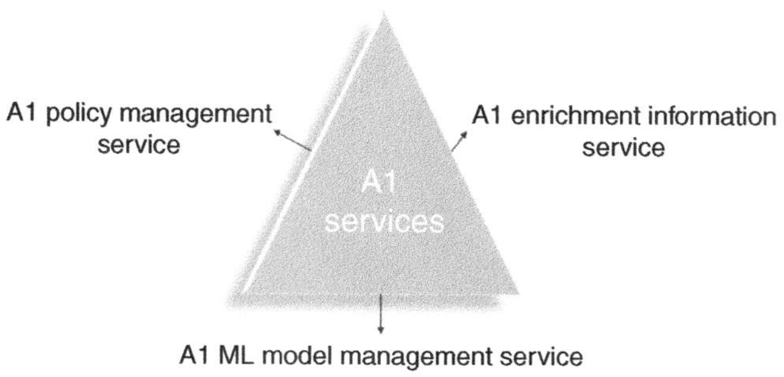

Figure 2.17 A1 Services.

2.3.3.2 A1 Enrichment Information (EI) and A1 EI Service

The EI is the information provided to an entity, in addition to the information that is available to it, to enhance the performance of its tasks (O-RAN ALLIANCE 2024d). In the O-RAN architecture, the EI is the enrichment information that is collected or derived at the SMO/Non-RT RIC from (i) external nonnetwork data sources or (ii) internal O-RAN network functions.

The EI can be produced based on information from one or more sources using processing such as relaying, combining, refining, or analyzing information from the input sources. When systems external to the O-RAN provide enrichment data to the SMO, they are called "O-RAN external information sources." The network functions in the RAN providing enrichment data to the SMO are called "O-RAN internal information sources." Hence, two types of EI exist, A1 EI and external EI. Based on the EI collected by the SMO, the Non-RT RIC can derive information for the Non-RT RIC and/or Near-RT RIC internal functions or applications (e.g., an ML model) to improve the RAN performance.

The Non-RT RIC provides the EI to the Near-RT RIC via the A1 interface. The A1 interface is also used for the discovery of external EI. The SMO/Non-RT RIC handles the authentication of the source and the security of the direct connection. However, handling of direct connections and discovery or delivery of external EI using the A1 interface are beyond the scope of O-RAN specifications.

2.3.3.3 ML Model Management Service

The Non-RT RIC can facilitate the development and use of ML models. In one scenario, the ML model is trained in the SMO using O1 observables and deployed by the SMO via the O1 interface. The Non-RT RIC utilizes such trained model along with O1 observables for inference to enhance the monitoring and guidance of the RAN. In another scenario, the ML model is trained in the SMO using O1 observables and deployed by the SMO via the O1 interface. However, the Near-RT RIC utilizes such trained model along with O1 observables for inference to enhance the optimization of the RAN. The performance evaluation and the handling of explicit feedback from the ML model are carried out based on O1 observability. The Non-RT RIC can provide EI to the Near-RT RIC via the A1 interface to support the ML model inference in the Near-RT RIC.

2.3.3.4 A1 Interface Protocol Stack

Figure 2.18 specifies the protocol stack for the A1 interface (O-RAN ALLIANCE 2024d) (O-RAN ALLIANCE 2024e).

The transport network layer utilizes IP. IPv4 or IPv6 may be used. IP provides best-effort delivery. Hence, Transmission Control Protocol (TCP) and Hyper Text Transfer Protocol Secure (HTTPS) are added above the IP layer to enhance security and reliability of the transport of A1 messages.

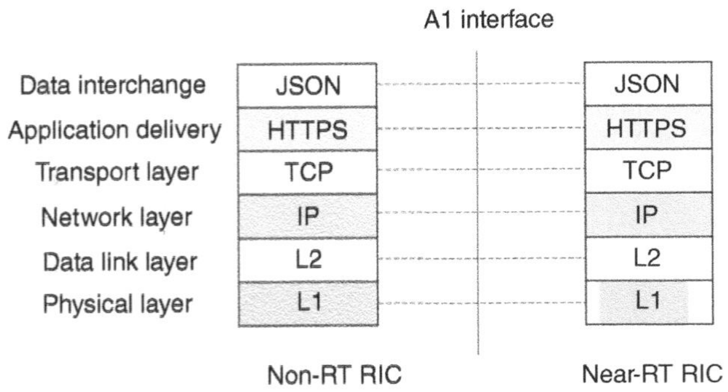

Figure 2.18 A1 Protocol Stack.

An HTTP connection is mapped to a TCP connection. Both the Non-RT RIC and the Near-RT RIC can act as an HTTP client and an HTTP server. Therefore, the Non-RT RIC and the Near-RT RIC establish a TCP connection for each direction. The application layer protocol, HTTPS, utilizes a RESTful approach and supports transfer of JavaScript Object Notation (JSON) formatted policy statements. Specifically, HTTP/1.1 is supported; HTTP/2 may also be optionally used. Default TCP port numbers for HTTP operations are used.

TLS is supported for security at both the transport layer and the application layer. Specifically, TLS v1.2, TLS v1.3, and OAuth2.0 with JSON Web Token are supported.[4] Furthermore, mutual TLS is used for mutual authentication. OAuth 2.0 is used for authorization at the application layer. HTTP over TLS is supported.

2.3.4 E2 Interface

The E2 interface is between the Near-RT RIC and an E2 node as shown in Figure 2.19 (O-RAN ALLIANCE 2024g). Examples of E2 nodes are O-CU-CP, O-CU-UP, O-DU, and O-eNB. The E2 nodes support all protocol layers and interfaces defined in 3GPP specifications related to (i) the eNB for Evolved Universal Terrestrial radio Access Network (E-UTRAN) and (ii) gNB/ng-eNB for the NG-RAN. The Near-RT RIC uses the E2 interface to collect near real-time information on UEs and cells and provide value-added services (e.g., optimization of a RAN operation such as handover). The E2 interface supports latency requirements for near-real-time optimization with the time scale ranging from 10 ms to 1 second. The Near-RT RIC and the E2 node can communicate with each other on the E2 interface using E2 Application protocol (E2AP). Even if there is a failure of the E2 interface or the Near-RT RIC, the E2 node can still provide services but there may be an outage for Near-RT RIC-provided value-added services.

A given E2 node is connected to only one Near-RT RIC so that it can receive consistent and harmonious guidance for its operations. In contrast, the Near-RT RIC can connect to multiple E2 nodes. The E2 interface utilizes Control Plane protocols.

The E2 agent routes the E2AP message toward the correct RAN Function. A RAN Function in Figure 2.19 is a specific Function in an E2 node. Example RAN functions include X2 Application Protocol (X2AP), F1 Application Protocol (F1AP), E1 Application Protocol (E1AP), S1AP, and NGAP for their respective interfaces and RAN internal functions that manage UEs and cells. The E2 node may have (and typically has) functions that do not have connectivity with the Near-RT RIC via an E2 agent.

Figure 2.19 Interworking Between the Near-RT RIC and E2 Nodes.

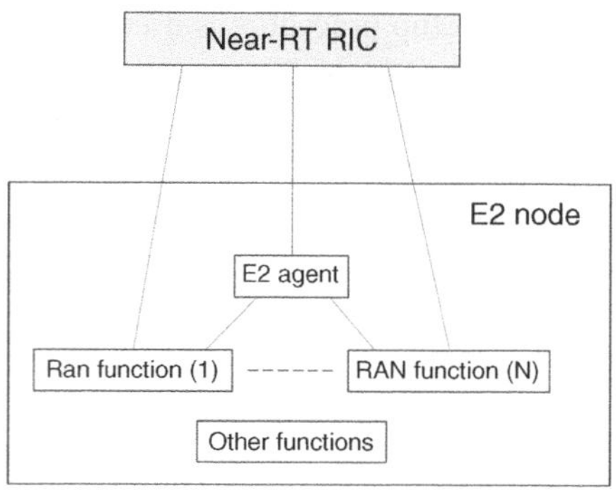

4 JWT defines a compact and self-contained mechanism to securely transmit information between two entities as a JSON object.

The Near-RT RIC and the xApps hosted in the Near-RT RIC use a set of services exposed by an E2 node. An E2SM defines what E2 node functionalities are accessible to the Near-RT RIC. The distribution of the RRM functional allocation between the Near-RT RIC and the E2 node depends on the capability of the E2 node exposed over the E2 interface by the E2SM to support a target use case. Specifically, the E2SM describes the functions in the E2 node that may be controlled by the near RT RIC, thus defining a function-specific RRM split between the E2 node and the near-RT RIC. For a function exposed in the E2SM, the Near-RT RIC may monitor, suspend, stop, override, or control the operations of the E2 node through policies.

As shown in Figure 2.20, E2 functions (i.e., functions of the E2 interface) can be divided into two categories, Near-RT RIC services and E2 support services (O-RAN ALLIANCE 2024g). A RIC service is a service provided on an E2 node to provide access to messages and measurements and/or enable control of the E2 node from the Near-RT RIC.

The Near-RT RIC services include REPORT, INSERT, CONTROL, and POLICY. The E2 support services include E2 interface management (e.g., E2 setup, E2 reset, and reporting of error situations) and Near-RT RIC service update (e.g., capability exchange related to the list of E2 node functions exposed over E2).

The RIC services of REPORT, INSERT, CONTROL, POLICY, and QUERY are supported by RIC functional procedures such as RIC Subscription, RIC Subscription Modification, RIC Subscription Modification Required, RIC Subscription Delete, RIC Subscription Delete Required, RIC Indication, RIC Control, and RIC Query. The E2 support services of interface management services are supported by global procedures such as E2 Setup, E2 Reset, E2 Node Configuration Update, E2 Removal, and Reporting of General Error Situations, while the E2 support services of RAN function services are supported by global procedures such as RIC service update and RIC service query.

In the REPORT service, upon request by the Near-RT RIC, an E2 node sends a REPORT message to Near-RT RIC, and the associated procedure continues in the E2 node after each occurrence of a defined RIC Subscription procedure Event Trigger. In the INSERT service, upon request by the Near-RT RIC, the E2 node sends an INSERT message to Near-RT RIC and suspends the associated procedure in the E2 node after each occurrence of a defined RIC Subscription procedure Event Trigger. In the CONTROL service, the Near-RT RIC sends a CONTROL message to E2 node to initiate a new associated procedure or resume a previously suspended associated procedure in the E2 node. As part of the POLICY service, upon request by the Near-RT RIC, the E2 node executes a specific POLICY during functioning of the E2 node after each occurrence of a defined RIC Subscription procedure Event Trigger. In the QUERY service, the Near-RT RIC sends a QUERY message to the E2 node to retrieve RAN-related and/or UE-related information from the E2 node.

The E2 setup procedure is used to establish the E2 interface between the Near-RT RIC and an E2 node. As part of this procedure, the E2 node provides a list of supported RIC services and

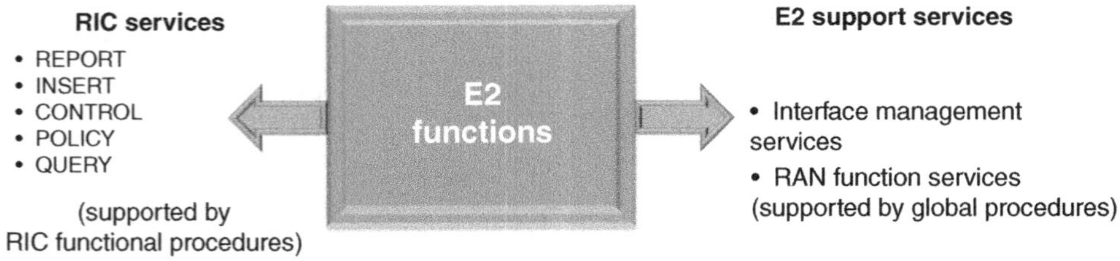

Figure 2.20 Categories of E2 Functions (O-RAN ALLIANCE 2024g.)

mapping of services to functions within the E2 node. Note that this information is specific to each RAN Function in the E2 node and is defined by a specific E2SM. The E2 node also provides E2 node configuration, which is specific to the E2 node type and defined by the E2 node system specifications.

The E2 node or the Near-RT RIC can use the E2 Reset procedure to reset the E2 interface. In this procedure, the outcome of all previous RIC Subscriptions is deleted from the E2 node, and the E2 node gracefully terminates any ongoing RIC Services. However, the information previously exchanged during E2 Setup, E2 Node Configuration Update, and RIC Service Update procedures are maintained.

The E2 node utilizes the RIC Service Update procedure to inform the Near-RT RIC of any change to the list of supported RIC services and mapping of services to RAN functions of the E2 node. Such information is specific to the RAN Function in the E2 node and is defined by a specific E2 SM.

The E2 node utilizes the E2 Node Configuration Update procedure to inform the Near-RT RIC of any change to the configuration of the E2 node and/or E2 node initiated changes to TNL associations of the E2 interface. Such information is specific to the E2 node type and defined by the E2 node system specifications.

The E2 Removal procedure is used by the E2 node or the Near-RT RIC to release the E2 signaling connection. The Near-RT RIC and E2 nodes releases all resources associated with the E2 interface.

Figure 2.21 specifies the Control Plane protocol stack for the E2 interface (O-RAN ALLIANCE 2024g).

The E2AP is an application layer protocol that governs the E2 interface. The E2AP utilizes IP transport. Since IP provides only best-effort delivery, SCTP is used to enhance reliability. Multiple SCTP associations are supported; in such case, the Near-RT RIC may request the E2 node to dynamically add or remove SCTP associations. Furthermore, the usage of SCTP association may be restricted for certain types of E2 signaling. The Payload Protocol Identifier assigned by IANA to be used by SCTP for E2AP is 70. A suitable port number for the SCTP Destination Port is selected via configuration.

The Near-RT RIC and E2 node support multiple TNL associations over the E2 interface. An initial TNL association is established during the E2 Setup procedure when the E2 node initiates the SCTP connection. Such TNL association is used for RIC services and E2 support services. TNL associations may be added, modified, or removed during subsequent E2 Connection Update and E2 Node Configuration Update procedures.

Figure 2.21 E2 Interface Protocol Stack.

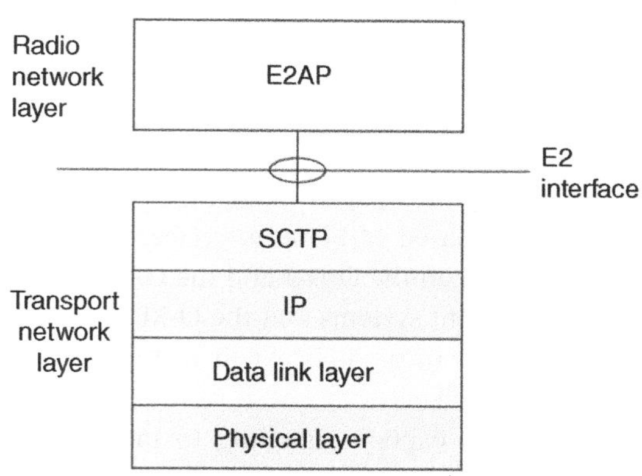

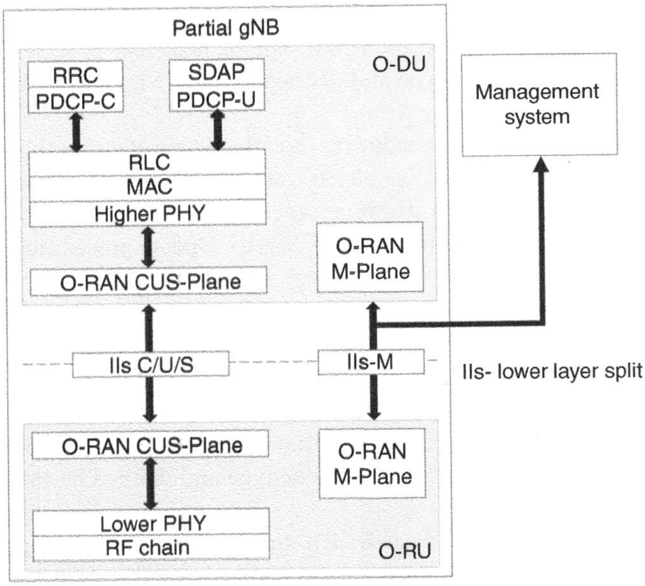

Figure 2.22 Functional Split with CUS-Plane and M-Plane of the Open FH Interface.

2.3.5 Open FH Interface

The open FH interface is between the O-DU and the O-RU (O-RAN ALLIANCE 2024k, 2024l). The open FH interface includes (i) the CUS Plane and (ii) the Management (M) Plane as shown in Figure 2.22 (O-RAN ALLIANCE 2024l). Both LTE and NR are supported on the open FH interface.

The Lower-Layer Split M-Plane (LLS-M) facilitates the initialization, configuration, and management of the O-RU in support of the 7.2x functional split. The Lower-Layer Split C/U/S-plane (LLS-C/U/S) facilitates control of Low-PHY in the O-RU by the O-DU, transfer of I/Q samples between the O-DU and the O-RU, and time, frequency, and phase synchronization between the clocks of O-DUs and the O-RU in support of the 7.2x functional split.

Section 2.3.5.1 provides details of the M-Plane, while Section 2.3.5.2 discusses the CUS-Plane.

2.3.5.1 M-Plane: A Closer Look

The M-Plane interface is defined between the O-RU Controller and the O-RU. The O-RU Controller function may be performed by entities such as the O-DU and the SMO. O-RAN defines a Network Configuration Protocol (NETCONF)/Yet Another Next Generation (YANG)-based M-Plane for start-up installation, software management, configuration management, performance management, fault management, and file management toward the O-RU (O-RAN ALLIANCE 2024l). The M-Plane supports two architectural models, a hierarchical model and a hybrid model. In the hierarchical model, the O-RU is managed entirely by one or more O-DU(s) using a NETCONF-based M-Plane interface. In contrast, the hybrid model utilizes both (i) a logical interface between the O-DU and the O-RU and (ii) one or more direct logical interfaces between the management systems and the O-RU. In the hybrid model, the O-RU has logical connectivity with the SMO through IP. Such connectivity may use the O-DU with the O-DU forwarding IP/Ethernet packets.

There is no explicit signaling to indicate that an O-RU is operating in a hierarchical or hybrid configuration. All O-RUs support hierarchical model and hybrid model deployments. A Multi-Operator O-RU additionally supports M-Plane architecture models, which involve one or more O-DUs operated by one or more different Shared Resource Operators.

The NETCONF/YANG framework supports integration of O-DUs and O-RUs with differing capabilities via well-defined published data models. Furthermore, NETCONF allows multiple clients to subscribe and receive information originating at the NETCONF server in the O-RU.

Different modes of transport network connectivity are possible between O-RU and O-DU and SMO.

The M-Plane has end to end IP connectivity between the O-RU and the entity managing it such as the O-DU, the SMO, or the O-RU Controller. The O-RU supports IPv4 or IPv6 and optionally supports dual stack IPv4 and IPv6.

Major functionalities provided by the M-Plane to the O-RU include the start-up installation, software management, configuration management, performance management, fault management, and file management (O-RAN ALLIANCE 2024l).

As part of the start-up installation functionality, the O-RU obtains network layer parameters through static configuration (e.g., preconfigured in the O-RU) or dynamically using DHCP or DHCPv6. The O-RU may also obtain the IP address of the O-RU controller(s) by establishing the NETCONF connectivity using the "call home" feature. When the O-RU interfaces with the SMO, the O-RU may obtain the IP address of the event-collector(s).

For software (SW) management, the M-Plane manages software download, installation, validation, and activation of new SW upon a request by the O-RU Controller. The software download is triggered by NETCONF Remote Procedure Call (RPC) procedures. The SW download is performed using secure File Transfer Protocol (sFTP) with Secure Shell (SSH) or File Transfer Protocol Explicit-mode Secure (FTPES) with TLS (IETF 2005).

In Configuration management, NETCONF get-config, and edit-config RPCs are be used for configuration parameter retrieval and updates at the O-RU. Scenarios such as Retrieve Resource State, Modify Resource State, Modify Parameters, and Retrieve Parameters are considered.

Performance management aims to optimize the operation of the O-RU. It describes the measurements and counters used to collect data related to O-RU operations. The measurement results are reported using a YANG notification (e.g., get RPC and/or notification) or file upload (e.g., periodic saving of results to a data file).

Fault management sends alarm notifications to the NETCONF Client. Fault management allows alarm subscription and enabling or disabling of alarm notifications.

File management enables the O-RU Controller to trigger an O-RU to perform upload of files stored on the O-RU to the O-RU Controller. Simultaneous multiple file upload operations can be supported under the same sFTP or FTPES connection between the O-RU and the O-DU/SMO.

Figure 2.23 specifies the protocol stack for the M-Plane of the open FH interface. As mentioned earlier, the M-Plane interface is between the O-RU Controller and the O-RU.

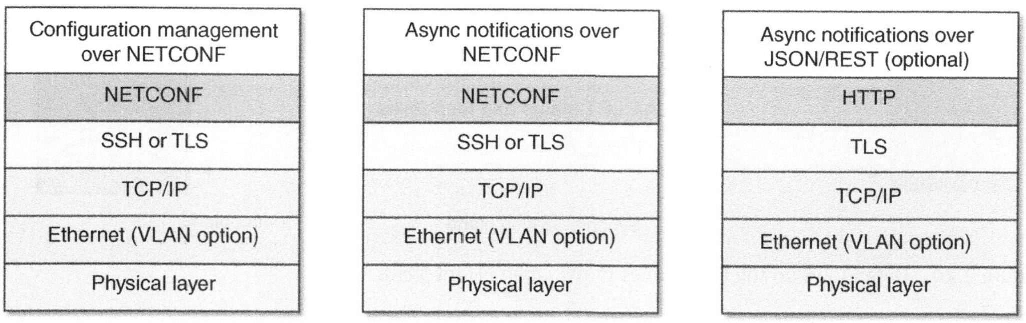

Figure 2.23 Protocol Stacks for the M-Plane of the Open FH Interface.

The M-Plane protocol stack utilizes TCP/IP protocol stack and adds protocols to enhance security and support NETCONF-based signaling. SSH or TLS is used above TCP. While NETCONF is typically used for configuration management and asynchronous notifications, asynchronous notifications can optionally be sent using HTTPS. The HTTPS-based option enables system optimization when the O-RU Controller function is performed by the SMO that is operating with a nonpersistent NETCONF session toward the O-RU. M-Plane security supports NETCONF/SSHv2 (IETF 2011) and NETCONF/TLS 1.2 (IETF 2015); support for TLS 1.3 is optional (IETF 2018).

2.3.5.2 CUS-Plane: A Closer Look

Recall that O-RAN utilizes the Functional Split 7.2x, where the protocol split occurs inside the PHY layer. Specifically, High-PHY refers to the portions of the PHY layer processing on the O-DU side of the FH interface including forward error correction encoding/decoding, scrambling, and modulation/demodulation. Low-PHY refers to the portions of the PHY layer processing on the O-RU side of the FH interface including Fast Fourier Transform (FFT)/Inverse Fast Fourier Transform (IFFT), digital beamforming, and PRACH extraction and filtering.

Figure 2.24 specifies the data flows on the CUS planes of the open FH interface (O-RAN ALLIANCE 2024k).

The User Plane on the open FH interface supports three types of data flows (O-RAN ALLIANCE 2024k). Data Flow 1a includes flows of IQ Data in FFT frequency domain on DL (e.g., DL user data on the Physical Downlink Shared Channel [PDSCH] and control channel data on the Physical Downlink Control channel [PDCCH]). Data Flow 1b includes flows of IQ Data in FFT frequency domain on UL (e.g., UL user data on the Physical Uplink Shared Channel [PUSCH] and control channel data on the PUCCH). Data Flow 1c includes flow of PRACH IQ data in FFT frequency domain (i.e., UL PRACH data).

The Control Plane on the open FH interface supports three types of data flows[5] (O-RAN ALLIANCE 2024k). Data Flow 2a carries scheduling commands for the DL and the UL (e.g., Scheduling information, FFT size, CP length, Subcarrier spacing, and UL PRACH scheduling) and beamforming commands (e.g., beam index). Data Flow 2b and 2c support Licensed-Assisted Access (LAA). Data Flow2b includes LAA Listen-Before-Talk (LBT) configuration commands (e.g., lbtHandle, lbtDeferFactor, lbtBackoffCounter, lbtOffset, MCOT, lbtMode,

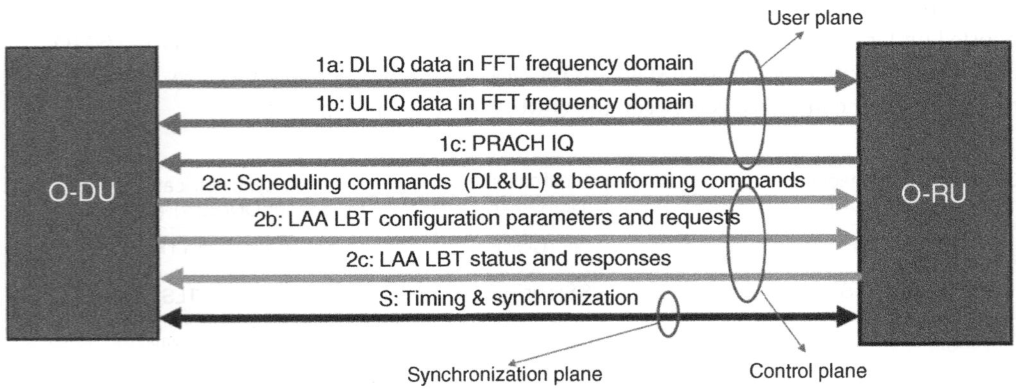

Figure 2.24 Data Flows on the CUS Planes of the Open FH Interface.

5 This Control Plane is specific to the open FH interface only and unrelated to the UE's Control Plane. The UE's Control Plane involves Access Stratum or RRC signaling and Non-Access Stratum signaling.

sfnSf, lbtCWconfig_H, lbtCWconfig_T, and lbtTrafficClass) and requests. Data Flow 2c carries LAA LBT status and response messages (e.g., LBT DL indication parameters such as lbtHandle, lbtResult, initialPartialSFs, bufferError, and lbtCWR_Result).

The Synchronization Plane on the open FH interface supports one data flow, Data Flow S (O-RAN ALLIANCE 2024k). This data flow carries timing and synchronization data (e.g., Synchronous Ethernet and IEEE 1588 packets and PTP packets).

The selection of the functional split 7.2x that defines the split within the baseband processing of the PHY layer leads to stringent bandwidth and tight latency requirements. The overall latency model for the open FH is based on eCPRI reference points (O-RAN ALLIANCE 2024k). The open FH interface is one of the most significant challenges faced in deploying O-RAN. The transport network needs to meet stringent bandwidth and latency requirements.

Figure 2.25 Shows how the CUS plane information can be carried by the transport network (O-RAN ALLIANCE 2024k). Two main transport mechanisms include (i) Ethernet encapsulation and (ii) IP/UDP encapsulation with IPv4 or IPv6.

Ethernet Encapsulation. Ethernet can be used as transport mechanism for the U-Plane and the C-Plane. The U-Plane and C-Plane messages are transmitted over standard Ethernet frames such as Ethernet II/DIX frame with type interpretation of the type length field. A unicast MAC address is assigned per Ethernet interface. The Ethernet termination feature supports MAC Client Data field sizes of up to 1,500 bytes for basic frames and 9,000 bytes for jumbo frames. In support of QoS, a 3-bit Priority Code Point (PCP) field is used, implying the capability for VLAN tagging for the traffic. While priority tagged frames or the frames tagged with VLAN ID=0 are always received, transmitting priority tagged frames is optional. Both the transport header and payload are contained within the Ethernet data field. The eCPRI Ethertype or the IEEE 1914.3 Ethertype is used.

IP/UDP Encapsulation. IP/UDP can also be used as the transport mechanism for the U-Plane and the C-Plane. IPv4 and/or IPv6 can be used. The IP/UDP encapsulation mechanism is identified by "IPv4" or "IPv6" Ethertype. IPv4 packets with any IPv4 header option field or IPv6 packets with "Next Header" value that is (i) an extension header or (ii) "59" are not supported. The C-Plane and U-Plane packets do not use IPv4 or IPv6 fragmentation due to stringent bandwidth and tight latency requirements on the open FH interface. The UDP destination port field identifies the encapsulated protocol. The UDP destination port field is set during initialization via M-Plane configuration. When the eCPRI Specification or IEEE 1914.3 (IEEE 2023) define their respective UDP destination ports, these will become the default for M-Plane configuration. UDP checksums are supported. Standard Ethernet packets or Jumbo frames up to 9,000 bytes in length can be supported.

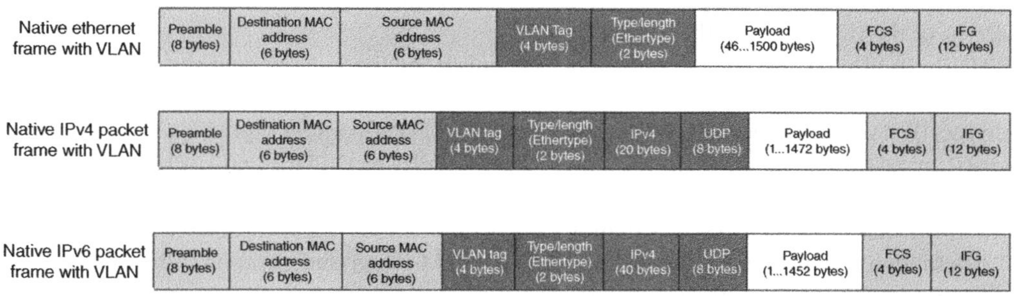

Figure 2.25 Transport Network Encapsulation for the CUS Planes of the Open FH Interface.

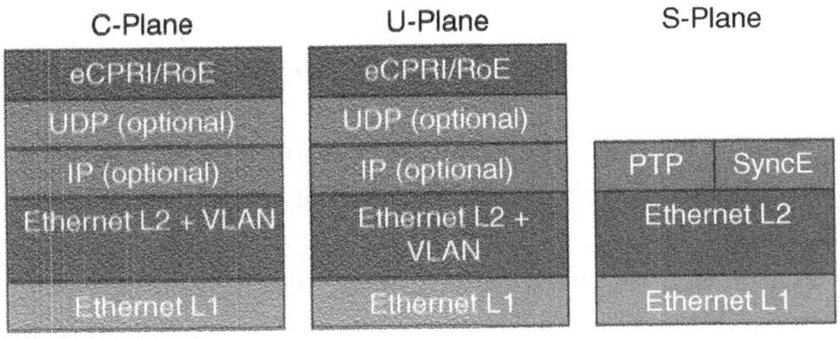

Figure 2.26 Protocol Stacks for the CUS Planes of the Open FH Interface.

Multiple transport headers within the Ethernet payload are supported to specify how the application data needs to be managed in the C-Plane and U-Plane. The 8-byte long transport header provides basic data routing capabilities and conveys the data flow type, sending and reception port identifiers, ability to support concatenation of multiple application messages in a single Ethernet packet, and sequence numbering.

Figure 2.26 specifies the protocol stacks for the C-Plane, U-Plane, and S-Plane for the open FH interface (O-RAN ALLIANCE 2024k). Ethernet is the primary transport mechanism for these protocol stacks.

The C-Plane utilizes eCPRI/Radio over Ethernet (RoE) over L2 Ethernet. The C-Plane optionally utilizes Layer 3 transport of UDP/IP if supported by the transmitting and receiving nodes.

The U-Plane also utilizes eCPRI/RoE over L2 Ethernet. Furthermore, like the C-Plane transport, the U-Plane optionally utilizes Layer 3 transport of UDP/IP if supported by the transmitting and receiving nodes.

The S-Plane utilizes PTP and SynE over L2 Ethernet to facilitate frequency and time synchronization of O-DUs and O-RUs. Transport of PTP directly over L2 Ethernet using Recommendation ITU-T G.8275.1 on full timing on-path support is considered as the typical mechanism. PTP over UDP/IP using Recommendation ITU-T G.8275.2 on partial timing support from the network may be used but synchronization performance may not be guaranteed. Security mechanisms are not used in the S-Plane.

Ethernet can be used as transport mechanism for the U-Plane and the C-Plane. The U-Plane and C-Plane messages are transmitted over standard Ethernet frames such as Ethernet II/DIX frame with type interpretation of the type length field. A unicast MAC address is assigned per Ethernet interface. The Ethernet termination feature supports MAC Client Data field sizes of up to 1,500 bytes for basic frames and 9,000 bytes for jumbo frames. In support of QoS, a 3-bit PCP field is used, implying the capability for VLAN tagging for the traffic. While priority tagged frames or the frames tagged with VLAN ID=0 are always received, transmitting priority tagged frames is optional. Both the transport header and payload are contained within the Ethernet data field. The eCPRI Ethertype or the IEEE 1914.3 Ethertype is used.

IP/UDP can also be used as the transport mechanism for the U-Plane and the C-Plane. IPv4 and/or IPv6 can be used. The IP/UDP encapsulation mechanism is identified by "IPv4" or "IPv6" Ethertype. IPv4 packets with any IPv4 header option field or IPv6 packets with "Next Header" value that is (i) an extension header or (ii) "59" are not supported. The C-Plane and U-Plane packets do not use IPv4 or IPv6 fragmentation due to stringent bandwidth and tight latency requirements on the open FH interface. The UDP destination port field identifies the encapsulated protocol. The UDP destination port field is set during initialization via M-Plane configuration. When the eCPRI Specification or IEEE 1914.3 (IEEE 2023) define their respective UDP destination ports, these will become the default for M-Plane configuration. UDP checksums are supported. Standard Ethernet packets or Jumbo frames up to 9,000 bytes in length can be supported.

Multiple transport headers within the Ethernet payload are supported to specify how the application data needs to be managed in the C-Plane and U-Plane. The 8-byte long transport header provides basic data routing capabilities and conveys the data flow type, sending and reception port identifiers, ability to support concatenation of multiple application messages in a single Ethernet packet, and sequence numbering.

2.3.6 O-Cloud Notification Interface

The O-Cloud Notification interface enables an EC such as an O-DU deployed on the O-Cloud to subscribe to events and status from the O-Cloud. The O-Cloud provides EPs to enable cloud workloads to receive events and status that might be known only to the O-Cloud infrastructure. Specifically, a REST API allows an EC to subscribe to events and/or status from the O-Cloud. Examples of ECs include a virtual O-DU (vO-DU) and CNF. This interface utilizes HTTP/2 and transports HTTP/2 over TCP.

An EC utilizes the REST API to subscribe to specific event types or categories of events by specifying the event/status producer address. The REST API acts as an integration point to an event and status framework that runs in the O-Cloud such as the IMS and/or the DMS. While intra-node delivery of events is the main focus, inter-node delivery of events is also allowed. The event framework interacts with northbound interfaces such as O2 through the IMS. The EC API complements the O2ims notifications. Furthermore, interfacing with external entities is necessary for communication with orchestrating entities and storing event information permanently for root-cause analysis. The events flow in one direction from the event framework to external entities.

2.3.7 CTI (Cooperative Transport Interface)

The CTI is an optional interface between O-DUs and Transport Nodes (TNs) of a packet-based transport network. Such transport network connects an O-DU to different O-RUs. Figure 2.27 illustrates the CTI reference architecture (O-RAN ALLIANCE 2024i).

The CTI is intended for the TNs that manage a shared point-to-multipoint network. The TNs working with point-to-point links do not exchange CTI messages with O-DUs. The CTI consists of a CTI Transport Control plane (TC-Plane) (O-RAN ALLIANCE 2024i) and a CTI Transport Management (TM) (O-RAN ALLIANCE 2024j).

The CTI supports real-time and non-real-time cooperation between the eNB/gNB and the resource allocation–based transport networks utilizing resource (e.g., bandwidth) allocation. The transport network enables transport of LLS-C/U/S traffic and LLS-M traffic between the O-DU and the O-RU. A TN is a master transport network node on the O-DU side and schedules one or multiple TUs to perform resource allocation for the upstream information. The TN connects to the O-DU via the CTI and decodes Transport-Plane (T-plane) messages from the O-DU to facilitate its scheduling.

The CTI use case involves packet-based transport networks with the TNs that terminate one or multiple shared distribution networks, where each distribution network (with one port on per TN) aggregates a multitude of TUs. The bandwidth of a distribution network is shared among multiple TUs. In the upstream direction, the TN manages bandwidth sharing by scheduling bandwidth allocations for multiple TUs statically or dynamically.

While static or continuous allocation of a fixed bandwidth results in the lowest latency, precious bandwidth is wasted when the bandwidth requirement is below the allocated bandwidth. In contrast, dynamic allocation of bandwidth involves observation of traffic and buffer status, increasing the efficiency of bandwidth utilization at the expense of latency due to the inherent processing delay of dynamic scheduling.

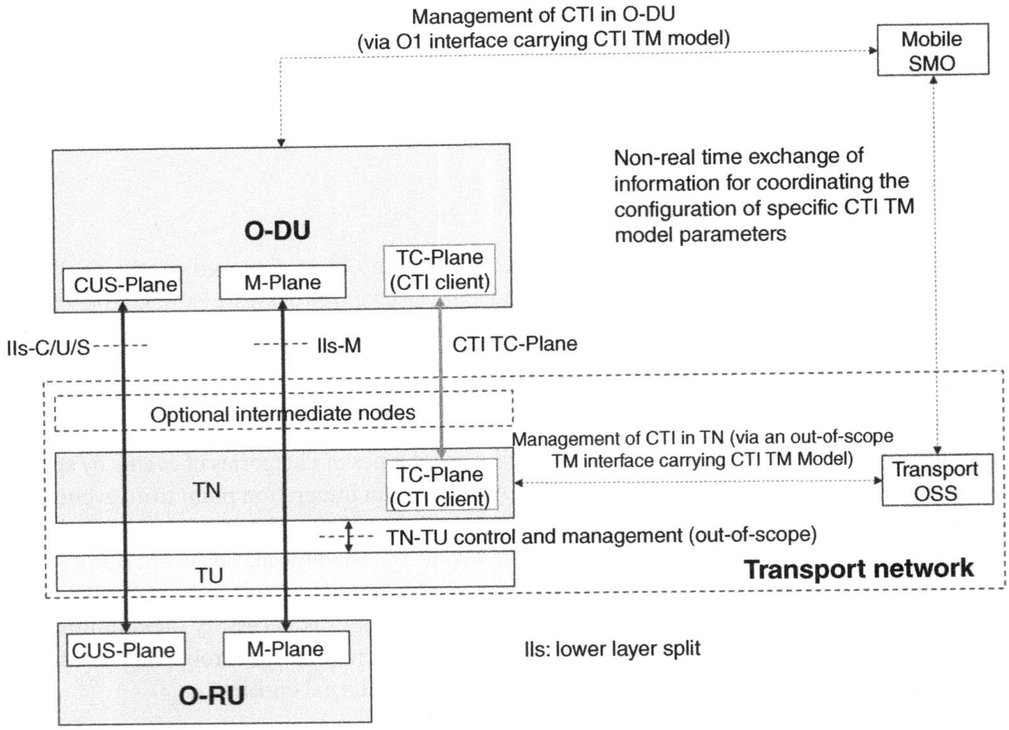

Figure 2.27 CTI Reference Architecture.

The CTI enables efficient upstream bandwidth allocation through dynamic scheduling performed by proactive TNs interacting with the O-DU. Specifically, the O-DU can notify the TNs about future bandwidth requirements based on the uplink grant requests that it has received from UEs. The TN can exploit statistical multiplexing while allocating bandwidths.

Figure 2.28 shows how the CTI is used as it relates to the transport network (O-RAN ALLIANCE 2024i).

The UEs communicate with the O-RU using lower layers of the NR radio protocol stack. The O-RU and the O-DU can communicate with each other using the transport network that makes use of the CTI.

The target uplink latency for the CTI is specified as T34max in O-RAN (O-RAN ALLIANCE 2024k). T34max is the delay between the R3 reference point of the O-RU and the R4 reference point of the O-DU. When intermediate nodes (e.g., switches or routers) exist between the O-DUs and the TNs, they introduce extra latency. The interface between the TN and the TU is transport network specific and is beyond the scope of O-RAN specifications.

The TU is considered to be a slave transport network node on the O-RU side. There are control and management interfaces between the TN and the TU. The TC-Plane is real-time transport control between the CTI Client in the O-DU and the CTI Server in the TN. The CTI TM Information Model/Data Model consists of (i) the configuration of parameters for the CTI functionality in the O-DU and (ii) the CTI configuration parameters that are coordinated between the mobile SMO and the transport network Operations Support System.

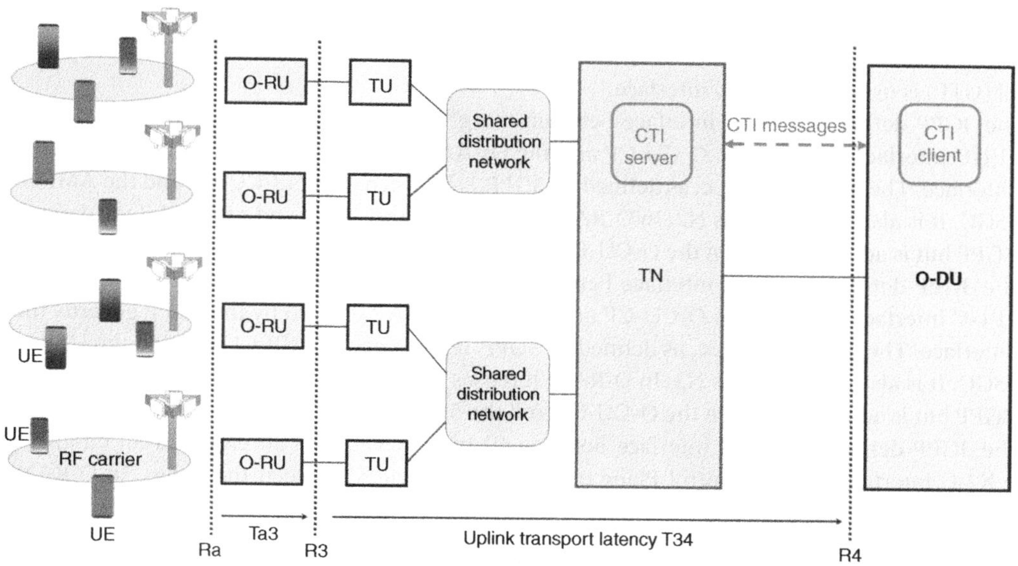

Figure 2.28 Interworking of CTI and the Transport Network.

2.3.8 3GPP Interfaces Relevant to O-RAN

As mentioned in Section 2.1, O-RAN reuses several interfaces defined by the 3GPP. Figure 2.29 shows such 3GPP interfaces (O-RAN ALLIANCE 2024b).

The 3GPP defines the E1 interface between the gNB-CU-CP and the gNB-CU-UP. In O-RAN, the E1 interface between the O-CU-CP and the O-CU-UP. E1AP defined by the 3GPP governs the E1 interface.

The 3GPP defines the F1-C interface between the gNB-CU-CP and the gNB-DU. In O-RAN, the F1-C interface is between the O-CU-CP and the O-DU. F1AP defined by the 3GPP governs the F1 interface.

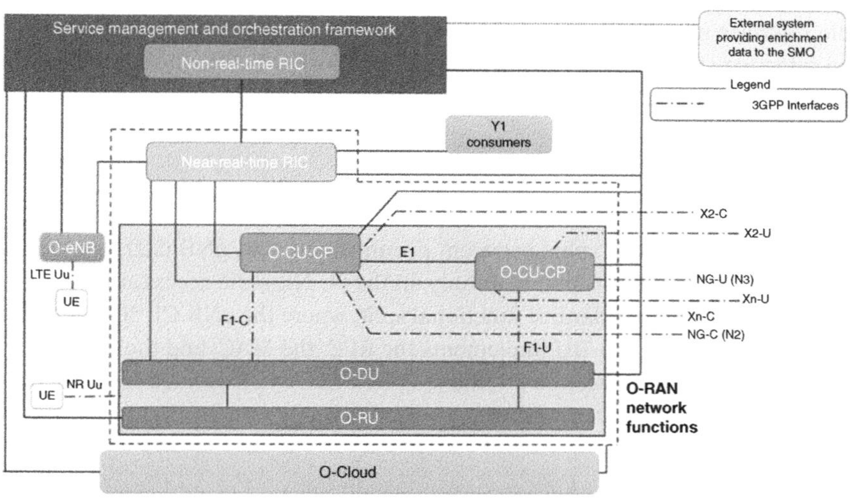

Figure 2.29 3GPP-Defined Interfaces in O-RAN.

The 3GPP defines the F1-U interface between the gNB-CU-UP and the gNB-DU. O-RAN defines the F1-U interface between the O-CU-UP and the O-DU. The 3GPP-defined GPRS[6] Tunneling Protocol (GTP) is used on the F1-U interface.

The 3GPP defines the F1-C interface between the gNB-CU-CP and the gNB-DU. In O-RAN, the F1-C interface between the O-CU-CP and the O-DU. F1AP defined by the 3GPP governs the F1 interface. The NG-C interface, as defined by 3GPP, is between the gNB-CU-CP and the AMF in the 5GC. It is also referred to as N2. In O-RAN, it reuses the principles and protocol stack defined by 3GPP but is adopted between the O-CU-CP and the 5GC.

The 3GPP defines the F1-C interface between the gNB-CU-CP and the gNB-DU. In O-RAN, the F1-C interface between the O-CU-CP and the O-DU. F1AP defined by the 3GPP governs the F1 interface. The NG-U interface, as defined by 3GPP, is between the gNB-CU-UP and the UPF in the 5GC. It is also referred to as N3. In O-RAN, it reuses the principles and protocol stack defined by 3GPP but is adopted between the O-CU-UP and the 5GC.

The 3GPP defines the X2-C interface between (i) two eNBs or (ii) an eNB and an en-gNB.[7] The X2-C interface carries Control Plane or signaling messages between the nodes. In O-RAN, the X2-C interface is between (i) two O-eNBs and (i) an O-eNB and an O-CU-CP. X2AP defined by the 3GPP governs the X2-C interface.

The 3GPP defines the X2-U interface between (i) two eNBs or (ii) an eNB and an en-gNB. The X2-U interface carries user traffic between the nodes. In O-RAN, the X2-U interface is between (i) two O-eNBs and (ii) an O-eNB and an O-CU-CP. The 3GPP-defined GTP is used on the X2-U interface.

The 3GPP defines the Xn-C interface between (i) two gNBs, (ii) two ng-eNBs, or (iii) an ng-eNB and a gNB. In O-RAN, the Xn-C interface is between (i) two O-CU-CPs, (ii) two O-eNBs, or (iii) an O-eNB and an O-CU-CP. Xn Application Protocol defined by the 3GPP governs the Xn interface.

The 3GPP defines the Xn-U interface between (i) two gNBs, (ii) two ng-eNBs, or (iii) an ng-eNB and a gNB. In O-RAN, the Xn-U interface is between (i) two O-CU-CPs, (ii) two O-eNBs, or (iii) an O-eNB and an O-CU-CP. The 3GPP-defined GTP is used on the X2-U interface.

The 3GPP defines the Uu interface between the UE and the eNB/gNB. More specifically, the UE communicates with the eNB using LTE-based Uu interface and the gNB using the 5G NR-based Uu interface. In O-RAN, all of the O-CU-CP, the O-CU-UP, the O-DU, and the O-RU together appear as the gNB to the UE, because these O-RAN nodes implement various aspects of the Nr-based radio interface protocol stack. Similarly, O-eNB appears as the eNB to the UE on the LTE-based radio interface. The UE can work with a gNB/eNB or O-RAN without any need to distinguish between the O-RAN-based or non-O-RAN-based configurations of the RAN.

2.4 Functional Split 7.2x

The 3GPP discussed multiple functional splits between a central unit (i.e., gNB-CU) and a distributed unit (i.e., gNB-DU) (3GPP 2005) as part of a study on the 5G NR radio access architecture and interfaces. The 3GPP decided to standardize functional split, where the gNB-CU implements RRC and PDCP (and SDAP), and the gNB-DU implements the RLC, the MAC, and the PHY layer (along with RF). As mentioned earlier in Section 2.1, the O-RAN ALLIANCE decided to utilize the functional split option 7 (or 7.2x to be more specific).

---

6 GPRS stands for General Packet Radio Service.

7 An en-gNB stands for EN-DC-gNB, which is a gNB that is operating in the E-UTRA-NR Dual Connectivity (EN-DC) configuration. EN-C is an example of dual connectivity, where a UE is communicating with an LTE eNB and a 5G gNB (=en-gNB).

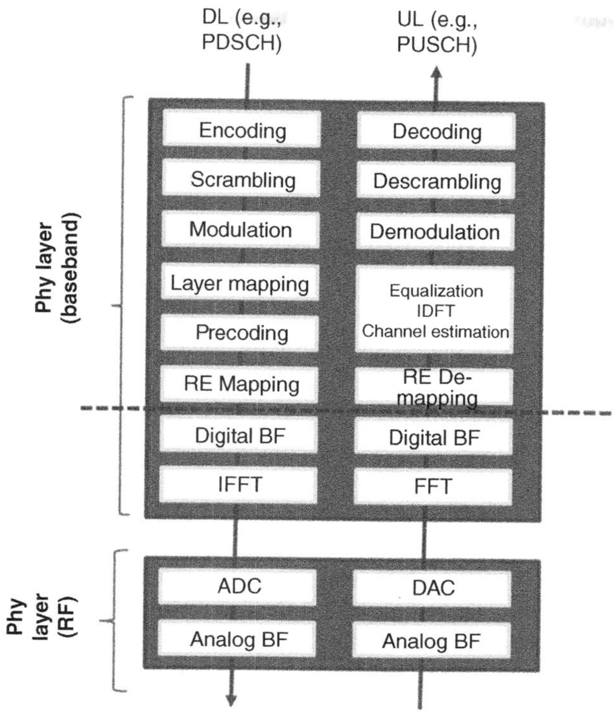

Figure 2.30 Functional Split 7.2x: A Closer Look.

Figure 2.30 shows the intra-PHY layer split selected by O-RAN (O-RAN ALLIANCE 2024k).

The Functional Split 7.2x involves the following physical layer functional split in the downlink (e.g., for PDSCH processing) (O-RAN ALLIANCE 2024k): (i) encoding, scrambling, digital modulation, layer mapping, precoding, and Resource Element (RE) mapping at the O-DU for the downlink and (ii) digital Beamforming (BF), IFFT, Analog-to-Digital Converter (ADC), and analog beamforming at the O-RU. Furthermore, the Functional Split 7.2x involves the following physical layer functional split in the uplink (e.g., for PUSCH processing): (i) analog beamforming, Digital-to-Analog Converter, FFT, and digital BF at the O-RU and (ii) RE de-mapping, equalization, Inverse Discrete Fourier Transform (if needed), demodulation, descrambling, and decoding at the O-DU.

There are two categories of the O-RU (O-RAN ALLIANCE 2024k). Category A O-RU does not have precoding and optionally supports digital and analog beamforming. Category B O-RU supports precoding and optionally supports digital and analog beamforming. Compression techniques may be used on the FH to compress the I/Q data, reducing the transport bandwidth requirements.

Per O-RAN ALLIANCE evaluation of various functional splits, Functional Split 7.2x provides a good balance among FH bandwidth requirements, RU implementation complexity, function extensibility, and multi-vendor interoperability. Functional Split simplifies the FH interface. The transfer of UP data is based on REs/Physical Resource Blocks (PRBs), which simplifies data mapping and limits the required control messages. The FH bandwidth is used efficiently, because the 7.2x interface is based on streams. Furthermore, only PRBs that contain user data need to be transported to reduce transport bandwidth. The Functional Split 7.2x supports different beamforming techniques such as digital, analog, and hybrid beamforming. Fewer user-specific parameters are used at split 7.2x compared to higher split options, simplifying specifications.

Functional Split 7.2x facilitates implementation of advanced receivers and coordination features (e.g., UL Coordinated Multi-Point), because many functions reside at the O-DU. Furthermore, this split option simplifies the O-RU design. Since more functions are at the O-DU and fewer functions are at the O-RU, new features can be introduced via software upgrades without requiring hardware changes at O-RU.

2.5 Inside O-RAN Domain: Functional Distribution Between O-RAN and Other SDOs

As discussed in Chapter 1, the design and deployment of O-RAN require the use of the work performed by non-O-RAN entities such as 3GPP, ETSI, IETF, CNCF, OpenStack, and ONAP. O-RAN specifications use the 3GPP specifications as baseline and introduces new O-RAN-specific functionalities in the 3GPP-defined NG-RAN Network Functions such as support for O-RAN-specific interfaces.

Before an operator can deploy O-RAN, it needs to select vendors to supply 3GPP-based O-RAN nodes such as the O-RU, O-DU, O-CU-CP, and O-CU-UP and non-3GPP-based O-RAN nodes such as the SMO (including the Non-RT-RIC) and the RT-RIC. Interoperability testing would need to be carried out for the selected set(s) of vendor products. One or more O-RAN Testing and Integration Centers (OTICs) can be used for such testing. Furthermore, testing in non-OTIC facilities such as the operator's test labs can also be helpful because the operator may have specific requirements beyond typical OTIC testing. Interoperability of O-RAN-based nodes with legacy RAN nodes would also need to be tested to ensure smooth interworking between them. Functional testing of xApps and/or rApps would also be needed. While challenging, performance testing of xApps and/or rApps individually and as a whole should also be carried out.

Since typical O-RAN nodes (except the O-RU) can be fully virtualized, suitable cloud infrastructure vendors and their systems would also need to be selected to verify the performance of O-RAN prior to the commercial deployment of O-RAN.

2.6 O-RAN Deployment Scenarios

This section presents an overview of O-RAN cloudification, hierarchical cloud architecture, and deployment scenarios. Few example mappings of O-RAN functions to the envisioned hierarchical cloud architecture are also illustrated.

2.6.1 Hierarchical Cloud Architecture

Figure 2.31 shows the hierarchical cloud architecture with cloud computing resources at various locations along with the cell sites (Dryjański 2023).

The Regional Cloud is placed at a central location, typically in a regional Data Center for an operator with large computing capabilities. The **Edge Cloud** is located close to the network edge, which facilitates placement of cloud computing resources close to the cell site to meet the latency requirements. The **Cell Site** is the location where the radio transmission and reception are carried out to provide cellular coverage for a desired area. Typical distances between elements for O-RAN architecture provided in Figure 2.31 are (i) less than 20 km between the cell site and closest Edge Cloud and (ii) greater than 200 km between the Regional Cloud and the closest Edge Cloud (O-RAN ALLIANCE 2024m).

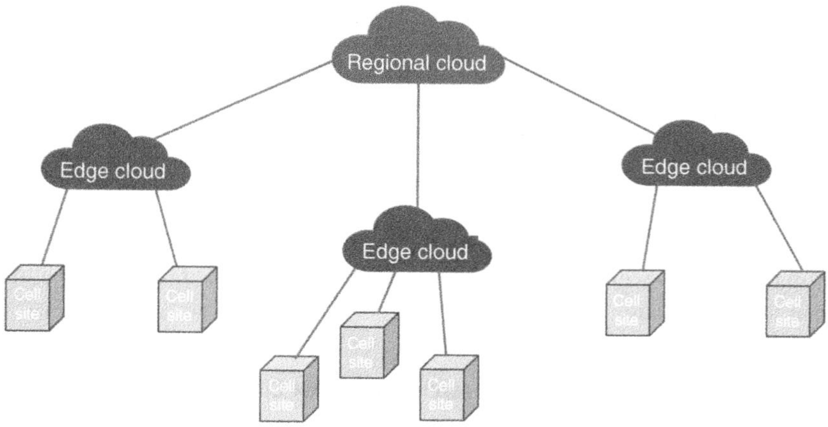

Figure 2.31 Cell Site, Edge Cloud, and Regional Cloud. Source: Dryjański, M. (2023). O-RAN Deployment Scenarios/Reprinted with permission from Rimedo Labs.

2.6.2 O-RAN Cloud Deployment Scenarios

Figure 2.32 identifies a list of O-Cloud deployment scenarios specified by the O-RAN ALLIANCE (O-RAN ALLIANCE 2024m). It displays six different O-RAN deployment scenarios, Scenarios A to F, explained below.

- **Scenario A.** The near-real-time RIC, the O-CU, and the O-DU are deployed at the edge cloud, while O-RUs are proprietary network elements or PNFs located at the cell site. This deployment provides the minimum delay, but the cost of such deployment will be the largest compared to other scenarios.
- **Scenario B.** The O-RU function is proprietary and hosted at the cell site as a PNF. The O-CU and the O-DU run on O-Cloud in an edge location, while the Near-RT RIC runs on a regional O-Cloud. This is the first focus of the O-RAN ALLIANCE for O-RAN deployments.

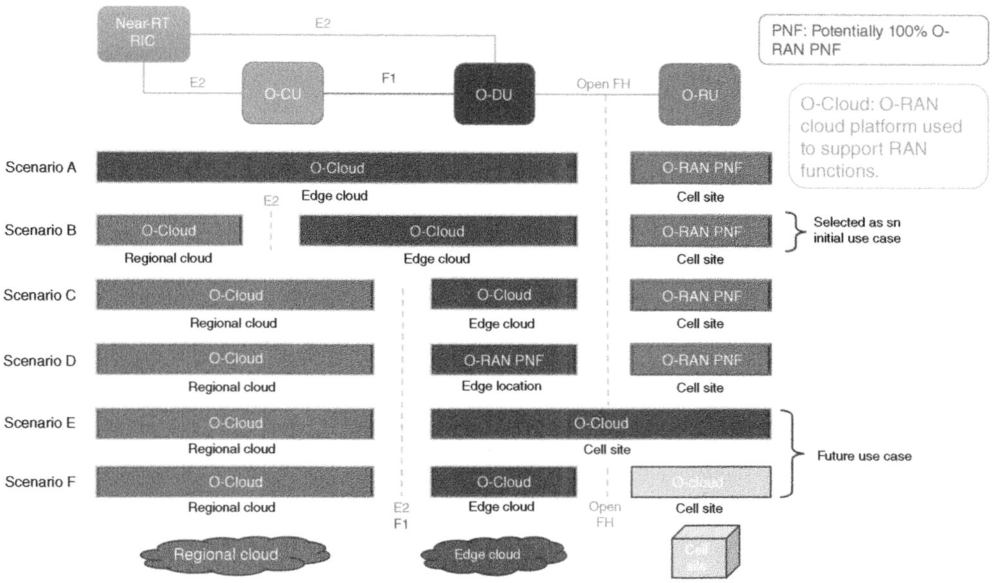

Figure 2.32 O-RAN Deployment Scenarios.

- **Scenario C.** Only the O-DU is at the edge, while the O-CU is moved to the regional O-Cloud. This saves on the cost and decreases energy consumption at the edge cloud.
- **Scenario D.** This setup is like Scenario C, except that the O-DU is realized as a PNF. This option enables the use of high-performing and optimized hardware for MAC/High-PHY layer processing.
- **Scenario E.** Everything is virtualized in Scenario E, including the O-RU (to the extent possible). In such a case, the O-DU and the O-RU can be put together at the cell site with the high-performing COTS hardware for PHY layer processing.
- **Scenario F.** The setup is similar to Scenario E, except that the O-RU is located at the cell site and the O-DU is located at the Edge cloud.

2.6.3 Example Mapping of O-RU Functions to Hierarchical O-RAN Cloud Architecture

Figure 2.33 shows an example mapping of O-RAN functions to O-Cloud architecture within the selected deployment options (Dryjański 2023).

The left part of Figure 2.32 shows the actual deployment of Scenario B, where the Near-RT RIC is placed in the regional cloud, the virtual or virtualized O-CU and the O-DU are placed at the edge cloud, and the O-RU is located at the cell site. This is appropriate, for example, for outdoor deployments in an urban scenario for low latency applications.

The right part of Figure 2.33 shows the deployment of Scenario C, where the O-DU sits at the edge cloud, while the O-CU is moved more centrally to a regional cloud. This is suitable, for example, for urban scenarios with enhanced Mobile Broadband (eMBB)/massive Machine Type Communication (mMTC) applications.

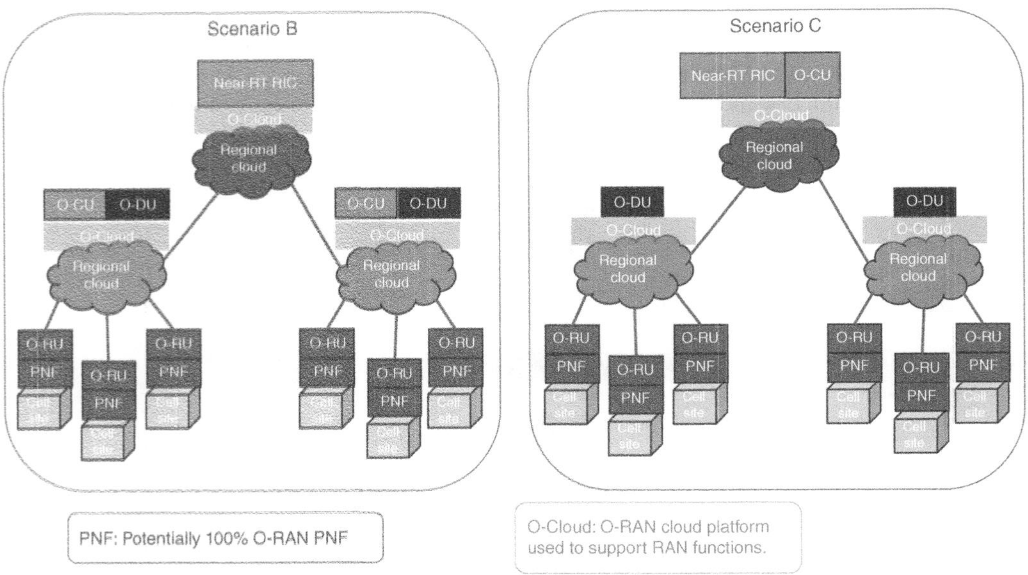

Figure 2.33 O-RAN Example Deployment Scenario Mapping. Source: Dryjański, M. (2023). O-RAN Deployment Scenarios/Reprinted with permission from Rimedo Labs.

2.7 Major Takeaways of the Chapter

The O-RAN ALLIANCE has defined a comprehensive RAN architecture that aims to achieve openness and intelligence. Here are the major takeaways of this chapter.

- Key nodes, network elements, functions, or components of O-RAN are SMO including Non-RT RIC, Near-RT RIC, O-CU-CP, O-DU, O-RU, O-eNB, and O-Cloud.
- O-RAN considers Near-RT-RIC, O-CU-CP, O-CU-UP, O-DU, and O-RU to be O-RAN Network Functions.
- Whether the 5G UE communicates with O-RAN-based RAN or non-O-RAN-based RAN is transparent to the UE, because the O-CU-CP, the O-CU-UP, the O-DU, and the O-RU together appear as the RAN to the UE by implementing suitable layers of the 3GPP-defined radio interface protocol stack.
- The SMO provides FCAPS interface to O-RAN Network Functions, houses the Non-RT RIC to facilitate RAN optimization, and carries out management, orchestration, and workflow management of the O-Cloud.
- The Non-RT RIC consists of the Non-RT RIC framework and rApps.
- The Non-RT RIC framework terminates the A1 interface and exposes a set of internal SMO services to rApps via the R1 interface. The Non-RT RIC framework provides rApps with AI/ML workflow that includes model training, inference, and relevant updates.
- rApps are Non-RT RIC applications that leverage the functionality exposed through the R1 interface to provide value-added services for RAN operations.
- The Near-RT RIC is a logical function that enables near-real-time control and optimization of RAN through suitable data collection and actions over the E2 interface.
- The O-CU-CP hosts the 3GPP-defined gNB-CU-CP containing RRC and PDCP-CP.
- The O-CU-UP hosts the 3GPP-defined gNB-CU-UP containing SDAP PDCP-UP.
- The O-DU hosts a large portion of the 3GPP-defined gNB-DU functions and contains RLC, MAC, and High-PHY layers based on the functional split 7.2x.
- The O-RU hosts lower PHY layer of the 3GPP-defined gNB-DU based on the functional split 7.2x.
- The O-eNB implements 3GPP-defined eNB or ng-eNB that supports O-RAN-defined interfaces such as the E2 interface toward the Near-RT RIC and the O1 interface toward the SMO.
- The O-Cloud is a cloud computing platform that hosts (i) O-RAN functions such as the Near-RT RIC, O-CU-CP, O-CU-UP, and O-DU, (ii) the supporting software components such as Operating System, VM Monitor and Container Runtime and (iii) suitable management and orchestration functions. It consists of physical infrastructure nodes.
- The O1 interface is between the management entity (e.g., SMO) and the O-RAN ME (e.g., the Near-RT RIC, The O-CU-CP, the O-CU-UP, the O-DU, the O-RU, and the O-eNB). The O1 interface supports numerous management services including provisioning management services, fault supervision MnSs, Performance Assurance MnSs, Trace MnSs, File MnSs, Heartbeat MnSs, PNF Startup and Registration MnSs, and PNF Software MnSs.
- The O2 interface is utilized by the SMO for the management of the O-Cloud infrastructure and the deployment LCM of O-RAN cloudified NFs implemented in the O-Cloud.
- The A1 interface is between the Non-RT-RIC residing in the SMO and the Near-RT RIC. The A1 interface supports three types of services to facilitate RAN optimization: A1 Policy MnS, A1 EI Service, and A1 ML Model MnS.

- The E2 interface is between the Near-RT RIC and an E2 node (e.g., the O-CU-CP, the O-CU-UP, the O-DU, and the O-eNB). The E2 nodes support all protocol layers and interfaces defined in 3GPP specifications related to (i) the eNB for E-UTRAN and (ii) gNB/ng-eNB for the NG-RAN. The Near-RT RIC uses the E2 interface to collect near real-time information on UEs and cells and provide value-added services (e.g., optimization of a RAN operation such as handover).
- The Open FH interface is between the O-DU and the O-RU. The Open FH interface includes (i) the CUS Plane and (ii) the M Plane. The M-Plane facilitates the initialization, configuration, and management of the O-RU in support of the 7.2x functional split. The CUS-Plane facilitates control of Low-PHY in the O-RU by the O-DU; transfer of I/Q samples between the O-DU and the O-RU; and time, frequency, and phase synchronization between the clocks of O-DUs and the O-RU in support of the 7.2x functional split.
- The O-Cloud Notification interface is between the O-Cloud infrastructure and a Network Function deployed in the O-Cloud. It enables an EC (e.g., an O-DU) deployed on the O-Cloud to subscribe to events and status from the O-Cloud. The O-Cloud provides EPs to enable cloud workloads to receive events and status that might be known only to the O-Cloud infrastructure.
- The CTI is an optional interface between O-DUs and TNs of a packet-based transport network. Such transport network connects an O-DU to different O-RUs.
- The design and deployment of O-RAN require the use of the work performed by non-O-RAN entities such as 3GPP, ETSI, IETF, CNCF, OpenStack, and ONAP and a suitable cloud infrastructure.
- O-RAN can be implemented using a hierarchical cloud using the regional cloud, the edge cloud, and the cell site. The Regional Cloud is placed at a central location, typically in a regional Data Center. The Edge Cloud is located close to the network edge, which facilitates placement of cloud computing resources close to the cell site to meet the latency requirements. The Cell Site is the location where the radio transmission and reception are carried out to provide cellular coverage for a desired area. O-RAN has studied six deployment scenarios with different placement of O-RAN nodes in the hierarchical cloud infrastructure.

References

3GPP (2005). *Study on New Radio Access Technology: Radio Access Architecture and Interfaces.* TR38.801. 3GPP, March 2017, https://portal.3gpp.org/desktopmodules/Specifications/SpecificationDetails.aspx?specificationId=3056 (accessed 24 July 2024).

Dryjański, M. (2023). *O-RAN Deployment Scenarios.* 2 July, https://rimedolabs.com/blog/o-ran-deployment-scenarios/ (accessed 24 July 2024).

IEEE (2023). *IEEE Standard for Radio over Ethernet Encapsulations and Mappings.* IEEE 1914.3-2023. IETF. https://standards.ieee.org/ieee/1914.3/10629/ (accessed 24 July 2024).

IETF (2005). *Securing FTP with TLS.* RFC 4217. IETF, October 2005, https://datatracker.ietf.org/doc/html/rfc4217 (accessed 24 July 2024).

IETF (2011). *Using the NETCONF Protocol over Secure Shell (SSH).* RFC 6242. IETF, June 2011, https://datatracker.ietf.org/doc/html/rfc6242 (accessed 24 July 2024).

IETF (2015). *Using the NETCONF Protocol over Transport Layer Security (TLS) with Mutual X.509 Authentication.* RFC 7589. IETF, June 2015, https://datatracker.ietf.org/doc/html/rfc7589 (accessed 24 July 2024).

IETF (2018). *Using the NETCONF Protocol over Transport Layer Security (TLS) with Mutual X.509 Authentication.* RFC 8446. IETF, August 2018, https://datatracker.ietf.org/doc/html/rfc8446 (accessed 24 July 2024).

O-RAN ALLIANCE (2024a). *O-RAN Operations and Maintenance Interface*. O-RAN.WG1.O1-Interface.0-v04.00. The O-RAN ALLIANCE. https://specifications.o-ran.org (accessed 24 July 2024).

O-RAN ALLIANCE (2024b). *O-RAN Architecture Description*. O-RAN.WG1.OAD-R003-v09.00. The O-RAN ALLIANCE. https://specifications.o-ran.org (accessed 24 July 2024).

O-RAN ALLIANCE (2024c). *O-RAN Operations and Maintenance Architecture*. O-RAN.WG1.OAM-Architecture-v04.00. The O-RAN ALLIANCE. https://specifications.o-ran.org (accessed 24 July 2024).

O-RAN ALLIANCE (2024d). *A1 Interface: General Aspects and Principles*. O-RAN.WG2.A1GAP-R003-v03.01. The O-RAN ALLIANCE. https://specifications.o-ran.org (accessed 24 July 2024).

O-RAN ALLIANCE (2024e). *A1 Interface: Transport Protocol*. O-RAN.WG2.A1TP-R003-v02.01. The O-RAN ALLIANCE. https://specifications.o-ran.org (accessed 24 July 2024).

O-RAN ALLIANCE (2024f). Non-RT RIC: Architecture. O-RAN.WG2.Non-RT-RIC-ARCH-R003-v05.00. The O-RAN ALLIANCE. https://specifications.o-ran.org (accessed 1 October 2024).

O-RAN ALLIANCE (2024g). *E2 General Aspects and Principles (E2GAP)*. O-RAN.WG3.E2GAP-R003-v04.01. The O-RAN ALLIANCE. https://specifications.o-ran.org (accessed 24 July 2024).

O-RAN ALLIANCE (2024h). *Near-RT RIC Architecture*. O-RAN.WG3.RICARCH-R003-v04.00. The O-RAN ALLIANCE. https://specifications.o-ran.org (accessed 24 July 2024).

O-RAN ALLIANCE (2024i). *Cooperative Transport Interface- Transport Control Plane Specification*. O-RAN.WG4.CTI-TCP.0-R003-v04.00. The O-RAN ALLIANCE. https://specifications.o-ran.org (accessed 24 July 2024).

O-RAN ALLIANCE (2024j). *Cooperative Transport Interface- Transport Management Procedures Specification*. O-RAN.WG4.CTI-TMP.0-R003-v04.00. The O-RAN ALLIANCE. https://specifications.o-ran.org (accessed 24 July 2024).

O-RAN ALLIANCE (2024k). *Control, User and Synchronization Plane Specification*. O-RAN.WG4.CUS.0-R003-v12.00. The O-RAN ALLIANCE. https://specifications.o-ran.org (accessed 24 July 2024).

O-RAN ALLIANCE (2024l). *Management Plane Specification*. O-RAN.WG4.MP.0-R003-v12.00. The O-RAN ALLIANCE. https://specifications.o-ran.org (accessed 24 July 2024).

O-RAN ALLIANCE (2024m). *Cloud Architecture and Deployment Scenarios for O-RAN Virtualized RAN*. O-RAN.WG6.CADS-v04.00. The O-RAN ALLIANCE. https://specifications.o-ran.org (accessed 24 July 2024).

O-RAN ALLIANCE (2024n). *O2 Interface General Aspects and Principles*. O-RAN.WG6.O2-GA&P-R003-v03.00. The O-RAN ALLIANCE. https://specifications.o-ran.org (accessed 1 October 2024).

O-RAN ALLIANCE (2024o). *O-RAN O-Cloud Notification API Specification for Event Consumers*. O-RAN.WG6.O-Cloud Notification API-v04.00. The O-RAN ALLIANCE. https://specifications.o-ran.org (accessed 24 July 2024).

3

O-RAN Applications: xApps and rApps

This chapter describes three key control loops within the Open Radio Access Network (O-RAN) architecture that play a crucial role in dynamically optimizing network performance at different time scales. The chapter delves into the details of each O-RAN control loop, its essential components, and microservices involved such as extended Applications (xApps) and remote Apps (rApps). These microservices enable the decomposition of complex RAN functionalities into modular and independently deployable units. This control loop and microservices paradigm shift not only empowers network operators to adapt swiftly to evolving demands but also promotes a more cost-effective and adaptable approach to building and maintaining telecommunications infrastructure. As part of control loop discussions, O-RAN RAN Intelligent Controllers (RICs) such as the non-real-time (non-RT) RIC and the Near-Real-Time (Near-RT) RIC and O-RAN applications such as xApps and rApps are also discussed extensively. Potential enhancements to the O-RAN architecture such as the real-time RIC and its microservices often referred to as micro applications (μApps) or distributed applications (dApps) are also summarized.

3.1 O-RAN Control Loops

The O-RAN architecture supports three major types of control loops that involve various O-RAN functions: (i) non-RT control loops, (ii) Near-RT control loops, and (iii) real-time (RT) control loops. Figure 3.1 shows time scales of these three O-RAN control loops in the O-RAN architecture (O-RAN ALLIANCE 2024a).

As seen in Figure 3.1, a typical execution time for the non-RT control loop is one second or longer. A Near-RT control loop executes in the order of 10 ms to 1 second. An RT control loop operates within 10 ms. For example, RT control loops in the E2 nodes operate quite fast to perform tasks such as radio scheduling in the O-RAN-Distributed Unit (O-DU). Similarly, a Near-Real-Time RAN intelligent controller (Near-RT RIC) enables near RT control and optimization of functions and resources of E2 nodes using Near-RT control loops that support fine-grained data collection and actions over the E2 interface.

A control loop host such as the Near-RT RIC, the non-RT RIC, and the O-DU acts as the controlling entity and interacts with other logical nodes or network functions. Note that the control loops exist at various levels and operate simultaneously. The three control loops may or may not interact with each other based on the target O-RAN use case. Examples of Non-RT and Near-RT control loops and the interaction between the RICs are defined in O-RAN ALLIANCE (2024b). For example, interactions among the O-RAN Central Unit-Control Plane (O-CU-CP), the O-DU, and the Service Management and Orchestration (SMO) related to call control and mobility, radio

Fundamentals of O-RAN, First Edition. Nishith D. Tripathi and Vijay K. Shah.

Companion website: www.wiley.com/go/tripathioran

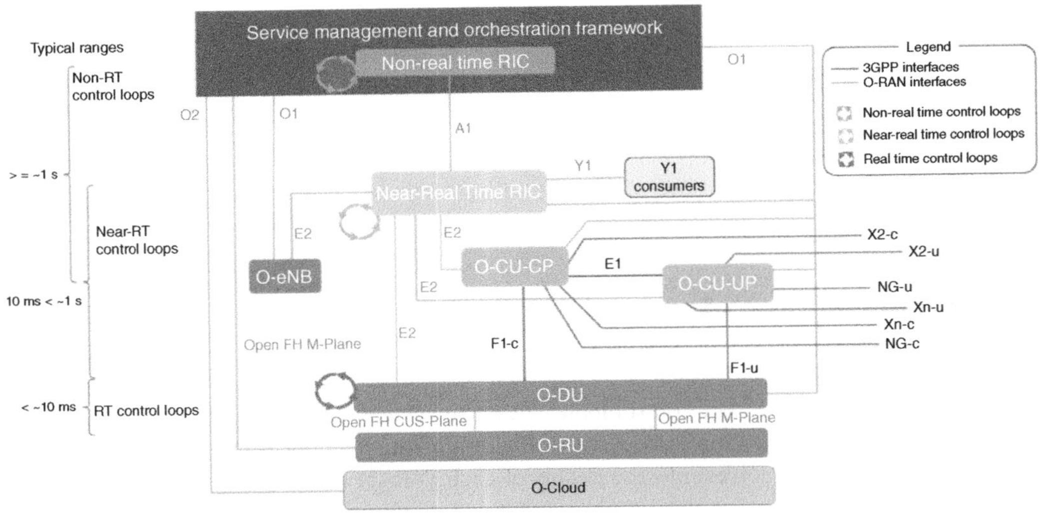

Figure 3.1 O-RAN Control Loops. Source: O-RAN WG1.OAD-2024. Reproduced with permission.

scheduling, Hybrid Automatic Repeat reQuest, and beamforming are also discussed in O-RAN ALLIANCE (2024b). Note that the exact timing of the control loops depends on the use case. A stable solution for a given use case requires the loop time in the non-RT RIC and the SMO management plane processes to be significantly longer than the loop time for such use case in the control entities such as the Near-RT RIC.

3.2 Near-RT Control Loop: Near-RT RIC and xApps

Let us now delve into the details of the Near-RT control loop host, Near-RT RIC along with its interplay with its microservices termed xApps and the relevant E2 interface. The Near-RT RIC plays a pivotal role in the control and optimization of the RAN, harnessing the powerful capabilities facilitated by the E2 interface. At its core, the Near-RT RIC platform serves as the hub for several critical elements. This includes hosting the terminations of three pivotal interfaces—O1, A1, and E2. In addition to supporting these interfaces, it also provides a home for the xApps themselves. Furthermore, the platform houses various components necessary for the seamless execution and management of these xApps, ensuring the efficient operation of the RAN optimization processes.

3.2.1 Near-RT Internal Components of the Near-RT RIC

Figure 3.2 provides an overview of a typical Near-RT RIC architecture. Additional details of the Near-RT RIC can be found in Section 2.2.2 in Chapter 2. Here, we will discuss the primary components that form the backbone of this architecture (O-RAN Working Group 3 2024).

- **Internal Messaging Infrastructure (IMI).** The IMI serves as the connective tissue, facilitating communication between xApps, platform services, and interface terminations. While the specifications do not mandate a specific technology although the O-RAN Software Community (OSC) employs a custom library known as RIC Message Router (RMR), they do specify the necessary requirements and functionalities. IMI must support endpoint registration,

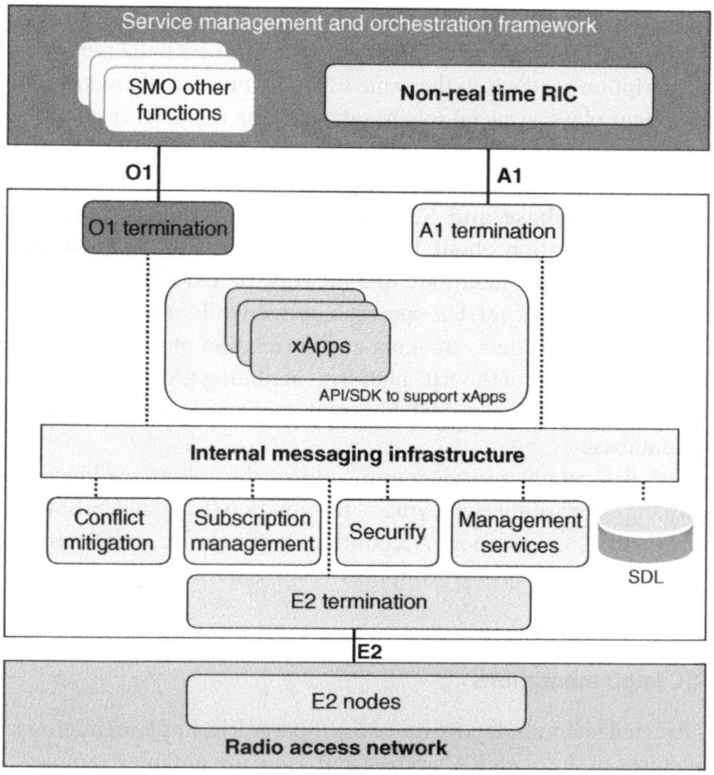

Figure 3.2 Near-RT RIC Architecture.

discovery, and deletion (covering both internal RIC components and xApps). The IMI provides application programming interfaces (APIs) for message transmission and reception, supporting point-to-point communication or publish/subscribe mechanisms. The IMI also offers routing and robustness mechanisms to prevent internal data loss.

- **Conflict Mitigation.** This crucial component addresses potential conflicts that can arise among different xApps. Diverse and independent xApps may apply conflicting configurations while pursuing separate optimization objectives, potentially leading to performance degradation. These conflicts can pertain to users, bearers, or cells and may involve any control actions executed by the RIC. The specifications highlight three classes of conflicts.
  - **Direct Conflicts.** These conflicts are readily detectable by the internal component. Examples include multiple xApps attempting to apply different settings for the same parameters within the same control target or xApps requesting more resources than are available. The conflict mitigation component resolves these issues, determining which xApp takes precedence or constrains the scope of the control action (preaction resolution).
  - **Indirect and Implicit Conflicts.** These conflicts cannot be directly observed and may or may not depend on relationships among various xApps. For instance, configurations that optimize the performance of one class of users may inadvertently degrade others in less obvious ways. These conflicts are detected and mitigated through postaction verification by monitoring system performance after applying different control policies. It is worth noting that as of the time of this writing, conflict mitigation is not included in the OSC Near-RT RIC.

- **Subscription Manager.** This functionality enables xApps to establish connections with functions accessible through the E2 interface. It also governs individual xApps' access to E2 messages and can consolidate identical subscription requests to the same E2 node into a single request.
- **Security Subsystem.** This component plays a crucial role in safeguarding the RAN from malicious xApps that could compromise sensitive RAN data or affect RAN performance. Detailed aspects of this component are reserved for future studies.
- **Network Information Base (NIB) Database and Shared Data Layer (SDL) API.** The RAN NIB (R-NIB) database stores information about E2 nodes, while the user equipment (UE)-NIB maintains entries for UEs and their identities. UE identity, or UE-ID, is a critical yet sensitive piece of information, as it allows for UE-specific control while also potentially exposing user-sensitive data. The UE-NIB facilitates tracking and correlation of user identities across different E2 nodes. Various components of the RIC platform, including xApps, can query the database using SDL APIs. The OSC RIC offers an SDL library that can be incorporated into xApps, along with a Redis-based database.
- **xApp Management.** The Near-RT RIC provides services and APIs for the automated lifecycle management of xApps, encompassing onboarding, deployment, termination (triggered by SMO), as well as tracing and logging for Fault, Configuration, Accounting, Performance, and Security (FCAPS). In the OSC RIC, this is accomplished through wrappers integrated with the Kubernetes infrastructure.

3.2.2 Open-Source Near-RT RIC Implementations

The open-source community has fostered a vibrant landscape of third-party RIC implementations, all of which contribute to the robustness of the open RAN ecosystem. Here are notable examples.

- **Open Software Community RIC** (Bimo et al. 2022). OSC is an open-source project, collaborated by O-RAN ALLIANCE and Linux Foundation, aims to develop reference software components based on Third Generation Partnership Project (3GPP) and O-RAN ALLIANCE specifications. Out of 12 projects, OSC has developed an O-RAN specifications compliant RIC platform that supports both Near-RT RIC and non-RT RIC and key interfaces, such as, E2 and A1.
- **Microservices-Driven Open Networking Operating System (ONOS)-Based RIC** (ONF White Paper 2020). The SD-RAN project, championed by the Open Networking Foundation (ONF), is actively crafting an open source and cloud-native embodiment of the O-RAN Near-RT RIC. This endeavor encompasses not only the RIC itself but also a robust collection of xApps, vital for orchestrating the RAN. Complementing these elements is a software development kit (SDK), designed to streamline the creation of novel xApps. These xApps seamlessly blend standard compliant E2 Service Models (E2SMs) with bespoke service models cultivated within the SD-RAN community. This RIC framework, underpinned by the ONOS controller, champions a microservices architecture. It boasts a rich array of services, including xApp subscription services, network- and user-centric information services, distributed data store services ensuring high availability, and operator services. This modular approach enhances flexibility and scalability while accommodating diverse RAN orchestration scenarios.
- **FlexRIC** (Schmidt et al. 2021). FlexRIC adopts a monolithic architecture for the Near-RT RIC and introduces an RAN agent to interface with the OpenAirInterface radio stack. This implementation is instrumental in facilitating the monitoring and slicing programmability use cases. Furthermore, it equips developers with an SDK, empowering the creation of tailored, service-oriented controllers.

- **5G-EmPOWER** (Coronado et al. 2019). Tailored for heterogeneous RAN environments, 5G-EmPOWER stands out with its unique, nonstandard compliant functionalities, including mobility management spanning Wi-Fi and cellular networks, support for multitenancy, and the ability to deploy custom resource allocation schemas within network slices. Such innovative capabilities broaden the spectrum of use cases that this Near-RT RIC can address.

These diverse implementations collectively amplify the O-RAN landscape, offering choice and innovation to developers and operators seeking to craft tailored solutions to meet the dynamic demands of modern telecommunications.

3.3 Near-RT RIC xApps: Enabling RAN Customization

At the heart of the Near-RT RIC lies its most vital component: the xApps. As shown in Figure 3.3, these xApps are the dynamic agents that empower the RIC with specialized capabilities, ranging from RT RAN data analysis to intricate RAN control maneuvers. Each xApp functions as a plug and play module, seamlessly integrating custom logic into the RAN orchestration.

xApps play a pivotal role in the orchestration of the RAN. They are designed to effortlessly receive and interpret data and telemetry from the RAN, allowing them to make informed decisions and execute control actions. These actions are communicated back to the RAN through the E2 interface, thereby influencing the network's behavior and performance.

3.3.1 Defining xApps

According to the O-RAN specifications (O-RAN Working Group 3 2024), an xApp is defined by two fundamental elements: the **xApp descriptor** and the **xApp software image**. The xApp descriptor, typically a YAML or JSON file, encapsulates essential information for xApp management.

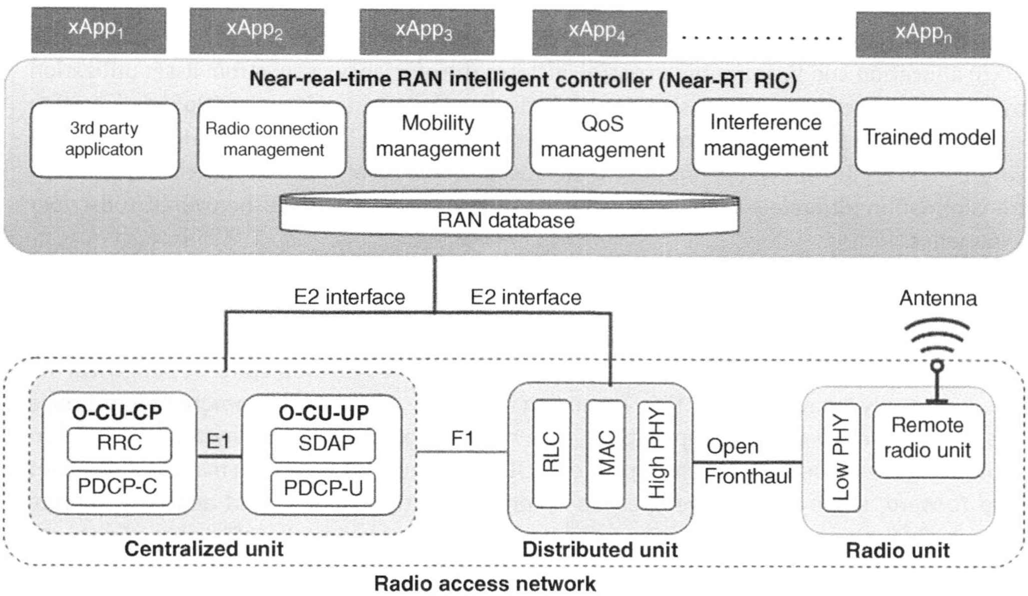

Figure 3.3 Extended Application (xApp).

It encompasses crucial aspects such as autoscaling policies, deployment instructions, deletion protocols, and upgrade guidelines. Additionally, this descriptor specifies the data types that the xApp processes and details its control capabilities. In the OSC RIC, an xApp is represented by a Docker image, which can be seamlessly deployed within a Kubernetes infrastructure. This deployment process is guided by the descriptor scheme, a file specifying container attributes. Refer to Section 3.8 for a detailed deployment and workflow of a typical xApp within Near-RT RIC.

3.3.2 Embracing Evolving Standards

As of Fall 2024, O-RAN specifications prescribe a limited set of APIs that the Near-RT RIC platform must furnish to xApps. This includes the provision of SDL APIs and APIs for registration, discovery, and subscription. However, there is a growing recognition that a broader spectrum of APIs could significantly enhance the potential of xApps. By consolidating these APIs into a comprehensive SDK, developers could create xApps that effortlessly adapt across various Near-RT RIC implementations. This promising direction has gained traction through the initiatives of the Telecom Infra Project (TIP) RAN Intelligence and Automation subgroup (Telecom Infra Project 2024). These efforts are poised to facilitate a more vibrant ecosystem of adaptable xApps, advancing RAN customization and performance optimization.

How xApp evolve?

The O-RAN architecture has undergone a transformative evolution with the integration of a software-defined networking (SDN) approach, reshaping the traditional separation between user plane (UP) and control plane (CP) functions. In a departure from the closed and vendor-specific control planes of traditional RAN vendors, O-RAN establishes a centralized network intelligence hub within a dedicated network element.

In the conventional RAN protocol stack, reliance on proprietary control algorithms, including critical functions such as radio resource control (RRC) for admission control and mobility management, as well as medium access control (MAC) functions like physical resource block (PRB) allocation and beamforming control, dictates the efficiency of spectrum asset utilization by operators. These algorithms significantly influence the quality of service (QoS) delivered to subscribers. The SDN approach of open RAN revolutionizes this landscape, providing operators with increased flexibility, openness, and advanced control over network functionalities. This transformation ultimately optimizes resource utilization and enhances the overall subscriber experience.

The maturation of O-RAN has seen the O-RAN ALLIANCE embrace a software-defined approach, enabling external developers to program the Near-RT RAN control plane. This specific subset of control algorithms accommodates control loop delays ranging from 10 ms to 1 second. At the heart of this capability within the O-RAN architecture lies the Near-RT RIC, commonly known as the RIC. As shown in Figure 3.4, external programmers utilize this architecture to develop applications, referred to as **xApps**, which play a pivotal role in controlling the algorithms governing the open RAN system. This evolution marks a significant leap forward, empowering developers to contribute to the dynamic and adaptable nature of open RAN, thereby fostering innovation and further improving the efficiency of network operations.

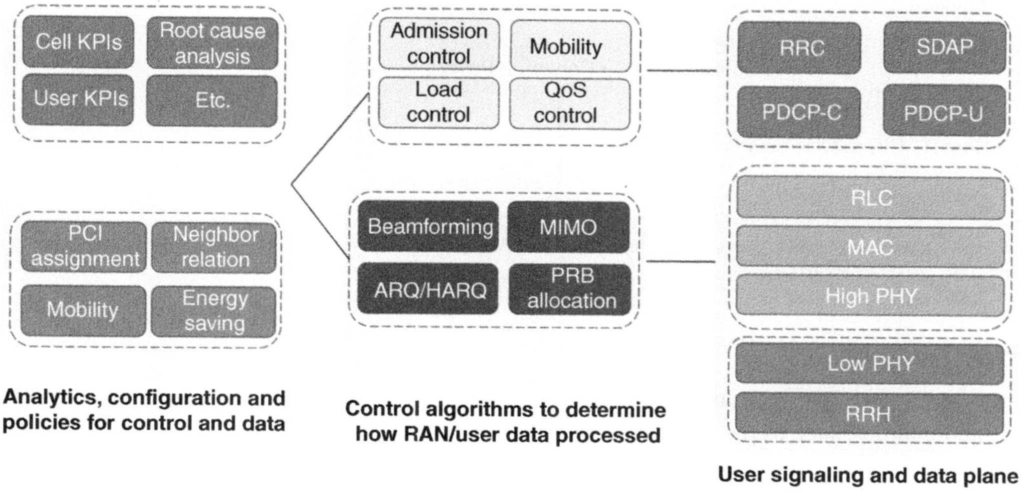

Figure 3.4 xApps: Enablers for Implementation of Control Algorithms.

> **xApp and SON Same?**
>
> **SON** is a Self Organizing Network. Functions of the SON and **xApps** are similar. xApps can be utilized as the next generation of the SON for O-RAN. xApps can be viewed as an enhanced SON that provide flexible, standards-based, interoperable, multivendor and programmable feature applications for **RAN optimization**.

3.4 E2 Interface: Empowering RAN Control and Data Exchange

The E2 interface serves as a critical conduit connecting two pivotal endpoints: the Near-RT RIC and the designated E2 nodes, which encompass Distributed Units (DUs), Central Units (CUs), and O-RAN compliant long-term evolution (LTE) evolved Node Bs (eNBs). This interface empowers the RIC with the authority to manage radio resource allocation and other essential functions within the E2 nodes. Furthermore, it facilitates the seamless collection of metrics from the RAN to the Near-RT RIC, whether through periodic updates or in response to predefined triggers. These control and data collection procedures can be applied to single or multiple cells, network slices, specific QoS classes, or individual UEs. The E2 interface operates atop the Stream Control Transmission Protocol (SCTP) protocol.

To enable these operations, the O-RAN ALLIANCE employs a range of unique identifiers. Specifically, O-RAN incorporates identifiers rooted in 3GPP specifications for next-generation-NodeBs (gNBs), network slices, and QoS classes. In the context of specific UEs, the O-RAN ALLIANCE introduces a common user identifier (i.e., UE-ID), ensuring a consistent and uniform user identity system wide without exposing sensitive user-related information.

The E2 interface is logically structured around two key protocols.

1) **E2 Application Protocol (E2AP)** (O-RAN Work Group 3 2024a). E2AP serves as the fundamental procedural protocol that orchestrates communication between the Near-RT RIC and E2 nodes. It provides a foundational set of services, illustrated in Figure 3.5. E2AP messages can encapsulate various E2SMs, which define specific functionalities.

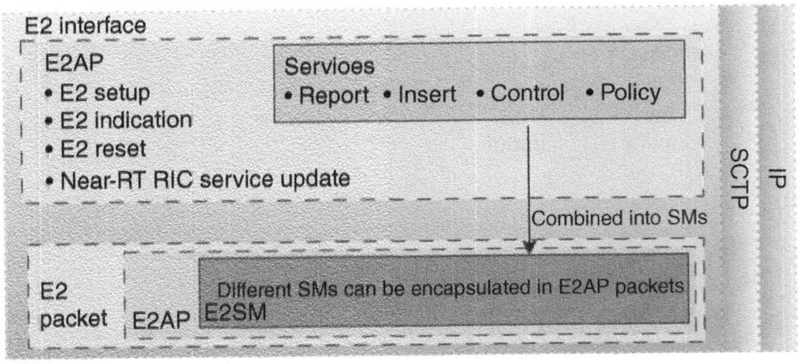

Figure 3.5 Illustration of E2AP Packet.

2) **E2SM** (O-RAN Work Group 3 2024b). E2SMs are instrumental in implementing specialized functionalities, such as the reporting of RAN metrics or the control of RAN parameters.

Each E2 node exposes a plethora of RAN functions, reflecting the services and capabilities it supports. For instance, DUs from different vendors may offer distinct control parameters and functionalities, as well as varied capabilities for collecting and reporting performance metrics. By adopting publish–subscribe mechanisms, E2 nodes can publish their data, while xApps within the Near-RT RIC can subscribe to one or more of these RAN functions via the E2 interface. This approach effectively segregates the capabilities of each node and dictates how xApps interact with the RAN.

3.4.1 E2 Application Protocol (E2AP)

At the foundational level, E2AP manages interface operations (establishment, reset, error reporting for the E2 interface itself) and updates regarding Near-RT RIC services. As an illustrative example, **the E2 setup procedure** is depicted in Figure 3.6. Initially, an SCTP connection is established between the Near-RT RIC and the E2 node, with the E2 node being aware of the IP address and port of the Near-RT RIC's E2 termination. Subsequently, the E2 node transmits an E2 setup request, listing the supported RAN functions, configurations, and identifiers. The Near-RT RIC processes this information and responds with an E2 setup response.

As defined in E2 General Aspects and Principles (O-RAN.WG3.E2GAP 2024), E2AP offers five services, which can be combined in various ways to create E2SMs (O-RAN Work Group 3 2024b).

a) **E2 Report.** This service is used to expose RAN control and UE context-related information. The E2 report service activates upon subscription by an xApp to a specific E2 node function. During the subscription negotiation, the Near-RT RIC's xApp can specify trigger events or the frequency with which the E2 node should transmit report messages. Based on this frequency, a timer is initiated within the E2 node, and a report is dispatched when the timer elapses. The RIC indication message is of the report type. For E2SM-RAN Control (E2SM-RC) discussed in Section 3.4.2, supported RIC REPORT services include message copy, call process outcome, E2 node information, and UE information (O-RAN Work Group 3 2024d).
b) **E2 Insert.** The insert service involves messages transmitted from an E2 node to an xApp within the Near-RT RIC, notifying the xApp about specific events within the E2 node (e.g., a UE signaling readiness for handover). Activation of this service occurs upon subscription by an xApp and involves an RIC indication message (of the insert type). In this scenario, the trigger is associated with an RAN radio resource management procedure, which is paused when the insert message is dispatched. A waiting timer is initiated as well, and if the RIC fails to respond

Figure 3.6 Setup Procedure for E2 Session Between the Near-RT RIC and an E2 Node.

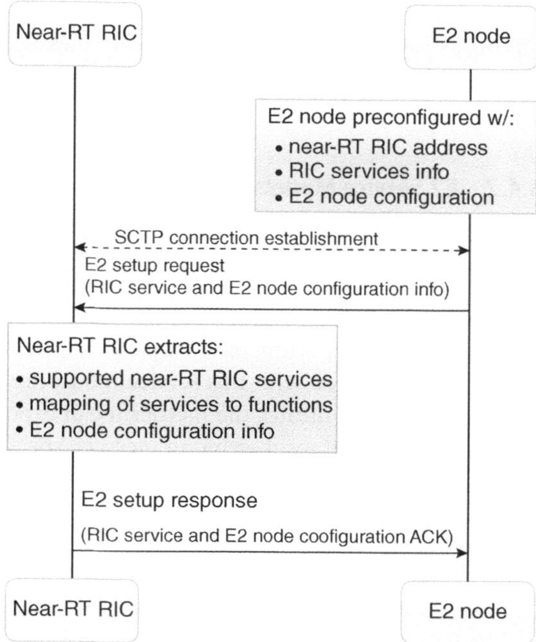

before the timer expires, the procedure within the E2 node may resume or be definitively halted. For E2SM-RC discussed in Section 3.4.2, supported E2 insert services include (i) Radio Bearer Control Request, (ii) Radio Resource Allocation Control Request, (iii) Connected Mode Mobility Request, (iv) Radio Access Control Request, (v) Dual Connectivity (DC) Control Request, (vi) Carrier Aggregation (CA) Control Request, (vii) Idle Mode Mobility Control Request, and (viii) Multiple Actions Control Request. Refer to O-RAN E2SM specification (O-RAN Work Group 3 2024b) for more details.

c) **E2 Control.** The control service can be initiated autonomously by the RIC or triggered by the receipt of an insert message at the Near-RT RIC. This service relies on a two-message procedure, comprising an RIC Control Request from the RIC to the E2 node and an RIC Control Acknowledge in the reverse direction. The control service holds the potential to influence parameters exposed by the RAN functions of the E2 node. For E2SM-RC discussed in Section 3.4.2, supported RIC CONTROL services include (i) Radio Bearer Control, (ii) Radio Resource Allocation Control, (iii) Connected Mode Mobility, (iv) Radio Access Control, (v) DC Control, (vi) CA Control, (vii) Idle Mode Mobility Control, (viii) UE Information and Assignment, (ix) Multiple Actions Control, (x) Measurement Reporting Configuration Control, and (xi) Beamforming Configuration Control.

d) **E2 Policy.** This service entails a subscription procedure specifying (i) an event trigger and (ii) a policy that the E2 node should autonomously follow to execute radio resource management. For E2SM-RC discussed in Section 3.4.2, supported RIC Policy services include (i) Radio Bearer Control, (ii) Radio Resource Allocation Control, (iii) Connected Mobility Control, (iv) Radio Access Control, (v) DC Control, (vi) CA Control, (vii) Idle Mode Mobility Control, (viii) Measurement Reporting Configuration Control, and (ix) Beamforming Configuration Control.

e) **E2 Query.** This service allows the RIC to request specific information from the E2 node. The information requested can be related to the E2 node's configuration, status, or any other relevant data. For E2SM-RC discussed in Section 3.4.2, support RIC Query services include (i) E2 node information query and (ii) UE information query.

These services are then combined to formulate a service model, with the service model message serving as payload within one of the E2AP messages, as depicted in Figure 3.5. The actual content is encoded using ASN.1 notation, employing well-defined types for numerical values and key-value pairs.

3.4.2 E2 Service Models

The O-RAN ALLIANCE Working Group 3 has standardized three pivotal service models, each playing a distinct role in enhancing RAN capabilities.

1) **E2SM Key Performance Metrics (KPM)** (O-RAN Work Group 3 2024c). E2SM KPM specializes in the reporting of RAN performance metrics through the E2 report service. The procedures associated with this service model are elucidated in Figure 3.7. During E2 setup, the E2 node communicates the metrics it can expose. Subsequently, an xApp within the Near-RT RIC can transmit a subscription message specifying the desired KPMs and whether reporting should be periodic or trigger-based. The E2 node then employs E2 Indication messages of the report type to stream the selected KPMs. Notably, distinct KPM messages are generated for different E2 nodes, encompassing performance metric containers with diverse fields for DUs, CU-UPs, and CU-CPs. Additionally, the latest Version 2 of the KPM specifications has introduced cell-specific and user-specific performance containers, rooted in metrics defined within 3GPP technical specifications for LTE and NR.
2) **E2SM Network Interfaces (NI)** (O-RAN Work Group 3 2024f). E2SM NI plays a pivotal role in routing messages received by the E2 node on specific network interfaces to the Near-RT RIC domain via E2 report messages over the E2 interface. During the subscription process, the E2

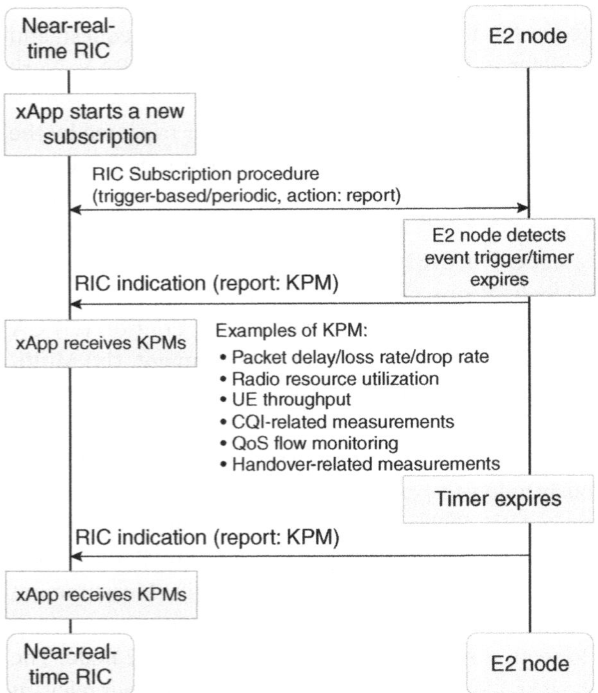

Figure 3.7 Procedures Related to Sending of KPMs from an E2 Node to the Near-RT RIC.

node advertises the interfaces it supports, encompassing X2 (connecting LTE eNBs), Xn (connecting various NR gNBs), and F1 (connecting DUs and CUs).

3) **E2SM Cell Configuration and Control (CCC)** (O-RAN Work Group 3 2024e). RAN function CCC performs the following functionalities: Exposure of node-level and cell-level configuration information, initiate control, and/or configuration of node-level and cell-level parameters.
4) **E2SM RAN Control (RC)** (O-RAN Work Group 3 2024d). One of the central goals of xApps is to optimize radio resource management within E2 nodes. E2SM RC, introduced in July 2021, embodies control functionalities through E2 control services. The initial release of the E2SM RC specifications encompasses a broad spectrum of control domains.
   - **Radio Bearer Control.** Modifying parameters for bearers of an E2 node or specific UEs, such as QoS parameters, bearer admission control, split bearers, and Packet Data Convergence Protocol (PDCP) duplication control.
   - **Radio Resource Allocation Control.** Adjusting elements like discontinuous reception and semipersistent scheduling for slice-level PRBs across E2 nodes, cells, slices, UEs, or QoS classes.
   - **Connected Mode Mobility Control.** Initiating mobility procedures for UEs in RRC connected states, including handovers to specific cells or conditional handovers to sets of cells.
   - **Radio Access Control.** Configuring parameters for random access back off, UE admission to cells, and more.
   - **DC Control.** Configuring and triggering UE handovers to selected target secondary cells, secondary cell updates, or DC releases.
   - **CA Control.** Initiating CA and altering component carriers for specific UEs.
   - **Idle Mobility Control.** Modifying functions related to mobility procedures for UEs in an RRC idle state, including cell reselection priorities and idle timers.

E2SM RC also equips the system with UE identification and UE information reporting capabilities. The control actions and policies specified by E2SM are linked to specific parameters standardized by 3GPP, with the control action often conveying the value of a 3GPP Information Element (IE) outlined in the E2SM specifications. For instance, the handover control IE includes, among other things, a target cell ID encoded as a 3GPP Target Cell GlobalID IE.

Effectively implementing control actions and policy enforcement, E2SM leverages a combination of the E2 services discussed in Section 3.4.2. Figure 3.8 provides an illustrative example of the E2 messages that can be exchanged between an E2 node and an xApp within the Near-RT RIC during a typical control procedure.

In this scenario, the xApp initiates a subscription to the E2 node, which must expose a specific RAN function referred to as "RAN control." During the subscription phase, the xApp can specify a triggering event or a timer. If the timer expires or the trigger condition is met, the E2 node dispatches an E2 insert message to the RIC. For instance, this might occur when a monitored gNB metric surpasses a certain threshold, such as when the Reference Signal Received Power (RSRP) for a neighboring cell (or target cell) exceeds that of the current cell plus an offset (referred to as an "A3 event" in 3GPP 5G NR specifications). In response, the E2 node sends a handover control request insert message, including the possibility to specify the target cell designated by the E2 node. This action suspends the radio resource management procedure and activates a timer. Subsequently, the Near-RT RIC could respond with an E2 control message either denying or accepting the control request, along with a control action (e.g., a target cell ID designated by the xApp). Alternatively, if the E2 node's timer expires before receiving a response from the RIC, the procedure may proceed autonomously or be terminated, depending on the specific procedure and the E2 node configuration.

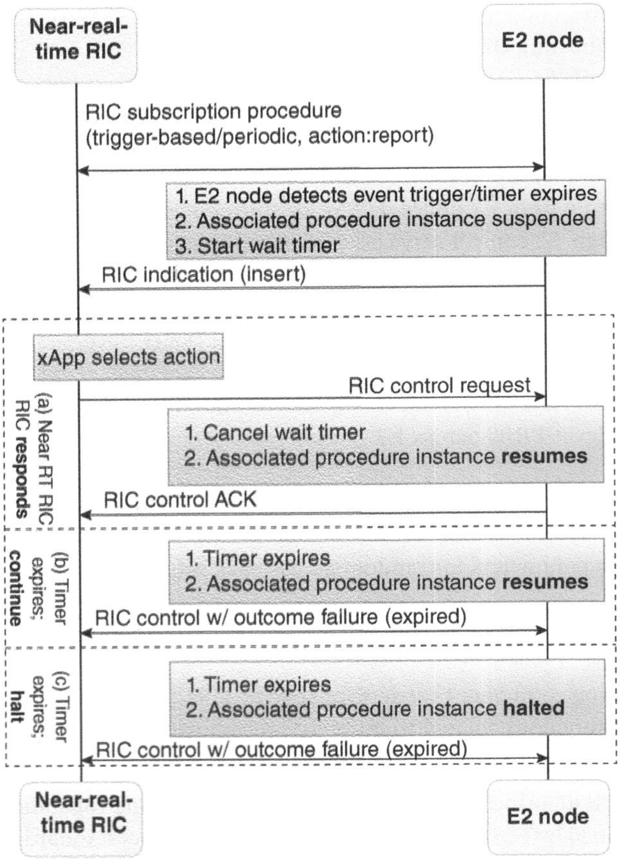

Figure 3.8 Illustration of Services Information Exchange Between the Near-RT RIC and an E2 Node.

Notably, the control action dispatched by the xApp can also be asynchronous, meaning it does not hinge on receiving an insert message from the E2 node. Furthermore, E2SM can specify policies, which fall into two categories: control policies enabling the E2 node to execute radio resource management actions autonomously when specific conditions are met and offset policies that modify 3GPP or vendor-defined thresholds by adding or removing offsets, thereby altering how the E2 node performs certain functions.

3.5 xApp Use Cases

Here, we discuss two example xApp usecases, namely, KPIMON xApp and RAN Slicing xApp.

3.5.1 KPIMON xApp

The KPIMON xApp is founded upon the E2SM-KPM v1.01 Service Model (SM) (O-RAN Work Group 3 2024c) and serves the purpose of gathering crucial Key Performance Indicators (KPIs) from the RAN.

Figure 3.9 details the end-to-end workflow and signaling procedure of the KPIMON xApp (Upadhyaya et al. 2022).

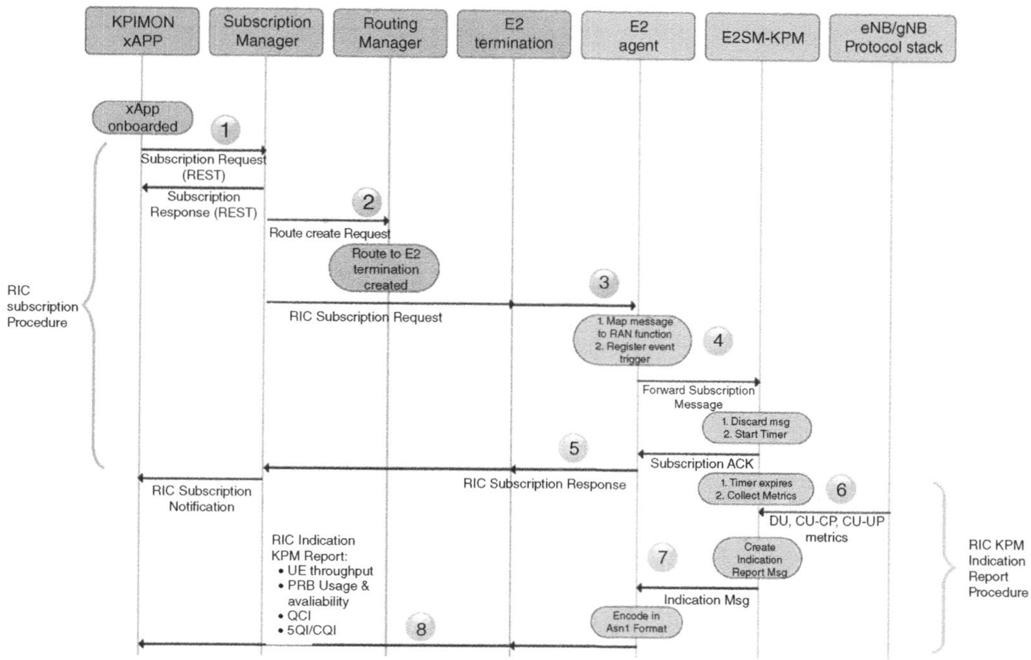

Figure 3.9 End-to-end Workflow and Signaling Procedure for the Collection of KPIs. The RIC Subscription and RIC Indication Procedures are Used by the KPIMON xApp.

1) **Subscription Procedure Initiation.** Upon deployment, the xApp engages with the subscription manager to instigate a subscription request procedure directed at all E2 nodes that support the specific functionality offered by the xApp. Notably, E2 nodes would have previously indicated the RAN functions they support, along with the corresponding RIC services and Service Models (SMs) during the E2 setup procedure.
2) **RIC Subscription Request Routing.** The RIC subscription request, initialized by the xApp, traverses through the E2 termination.
3) **Forwarding to E2 agent.** Subsequently, the E2 termination forwards the request to the E2 agent situated within the E2 node.
4) **E2 Agent Handling.** Within the E2 node, the E2 agent deciphers the RAN Function ID and service details from the message. It then maps this information to the KPM RAN function and forward the subscription message to the E2SM-KPM agent for further processing.
5) **E2SM-KPM Processing.** The E2SM-KPM agent interprets the subscription message, determining the periodicity of reports and initializing a timer. Following this, an RIC subscription response is dispatched to notify the xApp of a successful subscription.
6) **KPI Collection.** As the timer at the E2 node expires, the E2SM KPM agent collates KPI data, encompassing O-DU, O-CU-CP, and O-CU-UP metrics from the preceding time interval. These metrics are organized into containers, each bearing a unique ID. Additionally, every E2 message is equipped with a header containing a global RAN ID (PLMN ID) to identify the respective RAN node.
7) **KPI Report Formation.** The collected metrics are amalgamated to formulate an indication message, specifically a KPM Report. This process entails using the appropriate IEs defined in the Service Model and encoding the data through ASN.1 encoding.

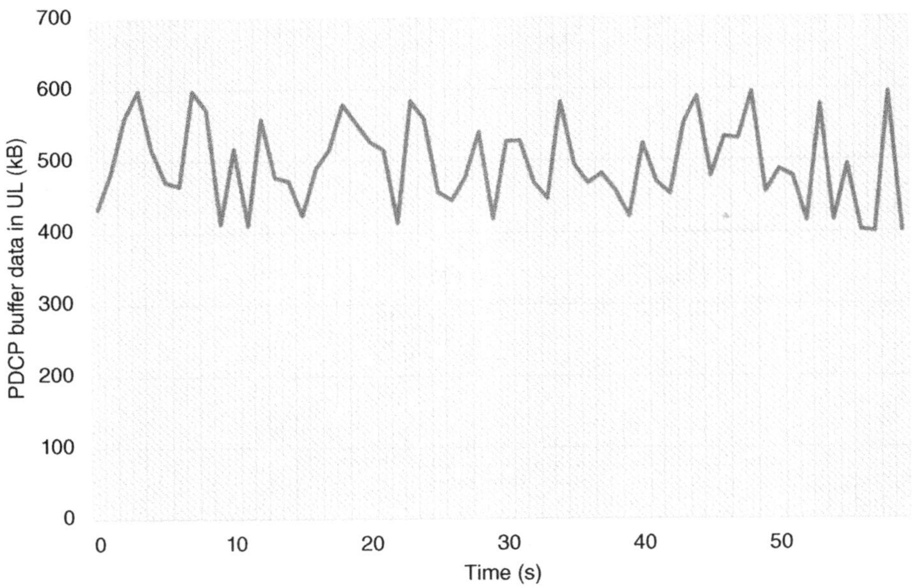

Figure 3.10 Total PDCP Bytes Received by the en-gNB in the Uplink from All UEs.

8) **Periodic Transmission.** The indication messages, comprising KPM Reports, are periodically dispatched to the KPIMON xApp. To decipher these messages and extract the metrics, the KPIMON xApp utilizes the same Service Model and ASN.1 definitions.

The KPIs gathered by the KPIMON xApp include the following:

- Number of utilized and available PRBs.
- Number of active UE connections.
- QoS Class Identifiers (QCIs) associated with active data transmissions.
- Downlink and uplink data transferred in bytes for each QCI session.

Data Received by the en-gNB. Figure 3.10 illustrates the data received in bytes by the en-gNB from all connected UEs for a specific QCI. In this scenario, two UEs are considered, each transmitting uplink data at rates of 4 and 7 Mbps, respectively. Both the eNB/gNB and the two UEs employ an X310 USRP alongside their respective compute nodes. The KPIMON SM accumulates the total PDCP Bytes for the entire session, while the UE remains connected to the en-gNB. The periodic event trigger timer is configured to one second.

3.5.2 RAN Slicing xApp

In this section, we illustrate the operation of the RAN Slicing xApp, showcasing its ability to combine two services for controlling and modifying RAN parameters from the Near-RT RIC. The RAN Slicing xApp is an extension of the NexRAN framework (Johnson et al. 2022), seamlessly integrated into the 5G NSA Testbed.

The SM supporting the RAN Slicing xApp encompasses both the REPORT and CONTROL services. The REPORT service serves the purpose of acquiring metrics from the RAN, while the CONTROL service is employed to dynamically adjust RAN parameters based on the collected metrics.

While the REPORT service configuration mirrors that of the KPIMON xApp, with a timer acting as an event trigger, the CONTROL service operates asynchronously, initiated by the Near-RT RIC xApp.

3.5.2.1 RAN Slicing Configuration and Resource Allocation

This xApp undertakes the configuration of the en-gNB into distinct slices and allocates resources to these slices through asynchronous E2 Control Messages. In this context, a "slice" refers to the bandwidth allocated to an operator to serve its users, with multiple operators sharing a single cell. For our scenario, we assume the presence of three operators, each with one active user. Consequently, each UE can only utilize the resources assigned within the network slice allocated to its respective operator.

In terms of hardware, we employ one X310 USRP for the base station and one B210 for each of the three UEs, with one per operator or slice.

3.5.2.2 Throughput Performance

Figure 3.11 provides insights into the downlink throughput performance of UEs associated with three distinct slices, each featuring different resource allocations. These slices collectively share a 10 MHz LTE cell.

Here is a breakdown of the performance observations.

- **Initial Allocation (First 60 Seconds).** During the initial 60 seconds, Slice 1 exclusively utilizes all available resources. This implies that Slice 1 monopolizes all subframes for data transfers between the RAN and the corresponding UE, leaving no subframe resources for the other slices.
- **Allocation Shift (60–120 Seconds).** After the first minute, Slice 2 is allocated 25% of the resources, while Slice 1 retains the remaining 75%. Consequently, the UE's throughput in Slice 1 decreases as expected.

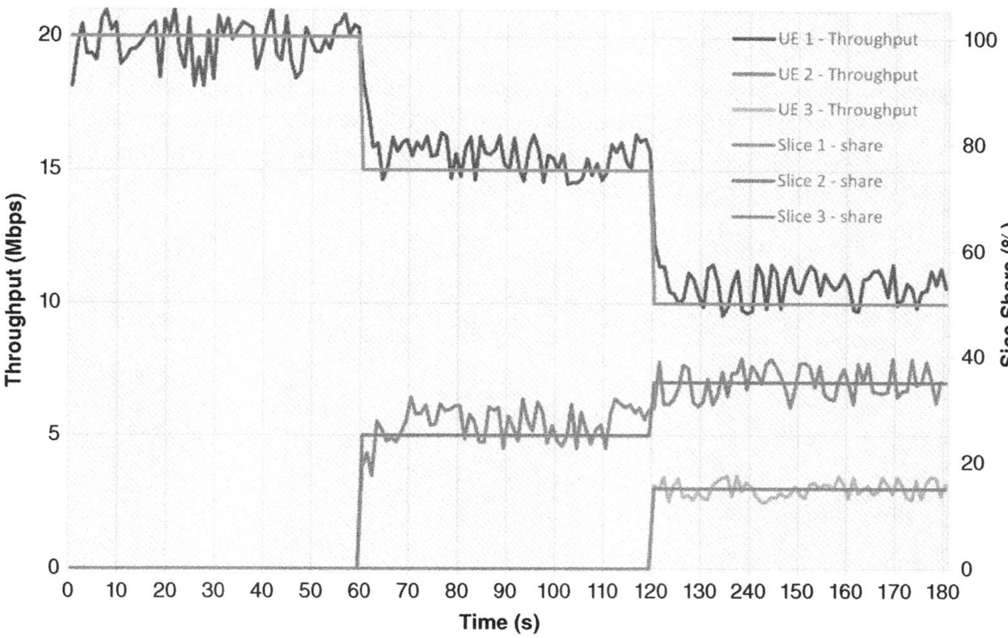

Figure 3.11 UE Bandwidth Variation When Associated with Different Slices.

- **Further Allocation Adjustment (After 120 Seconds).** Beyond the 120-second mark, Slice 2's resource share increases to 35%, with Slice 1's share further reducing to 50%. The remaining 15% of resources are allocated to Slice 3. These alterations manifest in the throughput plot.

Dynamic Subframe Allocation. The subframe allocation for each slice is dynamically managed through Control Messages. Simultaneously, E2 Indication messages, containing per-slice measurements, are reported to the RIC by the E2 node, enhancing the overall visibility and control over the network's performance.

3.6 Machine Learning (ML)-Based xApps

O-RAN ALLIANCE Report "O-RAN Use cases Detailed Specification" (O-RAN ALLIANCE 2024b) has identified various ML-based O-RAN xApp use cases for the RAN control, deployment, and self-optimization. Before we delve into the example ML-based xApp usecases in Section 3.6.2, let us discuss the AI/ML workflow in the O-RAN architecture in Section 3.6.1.

3.6.1 AI/ML Workflow

The AI/ML workflow is a critical component standardized by O-RAN WG2, with specifications detailed in O-RAN Work Group 1 (2024). While some aspects of this workflow are still under development, it primarily consists of six main steps:

1) Data Collection and Processing
2) Training
3) Validation and Publishing
4) Deployment
5) AI/ML Execution and Inference
6) Continuous Operations

These steps are briefly discussed below and illustrated in Figure 3.12. For clarity and illustration, we will describe the AI/ML lifecycle procedures within O-RAN systems using the example of developing an xApp. This xApp dynamically controls RAN slicing policies based on real-time network load and traffic demand using AI algorithms.

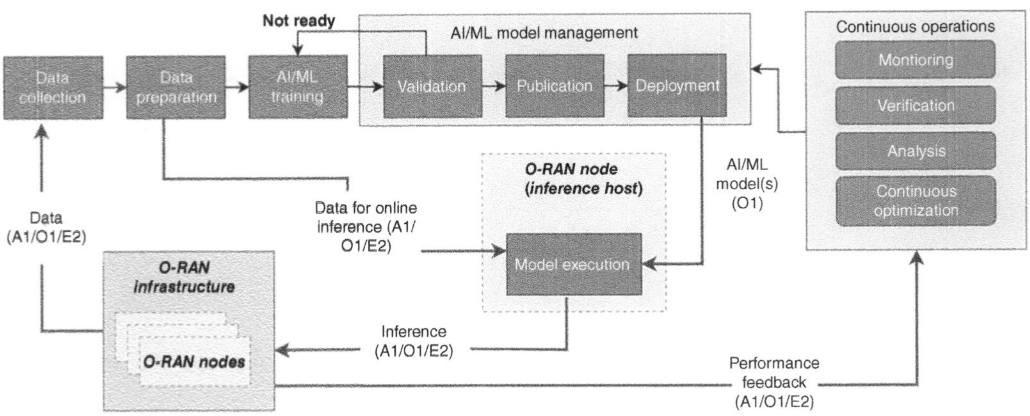

Figure 3.12 AI/ML Workflow Process for the O-RAN Architecture.

3.6.1.1 Data Collection and Processing

- Data are gathered from O1, A1, and E2 interfaces and stored in large datasets.
- Preliminary data preprocessing is applied to shape and format the data according to the AI/ML model's input requirements.
- For example, for an xApp controlling RAN slicing, data collection includes metrics over the O1 interface, such as PRB requirements for each user slice, throughput (enhanced Mobile Broadband [eMBB]), transmitted packets (massive machine-type communications [mMTC]), and latency (ultrareliable low-latency communications [URLLC]) measurements.
- Data processing includes techniques like autoencoders for dimensionality reduction, normalization, scaling, and reshaping.

3.6.1.2 Training

- Offline training of AI/ML models is mandated to ensure reliable intelligence and prevent inaccuracies in predictions or actions that could disrupt the network.
- Online training is allowed but solely for fine-tuning and updating previously offline-trained models.
- For example, multiple AI algorithms can be trained, such as Deep Reinforcement Learning (DRL) agents and decision trees, with variations in parameters, architecture, and input formats to identify the most suitable model for a specific task.

3.6.1.3 Validation and Publishing

- Trained models undergo validation to ensure reliability, robustness, and effectiveness.
- Successful models are published and stored in an AI/ML catalog on the SMO/non-RT RIC, while others are iteratively refined until they pass validation.
- **Example.** Models are compared against diverse validation datasets under varying network conditions, helping identify the most effective ones. This process also provides insights into ideal deployment conditions.

3.6.1.4 Deployment

- Models in the AI/ML catalog can be deployed through image-based or file-based methods.
- Image-based deployment involves containerized execution within O-RAN applications, typically at RICs.
- File-based deployment employs standalone files executed in an inference environment, forwarding outputs to O-RAN applications.
- **Example.** Pretrained AI-based RAN slicing models are selected and deployed as xApps within the Near-RT RIC.

3.6.1.5 AI/ML Execution and Inference

- Deployed models perform online inference tasks, including classification, prediction, policy derivation, and management/control actions.
- **Example.** The xApp for RAN slicing control uses KPMs collected over the E2 interface to compute control actions for PRB allocation in Near-RT.

3.6.1.6 Continuous Operations

- Monitoring and analysis of deployed AI/ML models ensure effective and accurate inference outputs without degrading network performance.
- Poor-performing models can be refined and re-trained for improvement.

- **Example.** Constant monitoring of the RAN slicing xApp allows for retraining of the embedded AI/ML model using new data from O1 and E2 interfaces in case of anomalies or inefficiencies.

This comprehensive AI/ML workflow in O-RAN systems (Polese et al. 2023) enables operators to harness intelligent solutions for optimizing network performance and resource allocation dynamically.

3.6.2 Example ML-Based xApp Use Cases

Recent years have seen an increase in ML-based applications and use cases (e.g., for the optimal configuration of O-RAN networks). At a high level, the scenarios and use cases can be classified in different ways: by considering the control knob or inference target or the domain that is being controlled or optimized (e.g., a UE, a slice, an RAN node, or the whole network). It is possible to associate an xApp with one of the following areas of interest:

- Mobility
- Resource allocation
- Quality of Experience (QoE)/QoS-based control
- RAN sharing
- Massive Multi-Input Multi-Output (MIMO)

Table 3.1 lists AI/ML models for example use cases. Refer to O-RAN ALLIANCE (2024b) for a detailed list of AI/ML use cases.

Table 3.1 Sample AI/ML Models in O-RAN Use Cases.

Use Case	AI/ML Model Functionality	Data Input	Data Output
QoE Optimization	Available bandwidth prediction	Network data, UE level radio channel info, mobility-related metrics, RAN protocol stack status, cell-level info	Available radio bandwidth
UE fingerprinting	Radio fingerprint prediction	Intra-frequency measurement report, PM counters (e.g., RSRP, RSRQ, MCS, CQI)	Inter-frequency MR data (e.g., RSRP, RSRQ, MCS, CQI)
Traffic Steering	Cell load prediction/traffic volume prediction	Load-related counters, e.g., UL/DL PRB occupation	Load-related counter, e.g., UL/DL PRB occupation
Dynamic Handover Management	Minimize handover anomalies	RSRP/RSRQ, CQI measurement data, UE connection, and mobility/handover statistics	Optimal handover sequences in RAN
Massive MIMO Beamforming	Optimal multicell M-MIMO beamforming performance	Cell site information, e.g., location, intersite distance, BS system configuration (e.g., frequency, bandwidth, frame structure, transmit power, default beam weight configuration), complete set of M-MIMO configurations	M-MIMO beam pattern configuration

3.7 xApp Workflow

The xApp workflow within the Near-RT RIC platform involves several key steps to ensure the seamless onboarding, configuration, deployment, and operational readiness of xApps. These steps leverage the Local Helm Repository, the App Manager (Appmgr), and the Reliable Message Routing (RMR) framework within the RIC cluster. Figure 3.13 illustrates the xApp workflow. The key steps are briefly explained below:

Step 1—Onboarding the xApp: The workflow begins with the onboarding of the xApp by submitting the xApp Descriptor Schema and the xApp Descriptor file. These files contain information such as the xApp's metadata, network requirements, Docker image location, and supported E2 messages. The onboarding process ensures that the necessary details are captured for the xApp to be integrated and deployed on the RIC platform.

Step 2—Uploading Helm Charts: The **Dms_cli** generates Helm charts for the xApp based on the descriptor file and uploads them to the Local Helm Repo. These Helm charts define how the xApp will be deployed in the Kubernates environment.

Step 3—Triggering Deployment: Once the Helm charts are uploaded, the operator triggers the deployment process for the xApp. This action initiates the deployment pipeline and informs the system to retrieve the required Helm charts.

Step 4—Fetching Helm Charts from the Repository: The Local Helm Repository fetches the uploaded Helm charts to prepare them for deployment within the RIC Cluster. This ensures that the deployment artifacts (helm chart files, configuration parameters, xApp metadata) are readily available to the cluster.

Step 5—Downloading Helm Charts to the RIC Cluster: The Helm charts are then downloaded to the RIC Cluster. At this stage, the deployment specifications are transferred to the Kubernetes cluster for further execution.

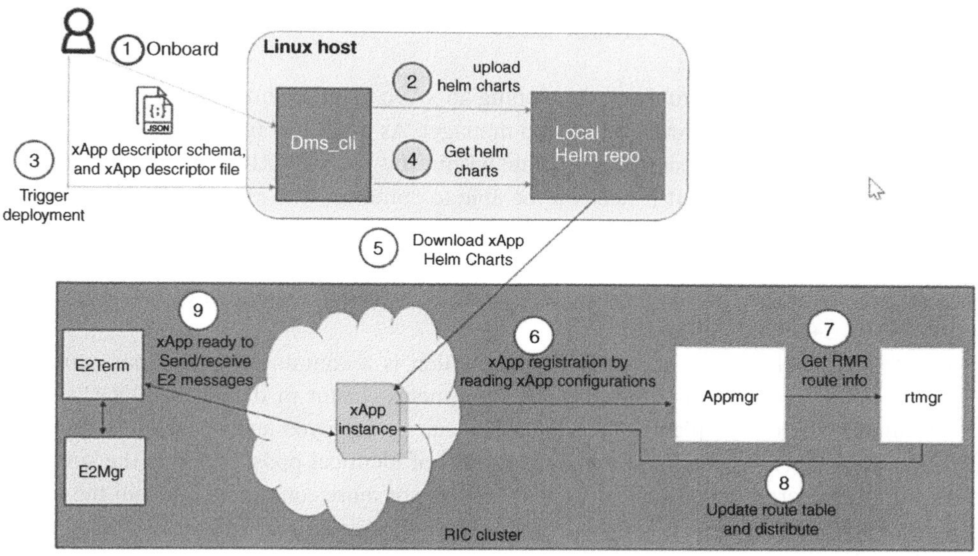

Figure 3.13 xApp Workflow Process. Source: OSC-RICApp-2020. Reproduced with permission.

Step 6—xApp Registration: Within the RIC Cluster, the xApp undergoes a registration process where it reads and applies the xApp configuration. This step ensures that the xApp is properly integrated into the RIC platform, with its configuration parameters aligned to the operational environment.

Once the Helm charts are downloaded and the xApp is registered in the RIC Cluster (Steps 5 and 6), the Kubernetes cluster creates the xApp instance based on the specifications provided in the Helm charts. This process involves setting up the necessary pods and containers to host the xApp. Each xApp deployment typically consists of at least one pod, which in turn contains one or more Docker containers. These containers encapsulate the xApp's logic and dependencies, ensuring it operates in a self-contained and scalable manner. For simple xApps, a single deployment may involve one pod hosting a single Docker container. For more complex xApps, multiple pods can be deployed, each with distinct resources and configurations. Kubernetes ensures fault tolerance by allowing individual pods to be restarted as needed, providing robustness to the xApp within the near-RT RIC environment. The creation of the xApp instance marks the point at which the xApp becomes operational and integrated with other RIC services.

Step 7—Fetching RMR Route Information: The App Manager communicates with the RMR manager (rtmgr) to retrieve the RMR route table. RMR is a messaging framework that enables reliable communication between xApps and other components in the RIC Cluster. This step establishes the necessary communication routes for the xApp.

Step 8—Updating and Distributing Route Table: The RMR route table is updated and distributed across the cluster to ensure that the xApp has access to the latest communication paths. This step ensures reliable and efficient message exchanges between the xApp and other RIC components.

Step 9—xApp Ready to Send/Receive E2 Messages: Finally, the xApp instance becomes operational and is ready to send and receive E2 messages. At this stage, the xApp can communicate with E2 nodes through E2Term and E2Mgr, enabling real-time interactions and decision-making within the Near-RT RIC platform.

3.8 xApp Development: A Closer Look

This section describes a detailed guide on developing an xApp using Python and the processes of deployment and undeployment through the App manager. As described in previous sections, an xApp is a software component designed to operate within the Near-RT RIC. It can be developed in any programming language, but it needs to be able to communicate over the E2 interface to E2 nodes. An E2 node here refers to a component of the RAN that can interface with the RIC via the E2 interface (e.g., the O-CU, the O-DU, the O-CU-CP, and the O-CU-UP). An xApp can be developed directly using RIC utility libraries such as RMR, SDL, and logging or by utilizing xApp frameworks in Go, C++, and Python.

The Near-RT RIC includes a Kubernetes cluster, which is a container system that manages the deployment and scaling of applications. The Kubernetes cluster in the Near-RT RIC serves as an environment where multiple instances of xApps can be deployed and managed. xApps are packaged into deployments, which are managed groups of identical pods. A pod is the smallest deployable unit in Kubernetes and typically consists of one or more containers that run the xApp application as shown in Figure 3.14. For this example, we will use a single Kubernetes deployment with a single pod, running a single Docker container. The example described in this section is based on the OAIC framework; the readers can refer to the OAIC GitHub page (Open AI Cellular (OAIC) Platform 2024) for more information.

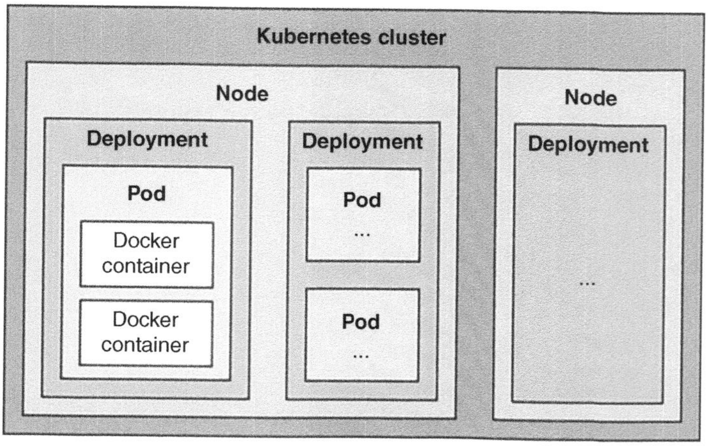

Figure 3.14 Kubernetes Cluster.

The xApp descriptor is a key component when developing an xApp, also known as the xApp config file, it serves as a blueprint for xApps, which are applications designed to run on the Near-RT RIC. This descriptor provides essential information about an xApp, facilitating its deployment, management, and maintenance in the Near-RT RIC. It includes details such as the xApp's name, version, vendor, resource requirements, and configuration parameters, alongside operational metadata about the xApp's capabilities. The information and configuration specified in the xApp descriptor or xApp config file will be used to uniquely identify the xApp and guide the Near-RT RIC components in understanding its functions and integration requirements.

A typical xApp descriptor encompasses several key components: metadata that helps in cataloging the xApp; dependencies that ensure all necessary components are available for the xApp's operation; resource requirements specifying needed computing resources like CPU and memory; a configuration template that outlines initial settings; lifecycle management processes for deployment and updates; and service endpoints and interfaces that list the APIs the xApp exposes for communication with the Near-RT RIC and other xApps. It is structured in the JSON format and contains the following key sections.

The **name** attribute within the descriptor serves as the primary identifier for the xApp, ensuring that each application can be uniquely identified within the Kubernetes deployment. This unique identification is important for managing multiple xApps and avoiding any potential conflicts during deployment or operation. Alongside the name, the **version** attribute specifies the semantic version number of the xApp, which is essential for tracking different versions of the application and managing updates. Together, the name and version create a unique identifier that helps in onboarding, distributing, and deploying the xApp artifacts, such as Helm charts, ensuring that the correct version of the application is always used. The **containers** section of the xApp descriptor is particularly significant as it details the containerized components of the xApp. Each entry in the containers section defines a container that the xApp requires, including its name, the Docker image to be used, and the specific version of the image. Additionally, it specifies the command and arguments that should be executed when the container starts, allowing for detailed customization of the container's runtime behavior. Figure 3.15 defines a single container.

The **controls** section of the xApp descriptor is used for setting internal configuration parameters that are specific to each xApp. It helps define arbitrary parameters that are crucial for the operation of the xApp applications. These parameters are consumed by the xApp by reading from a JSON file that is injected into the container at runtime. An environment variable

```
"containers": [
    {
        "name": "example_container_1",
        "image": {
            "registry": "example_image_registry_1",
            "name": "example_image_name_1",
            "tag": "example_image_tag_1"
        },
        "command": ["example_command_1"],
        "args": ["example_argument_1"]
        "resources": {
            "limits": {},
            "requests": {}
        }
    },
]
```

Figure 3.15 Container Attribute of xApp Descriptor.

```
"controls": {
    "active": True,
    "requestorId": 66,
    "ranFunctionId": 1,
    "ricActionId": 0,
    "interfaceId": {
        "globalGNBId": {
            "plmnId": "90170",
            "gNBId": 202451
        }
    }
}
```

Figure 3.16 Controls Attribute of xApp Descriptor.

"XAPP_DESCRIPTOR_PATH" indicates the directory where this JSON file is mounted inside the container. If this section is utilized, a schema file for these controls must be provided as shown in Figure 3.16.

The **metrics** section defines the metrics that the xApp will provide. This includes specifying the object name, instance, type, and a detailed description for each metric. These data are used for monitoring and is typically exposed to external systems like Prometheus through the VESPA manager in the Near-RT RIC platform. This setup facilitates real-time monitoring and analytics, enabling effective management and optimization of the RAN performance based on the xApp's activities.

In the **messaging** section, communication settings for the xApp are specified. This includes defining the communication ports for the containers, which are essential for handling incoming and outgoing messages. The section details RX and TX message types and the relevant A1 policies (A1-P) for RMR (Routing Manager) communications implemented by the xApp. Each defined port creates a Kubernetes (K8s) service port mapped to the container at the same port number, ensuring seamless data flow for the application.

The **liveness probes** and **readiness probes** sections are crucial for maintaining the health and efficiency of the xApp within the Kubernetes environment. These probes help determine the live and ready state of the xApp, ensuring that it remains operational and efficiently integrated within the Near-RT RIC. The probes can be defined either through HTTP or command-based checks, providing flexibility in how these checks are performed based on the xApp's architecture and needs. Figure 3.17 shows the additional messaging, liveness probes and readiness probes attributes of the xApp descriptor.

Following the development of the xApp's fundamental functionality, the subsequent steps involve the deployment of the xApp into the Near-RT RIC as specified below.

Step 1: Building the Docker Image. The xApp will be hosted in a Docker container. To create this container, it is necessary to provide a Docker file, which details the setup instructions for the virtual environment. In this case, Ubuntu setup with Python as the base for the Docker image will be used. An example of a Docker file is shown in Figure 3.18.

Once the Docker file is created, the Docker image can then be built and submitted to the xApp registry by running the following command:

```
Sudo docker build . -t xApp-registry.local:5000/ric-app-ml:1.0.0
```

This builds a Docker image labeled ric-app-ml with version 1.0.0 and submits it to the registry (Figure 3.19).

Step 2: Creating the xApp config (xApp Descriptor). In the xApp directory, there is a folder tagged ***init***, it contains the conFigurejson file, which signifies where the Docker image is located and provides the ports and capabilities of the E2 interface.

Step 3: Finding local IP address. Before proceeding, there is a need to set the local IP address of the system. On the Ubuntu operating system, the local IP addresses of the system can be obtained by running this command via the terminal *hostname–I*. The second IP address that appears should work. To set the *$HOST_IP* of the system, use this command to set the IP address *export HOST_IP=<ip address>*.

Step 4: Configuring the Nginx Web Server. The xApp descriptor files (conFigurejson) must be hosted on a web server when the **xApp onboarder** is used in deploying xApps. The xApp onboarder cannot access the local files. Hence, there is a need to upload them to the network where the xApp onboarder can find and download the files. Nginx is used as the web server to host the config files. To install the Nginx web server and check the status of the server after installation, the following commands will be used.

```
sudo apt install nginx

sudo systemctl status nginx
```

Create a few directories that will host the config files and be accessible by the server. This can be done using these commands:

```
sudo mkdir/var/www/xApp_conFigurelocal
```

```
"messaging": {
    "ports": [
        {
            "name": "http",
            "container": "demo-xapp",
            "port": 8080,
            "description": "http service"
        },
        {
            "name": "rmrdata",
            "container": "demo-xapp",
            "port": 4560,
            "txMessages": [
                "RIC_SUB_REQ",
                "RIC_SUB_DEL_REQ"
            ],
            "rxMessages": [
                "RIC_SUB_RESP",
                "RIC_SUB_FAILURE",
                "RIC_SUB_DEL_RESP",
                "RIC_INDICATION"
            ],
            "policies": [
                1,
                2
            ],
            "description": "rmr data port for demo-xapp"
        },
        {
            "name": "rmrroute",
            "container": "demo-xapp",
            "port": 4561,
            "description": "rmr route port for demo-xapp"
        }
    ]
},
"livenessProbe": {
    "httpGet": {
        "path": "near-rt-ric/v1/health/alive",
        "port": "8580"
    },
    "initialDelaySeconds": "5",
    "periodSeconds": "25"
},
"readinessProbe": {
    "exec": {
        "command": [
            "/usr/local/bin/rmr_probe"
        ]
    },
    "initialDelaySeconds": "5",
    "periodSeconds": "100"
},
```

Figure 3.17 Additional Attributes of the xApp Descriptor.

```
sudo mkdir/var/www/xApp_conFigurelocal/config_files
```

Create a custom configuration file and define the file location.

```
sudo nano /etc/nginx/conf.d/xApp_conFigurelocal.conf
```

Paste the following content in the conf file.

```
# Load a miniconda setup for our base Docker image which contains Python
FROM continuumio/miniconda3

# Install all necessary libraries
RUN apt-get update && apt-get -y install build-essential musl-dev libjpeg-dev zlib1g-dev libgl1-mesa-dev wget dpkg

# Copy all the files in the current directory to /tmp/ml in our Docker image
COPY . /tmp/ml

# Go to /tmp/ml
WORKDIR /tmp/ml

# Install requirements.txt
RUN pip install --upgrade pip && pip install requirements.txt

# Set our xApp to run immediately when deployed
ENV PYTHONUNBUFFERED 1
CMD app.py
```

Figure 3.18 An Example of a Docker File.

```
nearrtric@nearrtric-VirtualBox:~/ric-app-ml$ sudo docker build . -t xApp-registry.local:5008/ric-app-ml:1.0.0
Sending build context to Docker daemon  15.36kB
Step 1/7 : FROM continuumio/miniconda3
 ---> fbedfde8d058
Step 2/7 : RUN apt-get update && apt-get -y install build-essential musl-dev libjpeg-dev zlib1g-dev libgl1-mesa-dev wget dpkg
 ---> Using cache
 ---> 88f75a15a710
Step 3/7 : COPY . /tmp/ml
 ---> Using cache
 ---> ca877fc2be67
Step 4/7 : WORKDIR /tmp/ml
 ---> Using cache
 ---> a813836b2bca
Step 5/7 : RUN pip install --upgrade pip && pip install -r requirements.txt
 ---> Using cache
 ---> e706a4b0758f
Step 6/7 : ENV PYTHONUNBUFFERED 1
 ---> Using cache
 ---> 4615f9896743
Step 7/7 : CMD ["python", "app.py"]
 ---> Using cache
 ---> 8eafe9c58fff
Successfully built 8eafe9c58fff
Successfully tagged xApp-registry.local:5008/ric-app-ml:1.0.0
nearrtric@nearrtric-VirtualBox:~/ric-app-ml$
```

Figure 3.19 Output of the Docker Image Build and Submission to the xApp Registry.

To save and update the configuration file, execute the following command. If the process completes without any output, this indicates a successful update. Always ensure to verify that there are no errors in the configuration file after updating.

```
sudo nginx -t
```

```
nearrtric@nearrtric-VirtualBox:~/ric-app-ml$ sudo nginx -t
nginx: the configuration file /etc/nginx/nginx.conf syntax is ok
nginx: configuration file /etc/nginx/nginx.conf test is successful
nearrtric@nearrtric-VirtualBox:~/ric-app-ml$
```

Step 5: Hosting the config files. From the xApp directory, copy the xApp configuration file into the config files directory that was created in previous steps using the sudo command, which will protect the file from being modified. Use the chmod command to enable read/write permissions. Apply the chmod command as follows.

```
sudo cp init/conFigurejson/var/www/xApp_conFigurelocal/
config_files/ml-config-file.json
```

```
nearrtric@nearrtric-VirtualBox:~/ric-app-ml$ curl http://$HOST_IP:5010/config_files/ml-config-file.json
{
    "json_url": "ric-app-ml",
    "xapp_name": "ric-app-ml",
    "version": "1.0.0",
    "containers": [
        {
            "name": "ric-app-ml",
            "image": {
                "registry": "xApp-registry.local:5008",
                "name": "ric-app-ml",
                "tag": "1.0.0"
            }
        }
    ],
    "messaging": {
        "ports": [
            {
                "name": "rmr-data",
                "container": "ric-app-ml",
                "port": 4560,
                "rxMessages": [ "RIC_SUB_RESP", "RIC_SUB_FAILURE", "RIC_INDICATION", "RIC_SUB_DEL_RESP", "RIC_SUB_DEL_FAILURE" ],
                "txMessages": [ "RIC_SUB_REQ", "RIC_SUB_DEL_REQ" ],
                "policies": [1],
                "description": "rmr receive data port for ric-app-ml"
            },
            {
                "name": "rmr-route",
                "container": "ric-app-ml",
                "port": 4561,
                "description": "rmr route port for ric-app-ml"
            }
        ]
    },
```

Figure 3.20 Asserting the nginx Server Can Access the xApp config File.

```
sudo chmod 755 /var/www/xApp_conFigurelocal/config_files/
ml-config-file.json
```

```
sudo systemctl restart nginx
```

After restarting the server, check if the config file can be accessed from the server and run the following command to do this curl http://$HOST_IP:5010/config_files/ml-config-file.json. The output of the command should be similar to Figure 3.20.

Create onboard URL file. There is a need to create a .url file that will point the xApp-onboarder to the Nginx web server, where it can retrieve the xApp descriptor file. This descriptor is then used to create the helm chart necessary for deploying the xApp. Update the ml-onboard.url file with the following:

```
{"config-file.json_url":"http://<machine_ip_address>:5010/
config_files/ml-config-file.json"}
```

Replace the machine_ip_address with the IP address of the machine, this can be obtained by running echo $HOST_IP or hostname -I.

Onboard and deploy the xApp. First, collect and store the IP address of the Kong proxy to a variable, which will enable it to connect to the different components of the Near-RT RIC through a single address. Then, submit the onboard URL file to the xApp onboarder which indicates to the onboarder where the xApp config file can be located. Execute the commands as shown below:

```
export KONG_PROXY=`sudo kubectl get svc -n ricplt -l app.kubernetes.io/name=kong -o jsonpath='{.items[0].spec.clusterIP}'`
curl -L -X POST "http://$KONG_PROXY:32080/onboard/api/v1/onboard/download" --header 'Content-Type: application/json' --data-binary "@ml-onboard.url"
```

The output of the second command should return a status of "created," meaning the helm chart is created successfully. The config file will be processed by the xApp onboarder which contains the instructions to deploy the xApp. The next step will be to request the Appmgr to deploy the xApp,

```
nearrtric@nearrtric-VirtualBox:~/ric-app-ml$ sudo kubectl get pods -A
NAMESPACE     NAME                                                        READY   STATUS      RESTARTS   AGE
kube-system   coredns-5644d7b6d9-gll7s                                    1/1     Running     5          15d
kube-system   coredns-5644d7b6d9-q7pzf                                    1/1     Running     5          15d
kube-system   etcd-nearrtric-virtualbox                                   1/1     Running     5          15d
kube-system   kube-apiserver-nearrtric-virtualbox                         1/1     Running     5          15d
kube-system   kube-controller-manager-nearrtric-virtualbox                1/1     Running     6          15d
kube-system   kube-flannel-ds-npg25                                       1/1     Running     5          15d
kube-system   kube-proxy-mqf6b                                            1/1     Running     5          15d
kube-system   kube-scheduler-nearrtric-virtualbox                         1/1     Running     5          15d
kube-system   tiller-deploy-7d7bc87bb-z9lps                               1/1     Running     0          6d19h
ricinfra      deployment-tiller-ricxapp-67f67c95f4-k7s9s                  1/1     Running     3          6d19h
ricinfra      tiller-secret-generator-mq2r5                               0/1     Completed   0          6d19h
ricplt        deployment-ricplt-a1mediator-6ccd8896d7-mrkkf               1/1     Running     1          6d19h
ricplt        deployment-ricplt-alarmmanager-56d79dc55-k5v4b              1/1     Running     0          6d19h
ricplt        deployment-ricplt-appmgr-6fd6664755-2s6bj                   1/1     Running     0          6d19h
ricplt        deployment-ricplt-e2mgr-66cdc4d6b6-m5l8t                    1/1     Running     0          6d19h
ricplt        deployment-ricplt-e2term-alpha-db9c45968-rk9rg              1/1     Running     1          6d19h
ricplt        deployment-ricplt-jaegeradapter-76ddbf9c9-48g9z             1/1     Running     0          6d19h
ricplt        deployment-ricplt-o1mediator-677ff764d7-q2mz2               1/1     Running     0          6d19h
ricplt        deployment-ricplt-rtmgr-578c64f5cf-rgqbk                    1/1     Running     0          6d19h
ricplt        deployment-ricplt-submgr-7f6499555d-p2vp4                   1/1     Running     0          6d19h
ricplt        deployment-ricplt-vespamgr-84f7d87dfb-bgfcq                 1/1     Running     0          6d19h
ricplt        deployment-ricplt-xapp-onboarder-5958856fc8-z6nh8           2/2     Running     0          6d19h
ricplt        r4-infrastructure-kong-646b68bd88-tqvqc                     2/2     Running     1783       6d19h
ricplt        r4-infrastructure-prometheus-alertmanager-75dff54776-gwv7q  2/2     Running     0          6d19h
ricplt        r4-infrastructure-prometheus-server-5fd7695-ccjmh           1/1     Running     0          6d19h
ricplt        ricplt-influxdb-meta-0                                      0/1     Pending     0          6d19h
ricplt        statefulset-ricplt-dbaas-server-0                           1/1     Running     0          6d19h
ricxapp       ricxapp-ric-app-ic-795784bfb6-wl24j                         1/1     Running     1          3d18h
ricxapp       ricxapp-ric-app-ml-859c75c587-kv2rh                         1/1     Running     0          28s
nearrtric@nearrtric-VirtualBox:~/ric-app-ml$
```

Figure 3.21 Verifying Successful Deployment of the xApp.

which is the ric-app-ml in this example; it will use the chart owned by the xApp onboarder to deploy the xApp. Execute the following command to deploy the xApp:

```
curl -L -X POST 'http://$KONG_PROXY:32080/appmgr/ric/v1/xapps' --header 'Content-Type: application/json' --data-raw '{"xappName": "ric-app-ml"}'
```

To verify that the xApp is deployed successfully, run this command on the terminal sudo kubectl get pods –A, if successful, there should be a ric-app-ml pod visible in the "ricxapp" namespace as shown in Figure 3.21.

Step 6: Verification of the overall procedure. The following commands are useful for troubleshooting and managing the xApp.

a) View Kubernetes pods: sudo kubectl get pods –A
b) View Kubernetes services: sudo kubectl get svc –A
c) Build Docker image: sudo docker build. -t xApp-registry.local:5008/ric-app-ml:1.0.0
d) Restart the xApp: sudo kubectl rollout restart deployment ricxapp-ric-app-ml –n ricxapp
e) View xApp logs (replace podname with the name of the specific xApp's pod): sudo kubectl logs –n ricxapp podname
f) Open additional port for E2-like interface: sudo kubectl expose deployment ricxapp-ric-app-ml –port 5000 –target-port 5000 –protocol SCTP –n ricxapp –type=NodePort

3.8.1 Undeployment/Removal of xApps

To undeploy or remove the xApp, remove the xApp's helm chart from the xApp onboarder and redeploy or undeploy the Near-RT RIC components in the Kubernetes cluster. Figure 3.22 shows different commands for these processes.

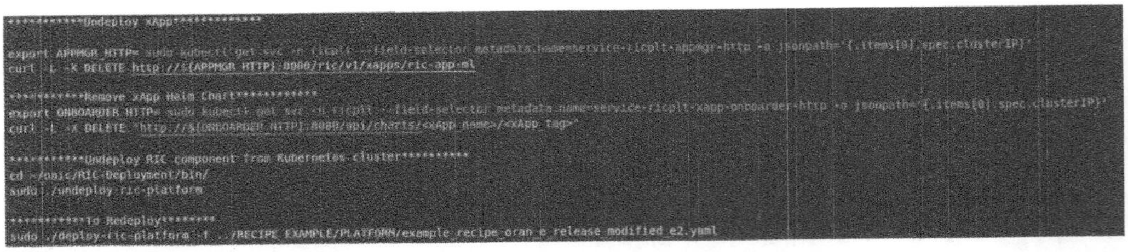

Figure 3.22 Various Commands to Remove xApp and RIC Components.

3.9 Non-RT Closed Loop: Non-RT RIC and rApps

The non-RT RIC is an essential component within the SMO framework of the O-RAN architecture, primarily designed to interact with the Near-RT RIC through the A1 interface. Its core function is to support intelligent optimization of the RAN by providing policy-based guidance, managing machine learning models, and supplying enrichment information (EI) to enhance the efficiency of operations such as Radio Resource Management (RRM) under specific conditions (O-RAN Work Group 2 2024a). The non-RT RIC operates on a non-RT interval, typically greater than one second, utilizing advanced data analytics and AI/ML training and inference to determine the best courses of action for RAN optimization. The architecture of the non-RT RIC includes two main subfunctions: the non-RT RIC Framework and non-RT RIC Applications called rApps (O-RAN Work Group 2 2024e). The Framework is integral to the SMO, managing the A1 interface and enabling the deployment of rApps by exposing necessary services through the R1 interface. These rApps are modular applications that use the functionalities provided by the Framework to execute RAN optimization tasks and other related functions. They interact with various interfaces such as A1, O1, O2, and open FH M-plane to obtain information and initiate actions like policy updates and system reconfigurations, thereby playing an important role in the intelligent management of RAN. Additional details of the non-RT RIC can be found in Section 2.2.1 in Chapter 2.

3.9.1 Requirements for the Non-RT RIC Framework

The non-RT RIC framework requirements are structured to support a cohesive workflow between (i) service registration, discovery, and management within the SMO and (ii) the non-RT RIC environments. Here is an overview of the core functionalities that the non-RT RIC supports.

- **Service Management.** The framework ensures robust mechanisms for service registration and discovery. It allows service producers to register their services while enabling consumers to discover and subscribe to updates regarding these services. This includes notifications about new registration, update, or deregistration of services.
- **Security and Access Control.** Essential to maintaining the integrity and security of the RAN, the framework supports functionalities to authenticate and authorize service consumers, ensuring that only authorized entities can access or modify the services.
- **Communication and Data Management.** It facilitates seamless communication between the non-RT RIC and the Near-RT RIC through the A1 interface, useful for effective RAN optimization. Additionally, it manages data interactions, allowing data consumers and producers (including rApps) to register, subscribe, and manage data types essential for RAN operations.
- **AI/ML Integration.** The framework is equipped to support AI/ML model training, deployment, and performance monitoring, which are required for improving RAN intelligence and adaptive

capabilities. This includes functionalities for managing the lifecycle of AI models and handling large-scale data analytics to inform model adjustments and RAN optimization strategies.
- **Telemetry and Logging.** This functionality supports the collection and storage of deployment telemetry, infrastructure telemetry, and log information from various entities within the network. These data are vital for ongoing monitoring, troubleshooting, and refining of network operations.
- **Performance and Fault Management.** The non-RT RIC framework provides tools for managing the performance and fault reporting of rApps, enhancing the reliability and efficacy of deployed applications and services.

These functionalities underscore the non-RT RIC framework's role as a central component in the O-RAN architecture, aimed at fostering a more responsive, secure, and efficiently managed RAN environment.

3.9.2 Non-RT RIC Architecture

Figure 3.23 describes the reference architecture of the non-RT RIC as an integral part of the SMO framework (O-RAN Work Group 2 2024a).

There are three main categories of logical functions within the non-RT RIC framework.

1) **Anchored Functions Within the Non-RT RIC Framework.** These functions are core to the non-RT RIC and are depicted in the architecture diagrams as solid blue boxes. They are essential for the internal operations of the non-RT RIC, including tasks such as service registration, notification, and AI/ML model management that directly interacts with the A1 interface for communication with the Near-RT RIC.
2) **Anchored Functions Outside the Non-RT RIC Framework.** Represented as solid orange boxes in the diagrams, these functions reside within the broader SMO framework but outside the direct responsibility of the non-RT RIC. They may include functionalities like network slicing OAM (Operations, Administration, and Maintenance) and other management tasks that

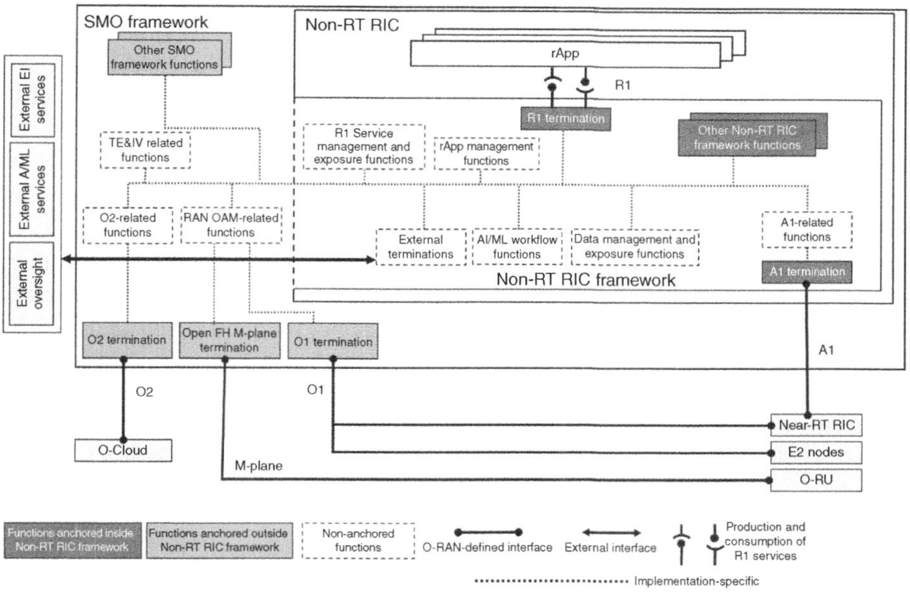

Figure 3.23 Non-RT RIC Architecture. Source: O-RAN Work Group 2 (2024a). Reproduced with permission.

support the broader network but are essential for the operation of the RAN.

3) **Nonanchored Functions.** Illustrated with dashed line boxes, these functions are flexible and can be implemented as needed based on specific requirements or vendor implementations. They are not strictly defined within the non-RT RIC or SMO frameworks and can include optional services that support the R1 interface or provide EI for the A1 interface.

The R1 services in this architecture facilitate the interactions between the non-RT RIC, the rApps, and other components of the SMO framework (O-RAN Work Group 2 2024d). These services enable operations such as service discovery, authorization, authentication, and communication support. They also manage the rApps, providing consistent access to configuration, fault information, performance data, and logging capabilities across different rApps, irrespective of the vendor. This flexible structured architecture ensures that the non-RT RIC can effectively manage and optimize RAN operations while supporting scalability and adaptability to meet the demands of 5G networks and beyond. The design allows for vendor-specific adaptations and is open to future expansions or integrations as defined by ongoing developments in O-RAN working groups.

To ensure clarity and organization, this section is divided into several subsections. These subsections explains the fundamental principles of the A1 interface, its role in the O-RAN architecture, and the specific services it supports, such as policy management, EI services, and A1/ML model management. This is designed to guide readers through each aspect of the non-RT RIC architecture, ensuring a comprehensive understanding of its integration and functionality within the broader O-RAN system.

3.9.2.1 A1 Interface: General Principles

The A1 interface within the O-RAN architecture is designed with specific principles to ensure flexible interactions between the non-RT RIC and the Near-RT RIC. Here is an overview of these principles as discussed in O-RAN Work Group 2 (2024b).

1) **Open Logical Interface.** The A1 interface acts as an open logical connection between the non-RT RIC and the Near-RT RIC. It is structured to support a multivendor environment, facilitating interoperability and independence from specific implementations within the SMO, non-RT RIC, and Near-RT RIC.
2) **Extensibility.** The design of the A1 interface is extensible, allowing for the addition of new services and data types without the need to modify the existing protocol stack or procedures. This flexibility supports evolving network needs and technological advancements.
3) **Policy-Based Guidance.** A key role of the A1 interface is to enable policy-based guidance for radio resource management. These policies guide internal functions or applications within the Near-RT RIC, ensuring efficient resource allocation and network management.
4) **Feedback Mechanism.** The A1 interface provides a fundamental feedback mechanism that allows the non-RT RIC to monitor the effectiveness and status of the policies implemented by the Near-RT RIC. This feedback is important for continuous policy refinement and adjustment.
5) **Standardized Policy Enforcement.** Policies transferred over the A1 interface are expressed using a standardized mechanism, which includes a defined language and syntax. This standardization ensures that policies are clear, consistent, and actionable by the Near-RT RIC.
6) **EI Transmission.** The A1 interface facilitates the transmission of EI from the non-RT RIC to the Near-RT RIC. This EI supports the Near-RT RIC by providing additional data needed to optimize RAN performance.
7) **Discovery and Secure Delivery.** The A1 interface allows the Near-RT RIC to discover available EI and request secure delivery. This capability ensures that the ear-RT RIC has access to the necessary information to perform optimally.

8) **Policy and EI Job Management.** The creation, modification, and deletion of A1-Ps are managed by the non-RT RIC, based on contextual information and RAN observability. Similarly, EI jobs related to the delivery of EI are managed by the Near-RT RIC based on its internal needs.

These principles underline the A1 interface's role in enhancing communication, management, and operational efficiency between the non-RT RIC and Near-RT RIC, making it a foundational component of the O-RAN architecture's aim to drive more intelligent and adaptable RAN solutions.

3.9.2.2 Role of the A1 Interface

In the O-RAN architecture, the A1 interface facilitates communication and operational synergy between the non-RT RIC and the Near-RT RIC within the SMO layer (O-RAN Work Group 2 2024a). This interface is essential for implementing policy-based guidance, managing machine learning models, and providing EI that enhances RAN optimization. The non-RT RIC, situated in the SMO layer, is involved in several key activities including the deployment, configuration, and data collection from RAN nodes. There are also functions within the SMO that handle the AI/ML workflow, which includes training and updating ML models and their deployment across the network. The SMO layer's capabilities extend beyond just managing the RAN network functions; it may access additional data sources that can provide more refined information. This information can further be used to enhance RAN guidance and optimization of network performance, especially in terms of meeting the specific Service-Level Agreements related to RAN intent. The A1 interface enables the non-RT RIC to deliver these comprehensive policies and manage ML models effectively, it supports the dynamic adjustment of network operations to accommodate varying conditions, particularly in optimizing the RRM. Through the A1 interface, the non-RT RIC provides the Near-RT RIC with the necessary tools and information to implement these optimizations effectively, thereby improving the overall network adaptation and responsiveness of the O-RAN system (O-RAN Work Group 2 2024b) (Figure 3.24).

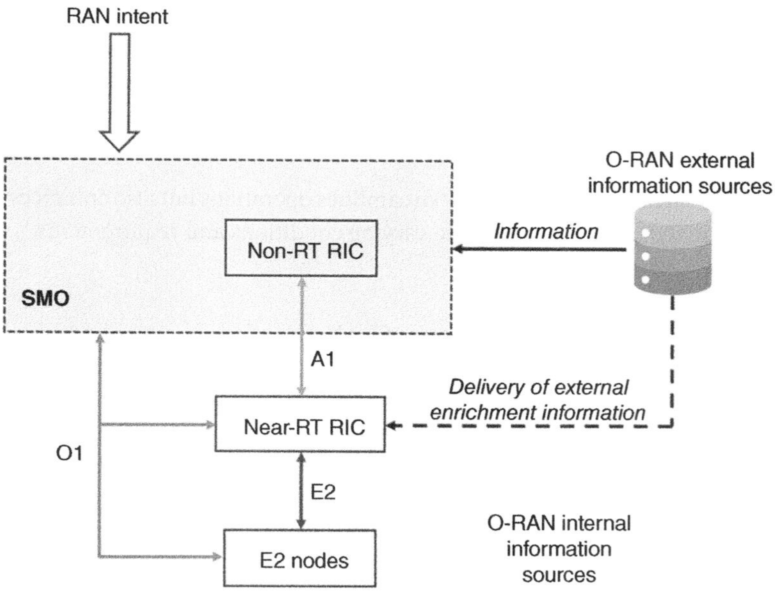

Figure 3.24 Illustration of A1 Interfaces and Other Related Interfaces.

3.9.2.3 A1-P Management Service (A1-P)

The A1-P management service in the O-RAN architecture leverages the capabilities of the non-RT RIC to guide and optimize RAN performance based on defined RAN intents. This service utilizes high-level system goals expressed in RAN intent, alongside observables (events and counters) obtained over the O1 interface, to formulate declarative A1-Ps (O-RAN Work Group 2 2024f). These policies, which are provided to the Near-RT RIC via the A1 interface, specify objectives and allocate resources effectively across the network, targeting elements like UEs and cell operations. The non-RT RIC dynamically manages these A1-Ps by continuously assessing the impact of implemented policies through feedback and network status updates received over the O1 interface. This continuous evaluation allows the non-RT RIC to adjust or update the A1-Ps to better align with the RAN intent, considering the real-time conditions within the network. Meanwhile, the Near-RT RIC utilizes these policies, along with its internal functions and configurations received over O1, to enforce the policy directives. The A1-P management services are designed to enhance the flexibility and control of policy management in a multivendor environment. These services allow for comprehensive management of A1-Ps through several key functionalities:

- **Discovery and Notification.** rApps can query the SMO/non-RT RIC framework to discover supported A1-P types and can subscribe to notifications for changes, such as additions or removals of policy types. This ensures that rApps remain updated on the available policy types they can utilize for network optimization.
- **Policy Lifecycle Management.** rApps have the capability to create, update, query, and delete A1-Ps within the SMO/non-RT RIC framework. When creating policies, the framework assigns unique identifiers to each policy, which rApps then use to manage these policies. This includes updating policies based on their identifiers, querying for specific policies applied to various network elements (e.g., UEs, cells, slices), and deleting policies.
- **Enforcement Status Queries.** rApps can query the enforcement status of A1-Ps using their identifiers. This functionality allows rApps to monitor how effectively the policies are being implemented in the network and adjust as needed to optimize performance.
- **Event Subscription for Notifications.** rApps can subscribe to notifications related to the creation, modification, or deletion of A1-Ps. They can also subscribe to notifications concerning changes in the enforcement status of these policies. These subscriptions can be tailored based on specific scopes or policy identifiers, ensuring that rApps receive relevant updates that are crucial for their operation.

The framework's flexibility in managing A1-Ps not only streamlines operations but also enhances the adaptability of the network to dynamically respond to varying conditions and requirements.

3.9.2.4 A1 Enrichment Information Service (A1-EI)

The A1-EI service enhances the functionality and efficiency of RAN operations by providing additional, valuable data to the SMO layer. This service involves gathering information from both internal and external sources within the O-RAN ecosystem to support the performance of network tasks, particularly through the non-RT RIC. *EI, as defined, is not just the raw data collected but also includes data that have been processed—relayed, combined, refined, or analyzed—to provide more insightful inputs* (O-RAN Work Group 2 2024f). Both the non-RT RIC and the Near-RT RIC use these refined data to improve internal functions or applications, such as machine learning models, enhancing their effectiveness in network management and optimization. The A1 interface plays an important role in this process, facilitating the discovery, request, and delivery of A1 EI. It also handles the discovery of external EI, ensuring the authentication of the source and the security of the connections.

However, the protocols for managing direct connections and the specific processes for the discovery or delivery of external EI involving the A1 interface are yet to be clearly defined in the current O-RAN specifications (O-RAN Work Group 2 2024f). This suggests an area of ongoing development within the O-RAN architecture, highlighting the evolving nature of the standards and practices as they adapt to new technological needs and challenges.

The A1-EI services primarily include the registration and deregistration of EI types.

- **Registration of EI Types.** rApps have the capability to register the types of EI data they can produce. This registration is essential for defining the data sources available for EI jobs, which are subsequently delivered over the A1 interface. This allows the SMO/non-RT RIC framework to leverage specific data types produced by rApps to enhance network performance and optimization tasks.
- **Deregistration of EI Types.** rApps can deregister previously registered EI types. This functionality is important for maintaining the relevance and efficiency of the data being utilized in the network. It ensures that only current and useful data types are active and contributing to network operations.

It is important to note that according to the "O-RAN.WG2.A1AP specification," EI types cannot be updated directly. If there is a need to modify the definition of an EI type, the rApp must deregister the existing type and register a new EI type. This approach maintains the integrity and accuracy of the data utilized across the network, ensuring that all EI remains current and correctly defined for optimal network performance.

3.9.2.5 A1 ML Model Management Service (A1-ML)

Within the O-RAN architecture, ML models can be trained and executed in different places. Two primary scenarios have been identified and described in O-RAN Work Group 1 (2024) that involve the non-RT RIC.

- **Scenario 1.1.** In this setup, the ML model is trained within the SMO layer, specifically to enhance the non-RT RIC's capabilities in monitoring and guiding the RAN. The model leverages observability data from the O1 interface, which are also the data used for the model's inference processes. The SMO's internal functions handle both the training and deployment of this ML model, ensuring that the non-RT RIC can effectively utilize the model to interpret and respond to real-time network conditions.
- **Scenario 1.2.** Here, the ML model is similarly trained within the SMO layer but is utilized by the Near-RT RIC to improve the precise optimization of the RAN. Like in the first scenario, the training data come from the O1 interface and are consistent with the data used by the Near-RT RIC for its operational inference. After training, the model is deployed to the Near-RT RIC through the O1 interface. This deployment includes mechanisms for evaluating the model's performance and integrating feedback from the model based on its execution results, thus allowing for adaptive and responsive network optimization.

Additionally, to aid the ML model inference process within the Near-RT RIC, the non-RT RIC can provide supplemental EI via the A1 interface. This EI helps in refining the model's predictive accuracy and effectiveness in real-time applications. The following A1/ML workflow services are defined.

- **AI/ML Training Services.** These allow for the specification of training requirements, including the necessary data, the model framework, and validation criteria. These services can be offered either directly by the SMO/non-RT RIC framework or through rApps, enhancing flexibility and accessibility for service consumers who wish to train AI/ML models tailored to specific network needs.

- **AI/ML Model Management and Exposure Services.** This category includes several of the following functionalities.
  - **Model Registration/Deregistration.** Enables rApps to register new AI/ML models, update existing registrations, or deregister models, ensuring that only current and effective models are active within the network.
  - **Model Discovery.** Allows rApps to discover available AI/ML models based on specified criteria, ensuring that they can access the most suitable models for their operational requirements.
  - **Model Change Subscription.** Allows rApps to subscribe to and receive notifications about changes to registered AI/ML models, keeping them updated with the latest developments and modifications.
  - **Model Storage.** Provides facilities for storing registered AI/ML models and associated data, securing valuable intellectual property and operational data.
  - **Model Training Capability Registration/Deregistration.** An optional service that allows service producers to showcase or update their capabilities in training AI/ML models or to remove their services from the registry if they are no longer available.
  - **Model Retrieve.** Enables rApps to retrieve location details for specific versions of registered AI/ML models, facilitating ease of access and integration.
- **AI/ML Model Performance Monitoring Services.** These services allow authorized consumers to monitor the performance of deployed AI/ML models, ensuring that the models are functioning optimally and providing expected outcomes within the network environment.

3.9.3 R1 Interface

The R1 interface within the non-RT RIC framework is designed to enhance interactions between the non-RT RIC framework and rApps, and these interactions enable effective management of various aspects of network functionality (O-RAN Work Group 2 2024c). The R1 interface is open and logical, positioned to ensure that it can accommodate multiple vendors, thereby promoting a multivendor environment that is independent of the specific implementations of the SMO and non-RT RIC framework. The design of the R1 interface is inherently extensible, allowing for the integration of new services and data types seamlessly, without the need for altering existing protocols or procedures. This flexibility is crucial for the evolving demands of telecommunications networks.

In its operational capacity, the R1 interface facilitates control signaling and the efficient exchange of data between the rApps and the non-RT RIC. This capability is essential for the dynamic management and optimization of the network, particularly in handling complex data flows and control commands necessary for the effective deployment of network services. The interface supports a comprehensive service management framework that includes the registration and deregistration of services and data types, which ensures that only relevant and up-to-date services are active at any given time. Furthermore, it includes security measures such as authentication and authorization processes to protect the network against unauthorized access and to ensure that only verified rApps can interact with the RIC framework.

Furthermore, the R1 interface aids in the discovery of available services and manages notifications about changes in these services, enhancing the responsiveness of network management practices. It also plays a role in the AI/ML workflow by supporting various functionalities related to artificial intelligence and machine learning applications, which are increasingly integral to modern RAN optimization strategies.

This section is organized into several subsections to systematically address the various service definitions and functionalities associated with the R1 interface. These include discussions on service management and exposure services, data management and exposure services, O2-related services, and rApp management services.

3.9.3.1 R1 Service Definitions

In the O-RAN architecture, services are defined as sets of interrelated capabilities that are specifically designed to be consumed by service consumers from service producers through well-defined service endpoints, typically APIs. These capabilities within a service usually share a common purpose and may depend on each other to function effectively. To enhance reusability and system efficiency, services are designed to minimize or avoid interdependencies across different services. This modular approach ensures that the interoperability of service producers and consumers is maintained through consistent support of service endpoints, independent of the underlying logical function decomposition. Within this framework, both rApps and logical functions in the SMO or the non-RT RIC can assume the roles of service producers or consumers.

The R1 interface supports a variety of services, which include but are not limited to service management and exposure services, data management and exposure services, A1-related services, rApp management services, O2-related services, AI/ML workflow services, and RAN OAM-related services (O-RAN Work Group 2 2024c).

Service Management and Exposure Service Service management and exposure services structured services, which provide a comprehensive list of functionalities to support the lifecycle management of rApps and their interactions with the network services framework. The following service management and exposure services have been defined in O-RAN Work Group 2 (2024c).

- **rApp Registration.** This service allows rApps to formally register with the service management system, making them recognized components within the O-RAN system. This is the initial step for an rApp to become an active participant in network operations.
- **Bootstrap Service (Optional).** An auxiliary service that helps rApps discover the endpoints of the service management and exposure services. This is particularly useful for new rApps integrating into the system, providing them with the necessary connectivity information to begin interactions.
- **Registration of Services.** This service goes beyond simple rApp registration by allowing the registration of specific services that rApps will offer or consume. This includes storing and maintaining a detailed profile for each service, which can be crucial for service discovery and management. It also facilitates the deregistration and deletion of services as needed.
- **Discovery of Services.** This service supports the ability of service consumers to find services based on specific criteria. This is complemented by notification mechanisms that alert service consumers about any changes such as newly registered, updated, or deregistered services.
- **Heartbeat Service (Optional).** This service ensures that an rApp maintains active and recognized registration with the service management system. This service helps in monitoring the presence and operational status of rApps over time.
- **Authentication and Authorization.** These are critical security services that manage the verification of service producers and consumers and ensure that only authorized entities can access or provide specific services. Authentication validates the identity of the participants, while authorization determines their access rights within the network.

- **Communication Support (Optional).** This service facilitates the interaction between rApp service consumers and producers, supporting robust communication across the service management framework.

These services collectively enhance the operational efficiency and security of rApps within the O-RAN architecture, ensuring that they can effectively register, discover, and interact with the network services they need to perform their functions.

Data Management and Exposure Services The data management and exposure services facilitate interaction between data producers and consumers, ensuring efficient data flow and accessibility within the SMO/non-RT RIC framework.

Data producers are entities (often applications or devices) that generate and provide data. In systems like O-RAN, these are typically rApps that produce various types of data based on their monitoring and operational functions. These data can include anything from performance metrics to operational logs, which are essential for network management and optimization. Data producers register the types of data they can supply and specify how these data should be collected, whether through periodic updates or event-triggered mechanisms.

Data consumers are entities that need to access and use the data produced by data producers for various purposes. In O-RAN, data consumers are also often rApps or other network management systems that require access to data to perform tasks such as network optimization, maintenance, and performance analysis. These consumers discover available data types via a management framework and can subscribe to or request specific datasets as needed. The consumption of these data is based on predefined criteria and authorization to ensure security and relevancy.

The data management and exposure services encompass several key functionalities (O-RAN Work Group 2 2024c).

- **Data Registration Service.** This service allows data producers, such as rApps, to register the data types they produce. Data registration is crucial because it communicates to the framework the specifics of the data types available, ensuring that only authorized data producers can register data. This foundational step sets the stage for subsequent data interactions.
- **Data Discovery Service.** Once data types are registered, data consumers can discover these data types based on their needs. This service is gated by authorization to ensure that only approved consumers can access the data, enhancing security and data integrity within the network.
- **Data Subscription and Request Services.** These services allow data consumers to either subscribe to continuously receive data or request data on a one-off basis from the SMO/non-RT RIC framework. Similarly, the framework is enabled to collect data continuously or once based on the needs from data producers. This flexibility in data consumption and collection methods supports dynamic network operations and management.
- **Data Delivery Services.** Critical for the operational efficiency of data exchange, these services ensure the seamless delivery of data from the SMO/non-RT RIC framework to data consumers and from data producers to the framework. This includes handling the logistics of how and when data are transmitted, whether on a scheduled basis or triggered by specific events.
- **Data Offer Service.** Data producers can initiate data collection by making a data offer to the SMO/non-RT RIC framework, which, upon confirmation, triggers the collection of the specified data instance. This proactive approach allows data producers to influence the timing and conditions under which their data are collected.

O2-Related Services O2-related services within the SMO/non-RT RIC framework provide comprehensive management and oversight of the O-Cloud infrastructure. These services enable rApps to interact effectively with the underlying cloud infrastructure, which is vital for the deployment, operation, and maintenance of network functions. These services are summarized below (O-RAN Work Group 6 2024).

- **Infrastructure Management Services.** These services provide rApps with capabilities to access and manage various aspects of the O-Cloud infrastructure:
  - **Infrastructure Inventory.** Accessing detailed listings of available infrastructure resources.
  - **Infrastructure Monitoring.** Overseeing the performance and status of infrastructure components to ensure they are functioning optimally.
  - **Infrastructure Provisioning.** Allocating and configuring infrastructure resources to meet specific network demands or operational requirements.
  - **Infrastructure Lifecycle Management.** Managing the stages of infrastructure components from deployment to decommissioning.
  - **Infrastructure Software Management.** Handling software updates, installations, and maintenance across the cloud infrastructure.
- **Deployment Management Services.** These services are centered around the deployment aspects of the O-Cloud and include the following:
  - **Deployment Inventory.** Keeping track of deployment specifics and statuses.
  - **Deployment Monitoring.** Continuously monitoring deployments to ensure they adhere to defined parameters and are delivering expected performance.
  - **Deployment Lifecycle Management.** Overseeing the entire lifecycle of deployments, from initiation and execution to completion and analysis.
- **Configuration Provisioning.** This service allows rApps to propose and implement configuration changes within the O-Cloud, facilitating adaptability and responsiveness to evolving network needs.
- **Additional Information Access.** Beyond standard management and monitoring functions, O2-related services also provide rApps with the ability to access supplementary information related to the O-Cloud. This might include advanced analytics, additional monitoring insights, or predictive data about infrastructure performance.

These O2-related services ensure that rApps have the necessary tools and information to effectively manage the O-Cloud infrastructure, contributing to optimized network operations and enhanced service delivery within the O-RAN system.

rApp Management Services In the non-RT RIC architecture, rApp management services are integral to ensuring that the app rApps operating within the network maintain optimal performance and reliability. These services provide functionalities that manage and monitor the rApps through various stages of their lifecycle. Here is an overview of the defined rApp management services (O-RAN Work Group 2 2024a).

- **rApp Configuration Service.** This service allows for the configuration settings of an rApp to be set and read. It provides the necessary tools for administrators or automated systems to adjust and fine-tune the settings of rApps to match the operational needs of the network or to adapt to changing conditions.
- **rApp Performance Service.** Through this service, an rApp can report on its own performance and its interactions with other network entities. This is crucial for monitoring the effectiveness

of the rApp and ensuring it is delivering the intended outcomes, such as optimized network performance or enhanced user experiences.

- **rApp Fault Service.** This service enables an rApp to report any faults related to itself or its communications with other entities within the network. Quick fault reporting and resolution are essential for maintaining network stability and reliability, particularly in complex environments where multiple rApps interact.
- **rApp Log Information Service.** This service allows rApps to convey log information, which can include a wide range of data from operational logs to error logs and more. The ability for rApps to report detailed logs is vital for troubleshooting, auditing, and optimizing the overall performance of the network.

These management services are designed to ensure that rApps function efficiently within the O-RAN system, providing essential data and control mechanisms to network operators.

3.9.4 Service Management and Orchestration

SMO in the O-RAN architecture leverages a Service-Based Architecture (SBA) to introduce streamlined roles such as service producers and consumers along with standardized interfaces for interoperability within the SMO. SBA, which does not dictate specific implementations, focuses on defining logical functions according to their roles in service production and consumption. This framework is used in validating services against consumer use cases, identifying service operations with a clear information model, and specifying APIs along with data models for various interfaces. It also pinpoints common services that multiple internal consumers can utilize, such as authentication, authorization, and data management. The SMO manages the RAN by providing a suite of functionalities supported through defined interfaces that interact with other components of the O-RAN architecture. These functionalities include the FCAPS interface for network management, the non-RT RIC for RAN optimization, and the management of the O-Cloud. Key interfaces facilitating these services include the A1 interface connecting the non-RT RIC to the Near-RT RIC for RAN optimization, the O1 interface for FCAPS support excluding the O-RAN-Radio Unit (O-RU), the open fronthaul M-plane interface for hybrid model support between SMO and O-RU, and the O2 interface for integrating platform resources and workload management (O-RAN Work Group 2 2024a). The SMO's flexibility in design allows for either a defined or an undefined boundary toward the non-RT RIC Framework, depending on specific deployment needs.

The SMO framework categorizes functionalities based on their association with the non-RT RIC Framework. First, there is the "Non-RT RIC Framework Anchored Functionality," which directly relates to the non-RT RIC and includes interfaces such as A1 and R1, managing specific interactions within the RIC. Secondly, the "O-RAN SMO Framework Anchored Functionality" includes functionalities that are independent of the non-RT RIC, such as the O1, open FH M-plane, and O2 interfaces, which handle broader network management and orchestration tasks. Lastly, "non-anchored Functionality" may be connected to the non-RT RIC Framework but is not limited to it, allowing flexibility in how services are deployed and utilized. This approach enables the SMO to extend certain services, like those facilitating rApps, beyond just the non-RT RIC. The relationships between the SMO services related to rApps are illustrated in Figure 3.25. Extending the "Services that enable rApps" outside the non-RT RIC (i.e., into the SMO Framework), denotes that the R1 services being exposed to rApps may either come from the non-RT RIC or the SMO. See Section 2.2.1 in Chapter 2 for additional information on the SMO.

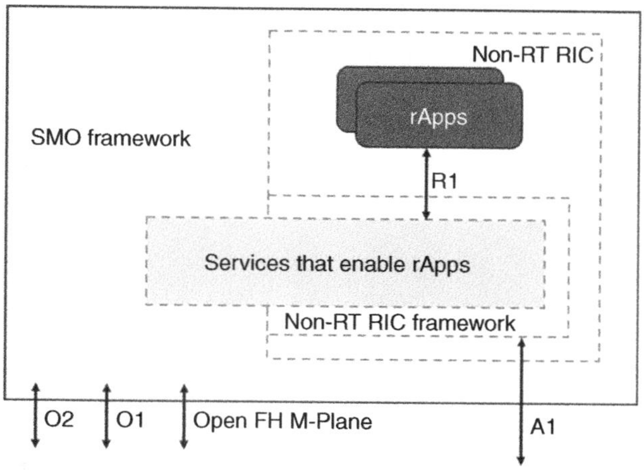

Figure 3.25 Exposure of SMO and Non-RT RIC Services.

3.9.5 rApp Use Cases

Here, we discuss two rApp use cases – traffic steering use case and quality of experience (QoE) use case.

3.9.5.1 Traffic Steering Use Case

As an example of how the non-RT RIC and the Near-RT RIC work with each other to exploit the intelligence aspect of the O-RAN, consider the use case of traffic steering. Note that the description below is one possible way of implementing traffic steering. The "Traffic Steering Use Case" focuses on optimizing network and UE performance across various access technologies in a multiaccess environment (O-RAN Work Group 2 2024e). The goal is to efficiently manage diverse traffic types and frequency bands to improve load balancing, latency, and throughput. This use case addresses the challenges of steering traffic in a balanced manner within networks that utilize multiple frequency bands and access technologies, such as LTE, NR, NR-U, and Wi-Fi. The complexities of QoS aspects, bearer selection, and the need for dynamic adjustment based on radio environment changes and application requirements are central to this use case. Traditional traffic management strategies, which are static, do not sufficiently address the dynamic nature of modern networks, leading to suboptimal performance, spectrum utilization, and increased handover failure.

The non-RT RIC plays a pivotal role in this use case. It retrieves performance and configuration data to define and update policies for traffic management, which are then communicated to the Near-RT RIC for execution. These policies guide decisions on aspects such as carrier and band preferences at a granular per-UE level or group of UEs, facilitating a more intelligent and proactive traffic management approach. The use of machine learning techniques to predict network and UE performance is a key feature, enabling the non-RT RIC to make informed decisions to optimize traffic flow across the network. High-level overview of key functionalities and actions involved are listed below. Detail of implementation and functionalities is provided in O-RAN Work Group 2 (2024e).

- **Data Collection and Policy Definition.** The non-RT RIC collects necessary performance, configuration, and other relevant data. These data aid in defining and regularly updating policies to direct the Near-RT RIC's traffic management functions.

- **Policy Communication and Enforcement.** Defined policies are communicated to the Near-RT RIC, which then interprets and enforces these policies to manage traffic effectively across different cells and access technologies.
- **Utilization of EI.** EI, such as radio fingerprints (e.g., RSRP, RSRQ, RSSI, MCS, and CQI) derived from UE measurement reports, is used by the Near-RT RIC to optimize control functions and accelerate traffic steering decisions, thus reducing signaling overhead and improving response times.

3.9.5.2 QoE Use Case

As another example of how the non-RT RIC and the Near-RT RIC work with each other to exploit the intelligence aspect of the O-RAN, consider the use case of QoE. Note that the description below is one possible way of implementing QoE. The QoE use case focuses on real-time optimization of QoE for 5G native applications that are bandwidth-intensive and latency-sensitive, such as Cloud VR. These applications demand a network that can dynamically adjust to fluctuating network conditions to maintain high service quality. The main goal here is to ensure effective QoE optimization within the O-RAN architecture using open interfaces. This use case is driven by the need to adapt the semistatic QoS frameworks to better meet the varied QoE requirements of applications. This is particularly challenging due to the potential significant fluctuations in radio transmission capabilities. The roles of the non-RT RIC, the Near-RT RIC, and the RAN are mentioned below (O-RAN Work Group 2 2024e).

- **Non-RT RIC.** It gathers QoE-related measurements from network-level reports and potentially from applications via the SMO. These data are used to construct and train AI/ML models that predict QoE, classify applications, and forecast available bandwidth. These models inform the creation of policies or intents that are sent to the Near-RT RIC to guide QoE optimization actions at the RAN level.
- **Near-RT RIC.** It updates and executes AI/ML models received from the non-RT RIC, applying them to assist in real-time QoE optimization.
- **RAN.** It supports QoS enforcement influenced by messages from the Near-RT RIC, derived from the A1 and E2 interfaces, impacting RRM behavior.

3.9.5.3 Operational Flow

There are three main steps.

- **Model Training and Distribution.** The non-RT RIC undertakes the training of ML models using the collected data and deploys these models to the Near-RT RIC. This training involves predictive models for application classification, QoE prediction, and bandwidth availability.
- **Policy Generation and Performance Evaluation.** Once models are deployed, the non-RT RIC generates QoE optimization policies based on current network data and model predictions. These policies are sent to the Near-RT RIC, which then translates them into specific control or policy commands for the RAN to execute.
- **Feedback and Adjustment.** Performance feedback from these operations is sent back to the non-RT RIC, which may adjust the policies and models in response to real-time conditions and performance outcomes.

This QoE use case exemplifies how O-RAN leverages advanced AI/ML capabilities and the dynamic interaction between the non-RT RIC and Near-RT RIC to optimize user experience in a flexible and responsive manner, ensuring that high-demand applications receive the necessary network resources and treatment to meet user expectations.

3.10 Real-Time RIC Closed Loop: RT RIC and μApps/dApps

The Real-Time RAN Intelligent Controller (RT RIC) and its microservices referred to as μApps or dApps represent an innovative approach toward more adaptive network management. Note that the O-RAN specifications do not support RT RIC, μApps/dApps at the time of writing. μApps/dApps can be viewed as candidate microservices that may be standardized in future O-RAN specifications. The μApps documentation, along with code and demos, can be found at EdgeRIC-Documentation (2024). While Northeastern is also developing RT RIC solution, referred to as RT RIC (S'Doro et al. 2022), the implementation detail is not available at the time of writing this chapter.

The RT RIC, especially when considering its real-time capabilities, functions as a component for managing RAN operations at time scales that are essential for applications that require immediate response, which is less than 10 ms.

In Ko et al. (2023), UC San Diego and Texas A&M University jointly developed EdgeRIC—an RT RIC platform located at the edge, near the RAN stack. EdgeRIC operates within a sub millisecond timescale, which is key for Transmission Time Interval (TTI)-level synchronized policies, optimizing PHY-MAC layer operations, and ensuring adherence to the application performance requirements. This real-time decision-making capability enables EdgeRIC to react swiftly to rapidly changing wireless channel conditions, ensuring that resources are allocated efficiently, which is vital for applications that depend on low-latency and high-reliability communication. The implementation details describe the system's ability to execute AI-optimized policies embedded within the RAN stack, demonstrating the potential for significant improvements in throughput and application-level performance benchmarks compared to the Near-RT-RIC solutions. For example, as shown in Figure 3.26, AI-optimized policies trained on DigitalTwin (an emulator hosted in non-RT RIC for training AI-based policies) can be adapted and executed on EdgeRIC, offering considerable performance improvements. These policies, once trained, can be rapidly

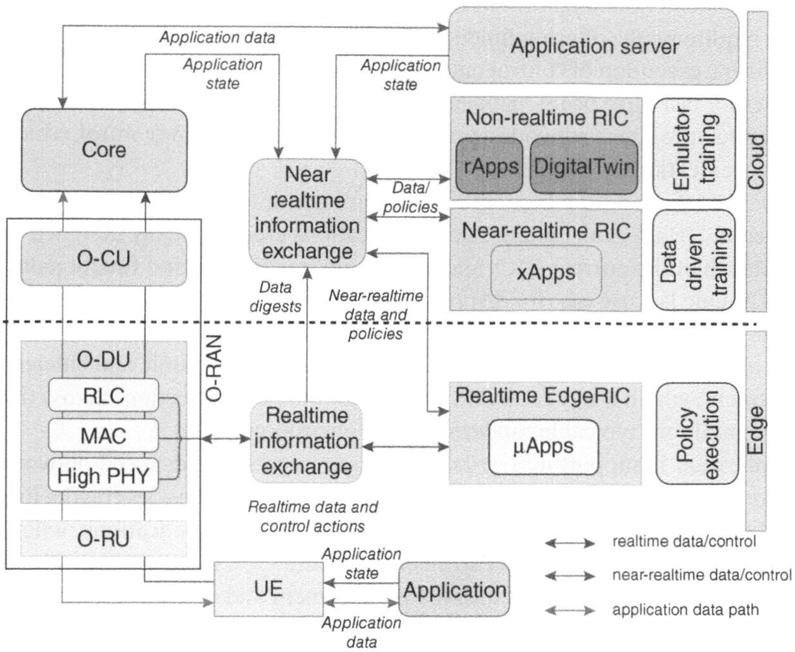

Figure 3.26 RICWorld Concept Architecture, Showing EdgeRIC and O-RAN Integration (Ko et al. 2023).

updated and executed on EdgeRIC, aligning with the goals of non-RT RIC for policy training and adaptation.

It is important to mention why such an RAN Intelligence Controller is being proposed. A RT RIC will address several challenges and requirements in the O-RAN system, particularly for high-demand, real-time applications like streaming media, AR/VR, and robotic control systems. The main motivations include but not limited to the following.

1) **Precision and Latency.** EdgeRIC is designed to operate on a submillisecond timescale, aligning with the TTI dynamics of the RAN. This fine-grained control is essential for applications requiring low latency and high reliability, as it allows for adjustments based on real-time channel variations and network conditions.
2) **Resource Efficiency.** Traditional approaches to network management, like resource slicing by non-RT and Near-RT RICs, often result in coarse and sometimes inefficient allocation of resources. EdgeRIC allows for more finite and dynamic resource management, enabling the RAN to adapt quickly and more efficiently without the heavy overhead of frequent reconfiguration.
3) **Integration and Fault Tolerance.** EdgeRIC operates as a decoupled microservice on edge compute platforms, which provides flexibility and scalability. It does not directly interfere with the primary RAN functions, thus maintaining the operational integrity of the RAN even if EdgeRIC fails or is offline. This design ensures that the RAN continues its normal operations without dependency on the EdgeRIC's state (Ko et al. 2023).
4) **Cross-layer Optimization.** By positioning EdgeRIC close to the RAN but maintaining decoupling, it can leverage real-time data for AI-driven optimizations without impacting the core latency-sensitive tasks of the RAN. This setup supports cross-layer decision-making that accounts for both application needs and network conditions, optimizing for factors like throughput and service quality dynamically.

On the other hand, Northeastern University discusses the concept of dApps (S'Doro et al. 2022) that enable fine-grained management and control at the CUs/DUs. These dApps can handle use cases with stricter timing requirements, offering functionalities that go beyond the capabilities of current RICs, such as real-time execution of control operations, access to lower-layer information, and processing I/Q samples for tasks like beam management. The introduction of dApps as a concept within the O-RAN architecture aims to provide a practical approach to real-time control, which is missing in the existing specifications.

dApps offer several use cases that address the real-time control and inference requirements in O-RAN systems, as detailed in S'Doro et al. (2022). These applications are specifically designed to operate in tighter timescales than the current RICs allow, bringing intelligence and functionality directly to the CUs and DUs. One key use case for dApps is enhancing beam management capabilities within the network. dApps can manage and optimize beamforming procedures directly at the edge, utilizing detailed, real-time data from the network. This enables more dynamic and efficient beam selection and adjustment, which is key for maintaining high-quality connections across the rapidly changing network conditions typical in modern cellular environments.

Another significant application is supporting low-latency requirements for advanced network services. dApps are positioned to handle RT RAN slicing and scheduling decisions, leveraging live data about network conditions and UE. This capability is vital for URLLC applications, where decisions need to be made in milliseconds to meet stringent QoS and QoE requirements. These use cases highlight the potential of dApps to transform the management and operational capabilities of next-generation networks by enabling real-time data processing and decision-making at the edge of the network. This advancement is key for supporting the evolving demands of high-speed, low-latency applications in 5G and beyond.

3.10.1 Resource Block (RB) Scheduling μApp

Scheduling decisions—such as how many PRBs to allocate to connected users or which user to serve in each slot—are made in real-time, typically on a per-slot or TTI basis. Scheduling is a critical network function, as optimized scheduling can significantly improve network performance in terms of throughput, latency, and overall efficiency. The formation of queues, which directly affects user Quality of Service (QoS), is closely tied to scheduling decisions. By intelligently prioritizing certain types of traffic demands, optimal QoS guarantees can be achieved, rather than using simple round-robin scheduling approaches.

Traditional scheduling algorithms offer several optimal approaches, but recent research has demonstrated the potential of AI-based schedulers to outperform conventional methods. For example, Ko et al. (2023) in their work on EdgeRIC have shown that Reinforcement Learning (RL)-based agents can effectively be used to make scheduling decisions within the network stack. These agents, trained with different reward maximization objectives, can adapt to various system performance goals, optimizing key metrics like throughput and latency in dynamic network environments.

3.10.1.1 Workflow for a RB Scheduling μApp

- Retrieve Network State: Since scheduling is an every-slot operation, the current network state must be continuously retrieved from the network. The network state typically includes metrics such as per-UE channel conditions, queue lengths, and other relevant parameters. These metrics are essential for making informed scheduling decisions regarding how many PRBs to allocate to each UE.
- Make a Decision: As the scheduling policy operates within a μApp, which functions at sub-millisecond timescales, it is crucial that the decision-making process is completed within this limited time frame. This imposes strict requirements on the efficiency of the scheduling algorithm, necessitating both algorithmic and system-level innovations to ensure real-time responsiveness.
- Scheduler Algorithm: The specific scheduling algorithm employed depends on the use case. For instance, the policy could aim to optimize QoS for individual users, or it could focus on maximizing overall system throughput. The design of the algorithm will vary based on the desired network performance objectives.
- Send the Decision Back: Once a scheduling decision has been made, the μApp sends the decision back to the network, ensuring timely resource allocation to the selected UEs.

3.10.1.2 Other Features and Future Directions

- Real Time synchronization: For efficient μApp functionality, maintaining TTI-by-TTI synchronization is crucial. This ensures that each decision is based on the latest network state and that actions are applied in real-time. Any delay or desynchronization can break the correlation between state and action, especially in policies trained for real-time reinforcement learning (RL) agents. To address potential lags, μApps must be capable of discarding outdated samples, ensuring that inferences are always made on the most recent data.
- Model adaptation: μApps gather, process, and infer time-series data, continuously collecting artifacts such as channel states and traffic patterns. This allows for more accurate and dynamic

control of network resources, adapting to real-time demands. By leveraging meta-learning techniques, μApps will be able to adapt models to new network conditions more rapidly. This capability allows for continuous learning and improvement, ensuring that policies remain optimal even as network environments evolve.

3.10.2 Beam Management dApp Use Case

The use case of enhanced beam management involves an application of real-time data processing and decision-making at the edge of the network. This will enhance beamforming optimization, one of the major aspects of modern cellular networks, especially in the new generation of wireless communication systems, where massive MIMO and beamforming are core technologies. Here is an overview of how dApps can enhance beam management.

3.10.2.1 Key Elements and Their Functions

- **dApps.** These are applications deployed directly on CUs and DUs within the O-RAN architecture (S'Doro et al. 2022). They are designed to process data in real-time, making immediate decisions required for managing dynamic network conditions.
- **Real-time Data Collection.** dApps collect real-time data from various sources within the network, including UE feedback, which provides information about signal quality, network conditions, and UE locations.
- **Beam Selection Algorithms.** These algorithms analyze the collected data to determine the optimal beam for each UE. The selection is based on several factors, including signal strength, interference patterns, and the mobility of the UE.
- **AI/ML Models.** AI/ML models are employed to predict changes in network conditions and UE behavior. These models can adapt beamforming strategies based on learned patterns and predictive analytics, enhancing the system's ability to respond to fluctuations in network demand and conditions.

3.10.2.2 Operational Flow for Beam Management

The sequence for enhancing beam management through dApps can be described as follows:

- **Data Acquisition.** dApps continuously gather and process real-time data from the network's edge. This includes signal measurements from UEs and other relevant telemetry data.
- **Data Analysis.** The collected data are analyzed using AI/ML models within the dApps. This analysis helps in understanding current network conditions and predicting future states. For example, if a group of UEs is moving from one cell site to another, the dApp can predict potential coverage issues or interference problems.
- **Decision Making.** Based on the analysis, dApps make decisions about beam allocation. This involves selecting the most appropriate beam for each UE to maximize signal quality and minimize interference. Decisions are made in real-time to reflect current network conditions.
- **Beam Adjustment.** Once a decision is made, commands are sent from the dApps to the RAN components to adjust the beams. This might involve switching a UE from one beam to another or altering the beam's characteristics (like its width or direction).
- **Closed Loop Control.** The system maintains a closed loop where the outcomes of beam adjustments are monitored. The performance data fed back into the dApps helps in refining the beam selection algorithms and AI/ML models, continuously improving the system's accuracy and efficiency.

- **Optimization and Updates.** As part of the closed loop implementation, the dApps periodically update their algorithms and models based on new data and emerging patterns. This ensures that the beam management system remains optimal and can handle evolving network scenarios.

At the time of writing, these two papers Ko et al. (2023) and S'Doro et al. (2022) discussed the architecture necessary to support real-time applications, proposing an extension of the O-RAN standards to include these applications. The proposed architecture underlines the advantages of real-time applications, such as reduced latency, AI at the edge, controlling MAC and PHY layer functionalities, and access to CU/DU data and functionalities in real time. The O-RAN ALLIANCE's nGRG's role in O-RAN is to drive research into these next-generation RAN architectures, focusing on the integration of AI and machine learning. They support the creation of an open and intelligent RAN system, which would benefit from RT RIC and μApps/dApps, enhancing network adaptability and efficiency. O-RAN ALLIANCE's nGRG group has recently approved RT RIC and its microservices (i.e., μApps/dApps) under research item 2 (RS-02), which presents a significant opportunity for potential adoption in the future versions of O-RAN architecture.

Overall, the proposal to enhance the RIC with real-time processing capabilities like EdgeRIC and μApps reflects a strategic response to the evolving demands of next-generation networks. It ensures that the network can handle the increasing complexity and stringent real-time requirements of modern telecommunications, thereby supporting a wide array of applications and services efficiently and effectively.

3.11 Multi-time-scale RIC Framework

The authors of EdgeRIC, in collaboration with the NextG Wireless Lab at NC State, introduce the Multi-Time-Scale RIC framework designed to optimize Radio Access Networks (RAN) across varying time scales. Figure 3.27 presents a high-level vision of multi-time-scale RIC framework and supported use cases. This framework facilitates the seamless interaction of Non-RT RIC, Near-RT

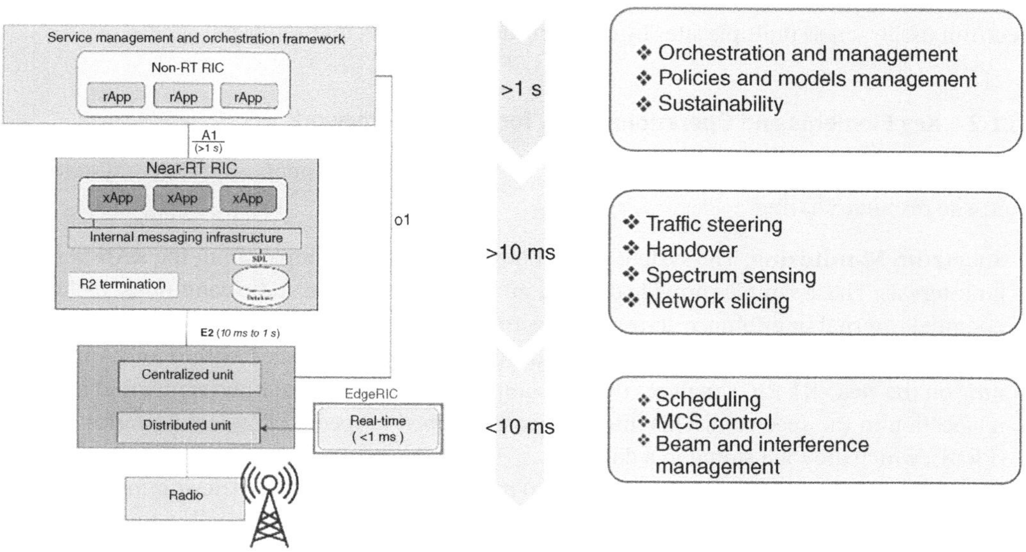

Figure 3.27 Multi-Time-Scale RIC Framework Vision and It's Use Cases.

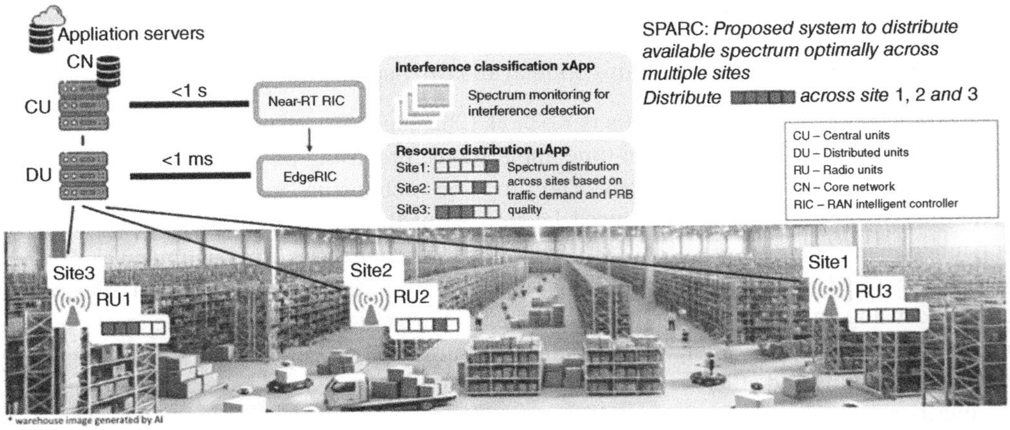

Figure 3.28 Overview of SPARC usecase.

RIC, and EdgeRIC, enabling them to share information and coordinate control actions to enhance RAN functionalities according to specific time requirements.

By adopting a multi-time-scale approach, this framework significantly improves RAN optimization, especially for applications such as spectrum sharing and multi-site spectrum management. By harnessing the distinct capabilities of each RIC operating at different time scales, it ensures that the limited spectrum is utilized more effectively and efficiently.

3.11.1 Use Case: SPARC for Multi-Site Spectrum Management

A notable example of this multi-time scale architecture in action is the **SPARC** system (Spatio-Temporal Adaptive Resource Control), which was developed for **multi-site spectrum management** in NextG cellular networks (Ghosh et al. 2024). As shown in Figure 3.28, SPARC effectively combines the Near-RT RIC with the Real-Time RIC (EdgeRIC) to optimize spectrum usage across multiple sites by detecting and mitigating interference.

3.11.2 Key Elements and Operational Flow for SPARC Framework

Figure 3.29 shows an overview/walkthrough of the main components of SPARC system for multi-site resource sharing.

1) **Spectrum Monitoring**: The system continuously collects I/Q samples from the RAN at regular intervals. These samples provide detailed information about the radio conditions, including potential external interference across the spectrum.
2) **Interference Detection and Localization**: An AI/ML model hosted within an xApp (running on the Near-RT RIC) analyzes the I/Q samples to detect external interference and identify its location in the spectrum. This interference is mapped to specific Physical Resource Blocks (PRBs), which are then stored in a database.
3) **Resource Allocation**: The EdgeRIC (μApp) queries the database to retrieve information on PRBs affected by interference. Based on this information, the μApp dynamically adjusts resource distribution and allocation across multiple sites to maximize spectral efficiency and minimize interference.

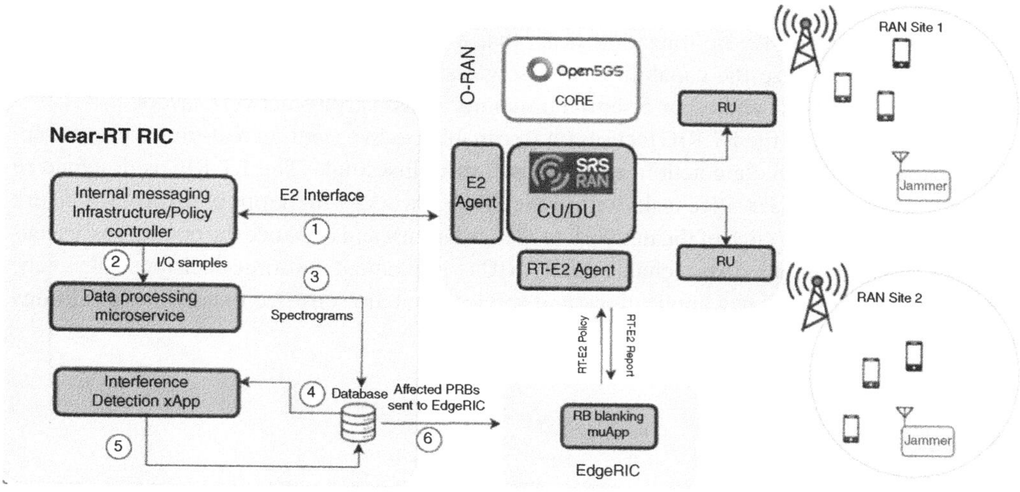

Figure 3.29 Operational Flow of SPARC System.

4) **Real-Time Adjustments**: The resource allocation decisions are implemented in real-time, adjusting user scheduling and resource assignment across the network every transmission time interval (TTI). This Near-real-time control loop operates with latency under 1 second, while real-time control at the EdgeRIC operates with sub-millisecond latency.

3.12 Major Takeaways of the Chapter

Control loops within O-RAN architecture play a crucial role in dynamically optimizing network performance at different time scales. This chapter discusses these control loops and associated microservices such as xApps, rApps, and also μApps/dApps. Here are the major takeaways of the chapter.

- There are three primary types of control loops—non-RT, Near-RT, and Real-Time. Each operates on different timescales to manage and optimize network functions efficiently. The non-RT loop operates for over 1 second, the Near-RT from 10 ms to 1 second, and the RT within milliseconds. These loops allow for the adaptive management of network resources.
- The internal architecture of Near-RT RIC includes several microservices like the IMI, Conflict Mitigation, Subscription Manager, Security Subsystem, and more. These services ensure efficient data handling, conflict resolution among xApps, secure operations, and effective communication within the RAN.
- The chapter highlights the Near-RT RIC) and its interaction with xApps. The Near-RT RIC serves as a component for RAN control and optimization, leveraging microservices architecture to host and manage xApps effectively. xApps are key enablers for customizing and enhancing the RAN operations. They interact with the Near-RT RIC to execute control actions based on real-time data analysis, directly influencing the RAN behavior to optimize the network performance and resource utilization.
- The non-RT RIC is positioned within the SMO framework, managing broader network optimization and policy management over timescales longer than those managed by the Near-RT

RIC. It provides high-level guidance and resources to the Near-RT RIC and rApps, helping to orchestrate more complex optimization tasks that do not require millisecond-level response times. The rApps utilize the capabilities and services exposed by the non-RT RIC to execute RAN optimization tasks and other essential functions across various network layers.

- The discussion around the RT RIC focuses on its capabilities in executing real-time control functions that require immediate action, typically within milliseconds. The RT RIC is designed to manage the most time-sensitive tasks within the RAN. It works with μαpps or dApps, which are deployed directly at the edge of the network to reduce latency and enhance the operations' execution speeds. These μApps/dApps enable the RT RIC to implement real-time decisions effectively, supporting ultra-low latency applications and services that are sensitive to network conditions and require instant response.

References

Bimo, F.A. et al. (2022, October). OSC community lab: The integration test bed for O-RAN software community. In: *2022 IEEE Future Networks World Forum (FNWF)*, 513–518. IEEE.

Coronado, E., Khan, S.N., and Riggio, R. (2019). 5G-EmPOWER: A software-defined networking platform for 5G radio access networks. *IEEE Transactions on Network and Service Management* 16 (2): 715–728.

'Doro, S., Polese, M., Bonati, L. et al. (2022). dApps: Distributed applications for real-time inference and control in O-RAN. *IEEE Communications Magazine* 60 (11): 52–58.

EdgeRIC-Documentation (2024). *EdgeRIC v0.1 Documentation*. https://edgeric.github.io/index.html (accessed July 24 2024).

Ghosh, U., Chiejina A., Stephenson N. H., Shah, V. K., Shakkottai S., and Bharadia, D. (2024). SPARC: Spatio-Temporal Adaptive Resource Control for Multi-site Spectrum Management in NextG Cellular Networks, *ACM Conference on Emerging Networking EXperiments and Technologies (CoNext)*.

Johnson, D., Maas, D. and Van Der Merwe, J. (2022). NexRAN: Closed-loop RAN slicing in POWDER-A top-to-bottom open-source open-RAN use case. In *Proceedings of the 15th ACM Workshop on Wireless Network Testbeds, Experimental Evaluation & Characterization* (pp. 17–23). (accessed July 24 2024).

Ko, W.H., Ghosh, U., Dinesha, U., Wu, R., Shakkottai, S. and Bharadia, D. (2023). *EdgeRIC: Empowering Realtime Intelligent Optimization and Control in NextG Networks*. arXiv preprint arXiv:2304.11199.

ONF White Paper (2020). *SD-RAN ONF's Software-Defined RAN Platform Consistent with the O-RAN Architecture*. https://opennetworking.org/wp-content/uploads/2020/08/SD-RAN-v2.0.pdf (accessed July 24 2024).

Open AI Cellular (OAIC) Platform (2024). https://openaicellular.github.io/oaic/xapp_python.html (accessed July 24 2024).

O-RAN ALLIANCE (2024a). *O-RAN Architecture Description*. O-RAN.WG1.OAD-R003-v09.00. The O-RAN ALLIANCE. https://orandownloadsweb.azurewebsites.net/specifications (accessed 24 July 2024).

O-RAN ALLIANCE (2024b). *Use Cases Analysis Report*. O-RAN.WG1.Use-Cases-Analysis-Report-R003-v14.00. https://specifications.o-ran.org/specifications (accessed 24 July 2024).

O-RAN Work Group 1 (2024). *O-RAN AI/ML Workflow Description and Requirements 1.03*. O-RAN.WG2.AIML-v01.03. https://specifications.o-ran.org/specifications (accessed 24 July 2024).

O-RAN Work Group 2 (2024a). Non-RT RIC: Architecture. O-RAN.WG2.Non-RT-RIC-ARCH-R003-v05.00. The O-RAN ALLIANCE. https://specifications.o-ran.org (accessed 1 October 2024).

O-RAN Work Group 2 (2024b). *O-RAN A1 Interface: General Aspects and Principles 3.03.* O-RAN.WG2.A1GAP-v03.03. The O-RAN ALLIANCE. https://orandownloadsweb.azurewebsites.net/specifications (accessed 24 July 2024).

O-RAN Work Group 2 (2024c). *O-RAN R1 Interface: General Aspects and Principles 8.0.* O-RAN.WG2.R1GAP-v08.00. https://specifications.o-ran.org/specifications (accessed 24 July 2024).

O-RAN Work Group 2 (2024d). *O-RAN R1 Interface: Use Cases Requirements.* O-RAN.WG2.R1UCR.v04.00. https://specifications.o-ran.org/specifications (accessed 29 July 2024).

O-RAN Work Group 2 (2024e). *O-RAN Non-RT RIC & A1 Interface: Use Cases and Requirements.* O-RAN.WG2.Use-Case-Requirements-R003.v10.00. https://specifications.o-ran.org/specifications (accessed 29 July 2024).

O-RAN Work Group 2 (2024f). O-RAN Working Group 2, "A1 interface: Application Protocol", *O-RAN.WG2.A1AP-R003-v04.02*, https://specifications.o-ran.org/specifications (accessed 4 Oct 2024).

O-RAN Work Group 3 (2024a). *O-RAN Near-Real-time RAN Intelligent Controller and E2 Interface: E2 Application Protocol (E2AP) 5.0.* O-RAN.WG3.E2AP-R003-v05.00. https://specifications.o-ran.org/specifications (accessed 24 July 2024).

O-RAN Work Group 3 (2024b). *Near-RT RIC and E2 Interface: E2 Service Model (E2SM) 5.0.* O-RAN.WG3.E2AP-R003-v05.00. O-RAN.WG3.E2SM-R003-v-05.00. https://specifications.o-ran.org/specifications (accessed 24 July 2024).

O-RAN Work Group 3 (2024c). *O-RAN E2 Service Model (E2SM) KPM 5.0.* O-RAN.WG3.E2SM-KPM-R003-v05.00. https://specifications.o-ran.org/specifications (accessed 24 July 2024).

O-RAN Work Group 3 (2024d). *O-RAN E2 Service Model (E2SM), RAN Control 6.0.* O-RAN.WG3.E2SM-RC-R003-v06.00. https://specifications.o-ran.org/specifications (accessed 24 July 2024).

O-RAN Work Group 3 (2024e). *O-RAN E2 Service Model (E2SM) Cell Configuration and Control 4.0.* O-RAN.WG3.E2SM-CCC-R003-v04.00. https://specifications.o-ran.org/specifications (accessed 24 July 2024).

O-RAN Work Group 3 (2024f). *O-RAN Near-Real-time RAN Intelligent Controller E2 Service Model (E2SM), RAN Function Network Interface (NI) 1.0.* ORAN-WG3.E2SM-NI-v01.00. https://specifications.o-ran.org/specifications (accessed 24 July 2024).

O-RAN Work Group 6 (2024). *O-RAN O2 Interface General Aspects and Principles.* O-RAN.WG6.O2-GA&P-R003-v07.00. https://specifications.o-ran.org/specifications (accessed 29 July 2024).

O-RAN Working Group 3 (2024). *O-RAN Near-RT RAN Intelligent Controller Near-RT RIC Architecture 6.00.* O-RAN.WG3.RICARCH-R003-v06.00. https://specifications.o-ran.org/specifications (accessed 24 July 2024).

O-RAN.WG3.E2GAP (2024). *O-RAN E2 General Aspects and Principles (E2GAP) 5.0.* O-RAN.WG3.E2GAP-R003-v05.00. https://specifications.o-ran.org/specifications (accessed 24 July 2024).

OSC-RICAPP (2020) *OSC RIC Applications: On-boarding and Deploying xApps.* https://wiki.o-ran-sc.org/display/RICA/On-boarding +and+Deploying+xApps (accessed 30 July 2024).

Polese, M., Bonati, L., D'oro, S. et al. (2023). Understanding O-RAN: Architecture, interfaces, algorithms, security, and research challenges. *IEEE Communications Surveys & Tutorials* 25 (2): 1376–1411.

Schmidt, R., Irazabal, M., and Nikaein, N. (2021, December). FlexRIC: An SDK for next-generation SD-RANs. In: *Proceedings of the 17th International Conference on emerging Networking EXperiments and Technologies*, 411–425. https://doi.org/10.1145/3485983.3494870.

Telecom Infra Project (2024). *OpenRAN: RAN Intelligence and Automation.* https://cdn.mediavalet.com/usva/telecominfraproject/Xhv2dK0k4E6fLYIktAv0AQ/dHvZIwafWkiP53YBcphg6A/Original/RIA_OpenRAN_ataglance_Glossy_v08_2021_06_16.pdf (accessed 24 July 2024).

Upadhyaya, P.S., Abdalla, A.S., Marojevic, V., Reed, J.H. and Shah, V.K. (2022). *Prototyping Next-generation O-RAN Research Testbeds with SDRs.* arXiv preprint arXiv:2205.13178.

4

O-RAN Security

In contrast to conventional cellular networks, Open Radio Access Network (O-RAN) represents a paradigm shift with its open, programmable, interoperable, and virtualized Radio Access Network (RAN) architecture. This innovative approach empowers network operators to seamlessly integrate hardware (HW) and software components from diverse vendors, fostering enhanced flexibility, scalability, and cost efficiency. Furthermore, it fosters competition among vendors. This chapter delves into the critical security implications stemming from these characteristics. It comprehensively examines the various security dimensions inherent in the O-RAN architecture, shedding light on the intricate interplay between openness and safeguarding network integrity.

4.1 Motivation for O-RAN Security

O-RAN is a significant advancement in the fifth-generation (5G) next-generation Radio Access Network (NG-RAN) that promotes a fundamental change toward open, interoperable, and intelligent network architectures. Through the standardization of interfaces between key RAN components such as the O-RAN Radio Unit (O-RU), O-RAN Distributed Unit (O-DU), and O-RAN Central Unit (O-CU), O-RAN facilitates the elimination of traditional vendor lock-ins, fostering an environment of interoperable solutions. This strategic openness aligns seamlessly with 5G's objectives of scalability, flexibility, and support for a diverse array of applications, including ultra-reliable low-latency communications (URLLC), massive machine-type communications (mMTC), and enhanced mobile broadband (eMBB). By disaggregating HW and software layers, O-RAN not only enables a modular network design (i.e., structural approach for network implementation) but also facilitates dynamic resource management and efficient network slicing, essential for meeting the dynamic demands of 5G networks.

At the core of O-RAN's innovation lies the RAN Intelligent Controller (RIC), which embodies the network's shift toward leveraging artificial intelligence (AI) and machine learning (ML) for real-time optimization. This focus on intelligence and automation aligns perfectly with 5G's need for optimized operational efficiency and peak performance across various service scenarios. Furthermore, O-RAN's emphasis on developing secure, energy-efficient network solutions reflects the priorities of modern telecommunications networks: ensuring trustworthiness in network applications supporting critical services and advancing sustainable technological development. Thus, O-RAN not only aligns with but also accelerates the fulfillment of 5G NG-RAN's goals, marking a significant step forward in the journey toward more open, flexible, and forward-thinking telecommunications networks.

Fundamentals of O-RAN, First Edition. Nishith D. Tripathi and Vijay K. Shah.

Companion website: www.wiley.com/go/tripathioran

4.1.1 Vulnerabilities of O-RAN Compared to Non-O-RAN Implementations

The O-RAN architecture (Figure 4.1), distinct from the traditional 3GPP RAN framework, introduces a broader attack surface due to its novel components, interfaces, and technologies. This transition from the conventional RAN framework to the state-of-the-art O-RAN implementation is characterized by the incorporation of new O-RAN-specific elements and the embrace of virtualization/containerization techniques, open-source code utilization, and AI/ML model support. Unlike the more contained environment of non-O-RAN implementation, O-RAN's design architecture significantly widens potential security vulnerabilities. Current research focuses on identifying these vulnerabilities and proposing mitigation strategies, yet it requires the necessity for further exploration. The open nature of O-RAN increases the number of potential threat actors, making comprehensive security solutions essential (Liyanage et al. 2023; Polese et al. 2023).

According to O-RAN WG11, these threat surfaces or vulnerabilities are categorized into six primary groups (O-RAN Working Group 11 2024a).

- **Additional functions**, including the Service Management and Orchestration (SMO), Non-Real-Time RIC (Non-RT RIC), and Near-Real-Time RIC (Near-RT RIC), introduce complex security challenges.
- **Additional open interfaces**, such as A1, E2, O1, O2, and open fronthaul, create new points of attack.
- **The modified architecture**, the lower layer split (LLS) 7.2x, creates a new attack surface.

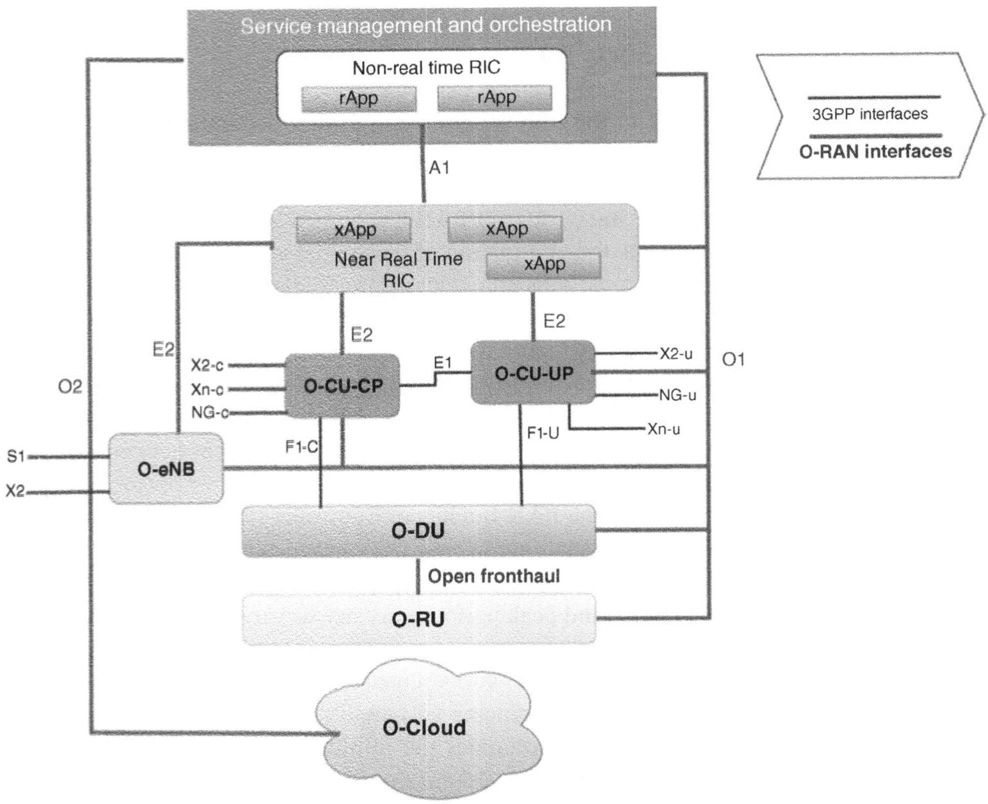

Figure 4.1 Logical Architecture of O-RAN System.

- **Decoupling** of HW and software amplifies threats to the trust chain, undermining the integrity and reliability of communications.
- **Containerization and virtualization,** representing disaggregation of software and HW, increases the complexity of securing network functions (NFs).
- The utilization of **open-source code** may elevate exposure to public exploits, resulting in more security breaches.

Furthermore, O-RAN WG11 identifies additional entry points that could facilitate the propagation of threats, including APIs between planes, internal threats originating within the O-RAN system itself, and external threats from actors outside the O-RAN system. This comprehensive identification of threat surfaces and entry points underscores the necessity for robust security measures tailored to the O-RAN architecture's unique vulnerabilities, ensuring the protection of this open network infrastructure against a wide array of potential cyber threats.

4.1.2 Scope of O-RAN Security Specifications

O-RAN has security specifications designed to address its unique architecture and components. These specifications cover a range of security aspects inherent to the O-RAN architecture, focusing on ensuring the O-RAN networks' confidentiality, integrity, and availability. However, some security mechanisms fall outside the O-RAN's scope, often relying on standards and specifications defined by other bodies, such as 3GPP (O-RAN Working Group 11 2024a).

4.1.2.1 Within O-RAN Security Scope

Within the O-RAN security framework, attention is devoted to protecting its distinctive NFs, applications, and cloud computing platform, collectively forming the backbone of O-RAN's innovative architecture. This comprehensive security scope covers the protection of unique interfaces such as A1, E2, O1, O2, and open fronthaul, alongside pivotal components including the SMO, the Non-RT RIC and remote applications (rApps), the Near-RT RIC and extended applications (xApps), as well as specific units like the O-CU-Control Plane/User Plane (O-CU-CP/UP), the O-DU, the O-RU, and the O-RAN next-generation-evolved Node B (O-eNB). The O-RAN cloud computing platform, or O-Cloud, further extends this protection to a collection of physical infrastructure nodes—designed to host essential functions like the Near-RT RIC and the O-CU—complemented by supporting software components (e.g., operating systems (OSs), virtual machine (VM) monitors, container runtimes) and crucial management and orchestration functions.

The security specifications within O-RAN ambitiously tackle the challenges posed by the heavy reliance on virtualized network functions (VNFs), containerized network functions (CNFs), and the use of open-source software, addressing risks through defining robust authentication, authorization, encryption, and integrity protection mechanisms. Additionally, they are geared toward ensuring the security of these virtualized environments, including the orchestration and management layers, isolation between virtual functions, and the secure integration and management of open-source components through practices such as vulnerability management, patching processes, and secure coding. Furthermore, O-RAN's security measures extend to protecting AI/ML models from adversarial attacks, thereby ensuring the integrity of AI-driven decisions and securing the data these models utilize. Relevant protocols used by the O-RAN system for enforcing security are mentioned below.

- **Transport Layer Security (TLS).** TLS is a cryptographic protocol designed to provide secure communication over a computer network. It is specified for protecting communication between the O-RAN system and other network elements. By establishing a secure channel, TLS ensures confidentiality, integrity, and authenticity of the transmitted data. For the O1 interface, TLS should be implemented to secure NETCONF and JSON/REST protocols, enhancing the security of these connections. Additionally, the application of TLS is advised for the A1 interface.
- **Secure Shell (SSH).** SSH is a protocol used to securely access and manage network devices over an unsecured network. It is specified to be used in O1 interface and NETCONF for the fronthaul management plane (M-plane) should be implemented over SSH.
- **Internet Protocol Security (IPsec).** IPsec is a suite of protocols developed to ensure CIA data communications over an IP network. It should be used to protect E2 traffic between the RAN and the Near-RT RIC.
- **File Transfer Protocol (FTP) and File Transfer Protocol Secure (FTPS).** FTP is a standard network protocol used to transfer files between a client and a server on a computer network. FTPS is an extension to FTP that adds support for the TLS and the Secure Socket Layer (SSL) cryptographic protocols. They are used in protecting file transfers over O1 and fronthaul M-plane interfaces.
- **Precision Timing Protocol (PTP)**, as defined by (IEEE-std-1588-2019 2023), is used to ensure precise synchronization of clocks throughout the RAN, which is essential for the coordinated and reliable operations in 5G networks. It supports secure and efficient data transmission, minimizing vulnerabilities related to timing discrepancies among network components.

4.1.2.2 Outside O-RAN Security Scope

While O-RAN focuses on the radio access network's openness and interoperability, it typically does not redefine security mechanisms that are already covered by 3GPP standards for core network components and end-to-end communication. This includes user data and signaling protection between the user equipment (UE) and the core network, which are handled by protocols like Evolved Packet System Authentication and Key Agreement (EPS AKA) for authentication and NAS encryption. Furthermore, the security of the underlying transport networks, crucial for backhauling data across Internet Protocol/Multiprotocol Label Switching (IP/MPLS) networks, and the provision of end-to-end encryption, particularly at the application layer, are areas typically considered outside O-RAN's scope. These components, alongside UE, multi-access edge computing (MEC), core network elements, and antennas, are guided by broader industry standards and practices or handled by application providers leveraging existing security protocols like TLS.

4.1.3 Critical Assets

In an O-RAN system, several critical assets warrant protection due to their importance in network operation, integrity, security, and performance. Protecting these assets is vital to ensure the confidentiality, integrity, and availability of the network and its services. In the O-RAN architecture, as delineated by the O-RAN WG11, a comprehensive list of 68 critical assets is identified, segmented into 38 data-critical assets and 30 component (logical, virtual, and physical) critical assets (O-RAN Working Group 11 2024a). This classification highlights the intricate nature of O-RAN systems and underscores the necessity of a multifaceted approach to security, tailored to protect these assets effectively.

4.1.3.1 Critical Data Assets

Among the data critical assets, several categories are recognized for their pivotal roles in ensuring the seamless operation and security of the O-RAN system. Some of these assets are given below.

- **Critical Synchronization Plane (S-Plane) Data.** This includes vital data flows for synchronization and timing information between nodes, which are essential for maintaining network coherence and reliability.
- **Critical Management Plane Data.** Such data, especially when transported over the fronthaul interface, include operational configurations and state information critical to network management.
- **Critical Control Plane (C-Plane) Data.** Information like scheduling details, Fast Fourier Transform (FFT) size, Cyclic Prefix (CP) length, subcarrier spacing, downlink and uplink (UL) PRACH scheduling is crucial for effective communication and resource allocation within the network.
- **Data Transported Over Interfaces (O1, A1, E2).** The integrity and confidentiality of data flowing through these open interfaces are paramount, given their role in controlling, optimizing, and managing the RAN.
- 3GPP application-related data such as subscription data, session data, and call control-related information.
- **Reference Signals and Synchronization Signals.** These signals are essential for network stability and connectivity, requiring stringent protection measures.
- **O-Cloud Provisioning Information and Cryptographic Keys.** These assets are critical for securing the virtualized infrastructure and encrypting communications, respectively.
- **Inter- and Intra-Slice UE Priority Information.** This data determines the prioritization of network resources among users, impacting both service quality and security.
- Security event log files generated by O-RAN components.

4.1.3.2 Critical Component Assets

The component critical assets include vital elements of the O-RAN architecture. These components can be physical, logical, or virtual. Some of these are highlighted below:

- **Service Management and Orchestration (SMO).** Acts as the central management hub, coordinating across the O-RAN components.
- **Near-RT RIC Software and xApps.** These software components are integral for real-time network optimization and decision-making.
- **O-CU, O-DU, O-RU, O-CU-CP, O-CU-UP Software.** The software driving these units is crucial for the processing and forwarding of data across the network.
- External interfaces termination at SMO framework, Non-RT RIC framework.
- **A1, R1, O1 Interfaces, Terminations, and Functions.** These include both the interfaces themselves and the functionalities they provide, crucial for intercomponent communication and network flexibility.

The highlighted critical assets within the O-RAN system, as detailed by O-RAN WG11, represent just a section of a broader list consisting of both data and component assets critical to the O-RAN network security. For those seeking comprehensive insights into all identified critical assets and their associated protection levels, O-RAN WG11 Threat Model and Remediation Analysis' documentation offers an in-depth explanation (O-RAN Working Group 11 2024a).

4.1.4 Protection Levels

O-RAN WG11 proposes specific protection levels for these critical assets, focusing on:

- **Confidentiality, Integrity, and Availability (CIA).** Ensuring that data is accessible only to authorized parties, remains unaltered during transit or storage, and is readily available when needed.
- **Replay Protection.** Guarding against the unauthorized retransmission of messages to prevent attacks such as session hijacking or message forgery.
- **Authenticity.** Verifying that data and communications are genuinely from the claimed source.
- **Point of Protection.** Distinguishing between the security measures needed "**at rest**" (for data stored within the system) versus "**in transit**" (for data being actively communicated between components).

These protection levels are meticulously defined to ensure that each critical asset within the O-RAN system is adequately protected against potential threats, thereby maintaining the network's integrity, performance, and trustworthiness. The comprehensive approach to classifying these assets and defining their required levels of protection exemplifies the depth of security considerations inherent in the O-RAN architecture.

This chapter is organized to provide a comprehensive overview of various aspects of security within the O-RAN systems. Section 4.2 explains the methodology of O-RAN security analysis, laying the groundwork for understanding the approaches and frameworks used to assess security. In Section 4.3, the focus shifts to the threat model, where potential security threats to O-RAN components and architecture are identified and analyzed. Section 4.4 provides an in-depth discussion on risk assessment, evaluating the likelihood and impact of different security threats. Section 4.5 outlines the security principles, security requirements, and security controls specific to O-RAN, offering a detailed guide on the necessary measures to protect the network. Following this, Section 4.6 explores the O-RAN security protocols, examining the protocols in place to ensure secure communication and operation within the network. Finally, Section 4.7 introduces the security testing framework, detailing the methods and tools used to test and validate the security measures implemented in O-RAN systems.

4.2 Methodology of O-RAN Security Analysis

As shown in Figure 4.2, the methodology of the O-RAN security analysis is based on three stages (O-RAN Working Group 11 2024a). These stages are explained below.

4.2.1 Stage 1: Risk Identification

The first stage in the methodology of the O-RAN security analysis, risk identification, lays the groundwork for a comprehensive security strategy by systematically addressing several key areas.

- **Identify Stakeholders.** This initial step involves identifying all parties involved with the O-RAN system across various capacities, including its implementation, management, operation, and maintenance. Understanding the roles and responsibilities of each stakeholder is critical for outlining accountability within the system's security framework. Some of the main stakeholders responsible for the different sections of the O-RAN system are listed below, the complete list can be found in O-RAN Working Group 11 (2024a).

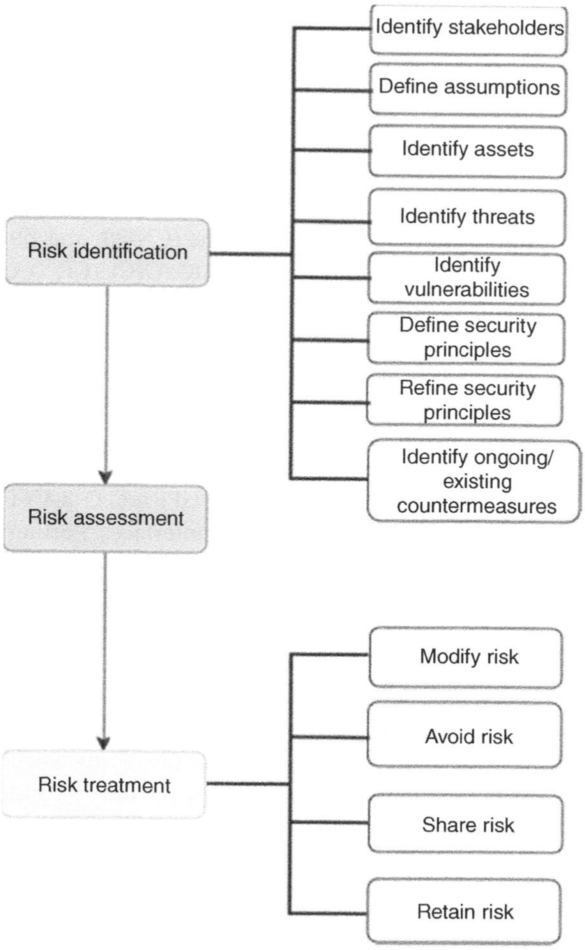

Figure 4.2 Methodology of O-RAN Security Analysis.

- **Mobile Network Operators (MNOs).** They are responsible for offering network services and possess a license to operate in an allocated spectrum.
- **HW/Network Vendors.** Network vendors or HW providers are tasked with supplying the foundational network infrastructure, such as servers, switches, routers, gateways, and radio HW, necessary for running a software-defined networking (SDN) controller. Their responsibilities extend to the installation, maintenance, or replacement of these HW devices, alongside ensuring that these devices can be securely configured, generate log events, and send log files to an MNO's externalized log analysis system. Additionally, they are responsible for facilitating a bug report submission process for users and security researchers, conducting security testing in alignment with 3GPP and O-RAN standards, managing vulnerabilities through continuous monitoring and evaluation, and maintaining firmware by providing necessary patches for identified bugs and vulnerabilities.
- **HW/Network Administrators.** Network administrators play a crucial role in managing and securing network infrastructure. Their responsibilities include configuring HW and network devices, enabling and overseeing the collection and analysis of log events generated by these devices, and deploying firmware patches as advised by HW or network vendors. They are also tasked with proactively monitoring for vulnerabilities, reporting any discoveries back to

the vendors, and conducting regular testing of the HW and network configurations to ensure security and efficiency. In some cases, the MNO may also assume the role of the HW and network administrator, centralizing the oversight and management of the network's operational integrity.

- **Orchestrators.** They oversee operations and orchestration of the services provided by the O-RAN system. The MNO could also serve as an orchestrator.
- **NF Vendors.** They are tasked with the development and provision of NFs, including VNFs, containerized network functions (CNFs), and physical network functions (PNF), essential for components like Near-RT RIC, O-CU-CP, O-CU-UP, O-DU, among others. They ensure that these NFs can be securely configured and manage vulnerabilities through comprehensive processes that include monitoring, identification, evaluation, treatment, and reporting of any security vulnerabilities. Additionally, NF vendors are responsible for the entire life cycle of patch management, from development and testing to delivery, and for maintaining the NFs by providing timely patches for any bugs and vulnerabilities identified. They equip NFs with the capability to generate and send log events to an externalized log analysis system, as specified by the MNO. Compliance with 3GPP and O-RAN test plans is mandatory, incorporating security testing of the NFs and their interfaces. Furthermore, they establish mechanisms for users and security researchers to report bugs, enhancing the continuous improvement and security of the NFs.
- **NF Administrators.** *They* are responsible for the operational and security management of NFs, including their deployment, configuration, and maintenance. Their key duties encompass deploying patches according to the guidelines provided by NF vendors to ensure that NFs remain secure and up to date. They actively monitor for vulnerabilities within the NFs, promptly notifying NF vendors of any issues discovered to facilitate swift remediation. Secure configuration of the NFs is a critical task, ensuring that all settings adhere to the best security practices. Administrators also conduct regular testing of NF configurations to validate their effectiveness and compliance with security standards. Additionally, they are responsible for enabling the collection of log events and analyzing these logs to detect anomalies or security incidents that may indicate potential threats. In many scenarios, the MNO may assume the role of the NF administrator, taking on these responsibilities to ensure the secure and efficient operation of the NFs within their domain.
- **Virtualization/Containerization Software Infrastructure Providers.** They are pivotal in crafting the foundational services essential for contemporary network frameworks, focusing on the design, construction, and management of infrastructures composed of critical software components like hypervisors, host OSs, and container runtime systems. Their responsibilities encompass not only the secure configuration of these infrastructures but also the establishment of a thorough vulnerability management cycle—monitoring, identifying, evaluating, addressing, and reporting vulnerabilities. Furthermore, they are charged with the development, testing, and dissemination of patches to rectify bugs and vulnerabilities, alongside ensuring the infrastructure's ability to log events and communicate these logs to external analysis systems provided by MNOs. They also facilitate a collaborative security improvement process by enabling users and security researchers to report issues through designated channels, thereby maintaining the security and operational integrity of the virtualization and containerization infrastructure.

• **Define Assumptions.** It is essential to outline the operational prerequisites and assumptions for the environment in which the O-RAN system operates. These assumptions help clarify the conditions deemed necessary for the system's successful operation, particularly those outside the direct control of O-RAN.

- **Identify Assets.** Identifying the O-RAN system's assets involves pinpointing and documenting all relevant data and components. This includes detailing their types (such as data or components), security properties based on the CIA triad (Confidentiality, Integrity, Availability) both at rest and in transit, and their locations. This process is foundational for understanding what needs protection and where these assets are situated.
- **Identify Threats.** This step requires identifying potential threats that could exploit vulnerabilities in the new O-RAN components, interfaces, and technologies. It also involves detailing the threat surface and identifying potential threat agents, providing a comprehensive view of security challenges.
- **Identify Vulnerabilities.** The O-RAN system's vulnerabilities, especially those related to its new components, interfaces, and technologies, must be systematically identified. These vulnerabilities represent areas of weakness that could be targeted by threats.
- **Define Security Principles.** Defining security principles involves establishing the fundamental goals and objectives that the O-RAN security strategy aims to achieve. These principles serve as the basis for developing a coherent approach to reducing risk exposure.
- **Elaborate and Refine Security Principles.** Following the definition of security principles, this step involves expanding each principle into specific security requirements, recommendations, and countermeasures. This refinement process ensures that principles are actionable and directly tied to mitigating identified risks.
- **Identify Existing/Ongoing Countermeasures.** Finally, it is crucial to inventory all existing or ongoing security measures within the O-RAN system. Recognizing these allows for a more informed approach to introducing new security measures, ensuring that they complement existing controls and address any gaps.

4.2.2 Stage 2: Risk Assessment

Following the identification of threats, the second stage delves into assessing the risks associated with each identified threat. This assessment evaluates the likelihood of each threat being exploited and its potential impact on the system's key security principles: confidentiality, integrity, and availability. Factors considered in this evaluation include the complexity of potential attacks, the level of access required to exploit each vulnerability, and the sensitivity of the affected assets. This risk assessment phase is crucial for prioritizing security efforts, focusing on the most significant threats to the system. Prioritizing threats can be based on the value of risks associated with the threats; the risk of each threat can be quantified as the product of the probability that the threat will exploit a vulnerability and the overall impact resulting from the exploitation of that vulnerability.

4.2.3 Stage 3: Risk Treatment

The third stage in the methodology of O-RAN security analysis revolves around risk treatment. Upon establishing the level of risk associated with each identified threat, the focus shifts to determining the most suitable strategy for addressing these risks. This stage offers four strategic options as briefly mentioned below.

(I) **Modify the Risk.** This approach involves implementing specific controls aimed at reducing the risk's likelihood. It is a proactive measure to alter the risk's conditions, making it less likely to occur or less impactful if it does.

(II) **Avoid the Risk.** Avoidance involves halting any operations or activities recognized as the sources of the identified risk. This strategy is particularly viable when the risk is deemed too significant to be effectively managed through conventional security controls, necessitating the cessation of actions that contribute to its presence.

(III) **Share the Risk.** Risk sharing involves transferring or distributing some of the risk to a third party. This can be achieved either by outsourcing security tasks to another organization that specializes in managing such risks or by purchasing cyber insurance to mitigate financial losses in the event of a security breach. While not entirely removing the responsibility from the original organization, these options provide a means to manage risks that exceed the organization's capacity to address internally.

(IV) **Retain the Risk.** In some cases, an organization may choose to accept the risk, acknowledging that the cost of implementing measures to mitigate it would outweigh the potential damage it could cause. This decision involves a calculated acceptance of the risk, usually after determining that the likelihood of occurrence or the impact of the risk is manageable within the organization's risk tolerance levels.

4.3 Threat Model

The threat model for O-RAN encompasses a comprehensive analysis of the architecture's expanded threat surface, diverse threat agents, specific vulnerabilities, and targeted threats to O-RAN components, structured through a coverage matrix of threats. The O-RAN's openness and integration of innovative technologies, including open interfaces like A1, E2, O1, O2, cloud-native functions, the RIC, and virtualization/containerization layers, significantly broaden the threat surface. This diversity introduces potential vulnerabilities, ranging from software bugs and misconfigurations to insecure APIs and encryption weaknesses, which could be exploited by a wide array of threat agents, including cybercriminals, state actors, and insiders, aiming for espionage, service disruption, or financial gain (O-RAN Working Group 11 2024a).

Threats against O-RAN components might involve denial-of-service (DoS) attacks to disrupt operations, man-in-the-middle (MITM) attacks on insecure communications, and supply chain attacks aiming to compromise network elements. The use of a coverage matrix facilitates a systematic approach to identifying, evaluating, and prioritizing security risks, focusing mitigation efforts on areas of higher vulnerability within the O-RAN framework. This comprehensive threat model underscores the importance of adopting effective security strategies tailored to the unique challenges posed by the O-RAN architecture, ensuring resilience against potential cyber threats and vulnerabilities.

4.3.1 Building Blocks of the Threat Model

The threat model in O-RAN encompasses several critical building blocks designed to systematically identify, assess, and mitigate security risks within its open architecture shown in Figure 4.3. The model begins with the identification of the threat surface, encompassing all potential areas of attack within O-RAN's components and interfaces, and threat agents, which include any entities that might exploit system vulnerabilities. It then assesses vulnerabilities inherent within the system, identifying specific weaknesses that could be targeted. Direct threats against O-RAN components are analyzed, leading to the development of a coverage matrix of threats that prioritizes risks based on their potential impact and likelihood. Finally, mitigation strategies are devised, outlining specific

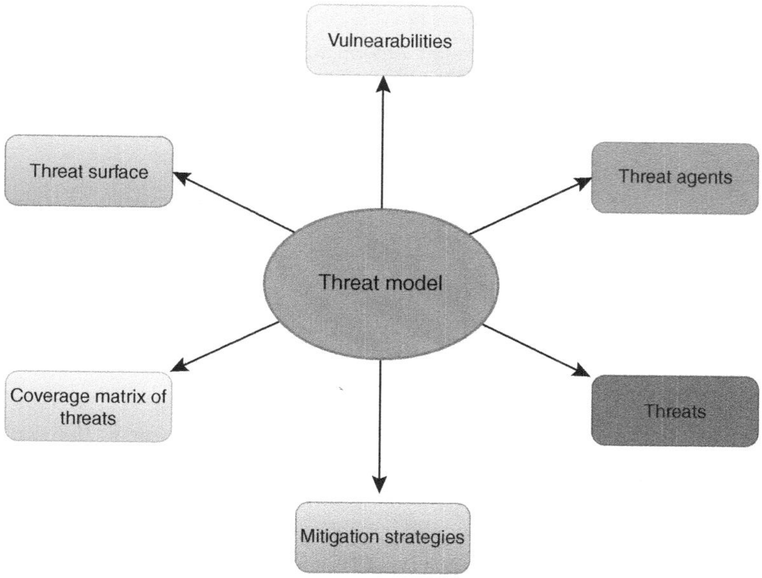

Figure 4.3 Building Blocks for O-RAN Threat Model.

countermeasures and security practices to protect against identified threats and vulnerabilities, ensuring the resilience and security of the O-RAN system.

4.3.2 An Overview of O-RAN Security Threats

In the O-RAN system, security threats encompass a wide range of vulnerabilities and attack vectors that exploit the network's open architecture, advanced technological integrations, and more reliance on software-based components. These threats, which are diverse in nature, span from exploiting cloud-based infrastructures and open-source code vulnerabilities to leveraging physical access for tampering with network devices. Additionally, they capitalize on the specificities of 5G radio networks, manipulate the AI and machine learning systems designed for network optimization, and exploit weaknesses within the communication protocols. The comprehensive threat landscape also includes the risk posed by insiders, who might misuse their access to the network for malicious purposes. Figure 4.4 shows the seven categories of these security threats in an O-RAN system. The O-RAN WG11 in O-RAN Working Group 11 (2024a) has developed a comprehensive threat inventory. This inventory categorizes the diverse threats into distinct groups. Each identified threat within this inventory is clearly documented, including a unique threat ID, a concise threat title, and a detailed description that outlines the nature and mechanics of the threat. Furthermore, the inventory specifies the threat agent responsible for the risk, the vulnerabilities that could be exploited, the assets that are under threat, and the components of the O-RAN system that are affected.

4.3.2.1 Threats Against O-RAN System

The openness of the O-RAN architecture introduces common threats among its components. Attackers can exploit insecure designs, or the lack of security measures adopted in O-RAN components to gain unauthorized access or inject malware, compromising the integrity and functionality of the network. Misconfigurations of O-RAN components further open vulnerabilities that malicious actors can leverage to disrupt services or steal sensitive information.

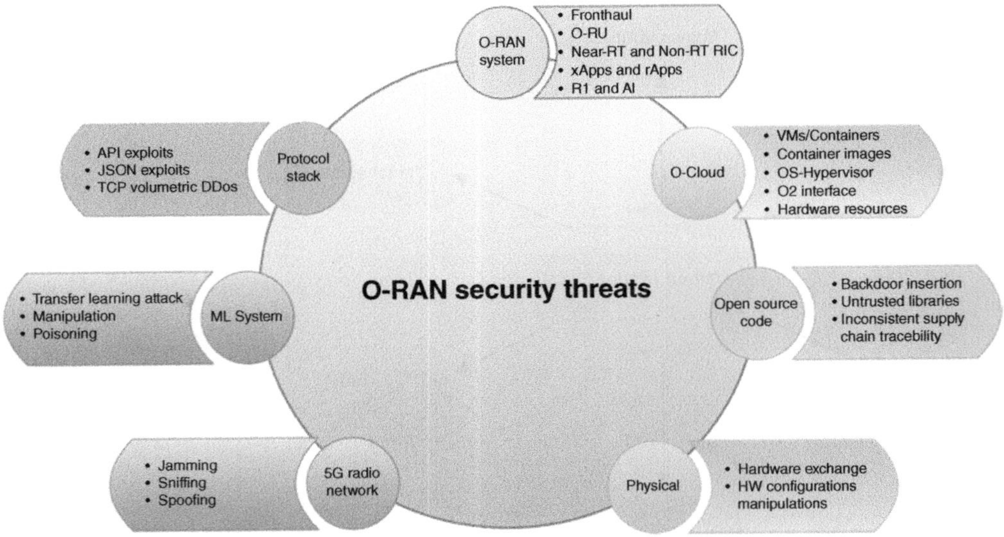

Figure 4.4 O-RAN Security Threats Groups.

Additionally, the internet-facing aspects of O-RAN are particularly vulnerable to attacks that exploit weak authentication and access control mechanisms, allowing attackers to penetrate the O-RAN network boundary. The integration of Internet of Things (IoT) devices introduces another vector for attack, where compromised devices can be used to jam airlink signals, severely affecting network availability. Furthermore, the O-RAN's reliance on open interfaces, such as fronthaul, O1, O2, A1, and E2, while facilitating flexibility and innovation, also poses significant risks if not secured properly, leading to potential penetration and compromise of the O-RAN system.

The fronthaul interface and the M-S-C-U planes within the O-RAN system play pivotal roles in ensuring network integrity, synchronization, and efficient data transmission. However, these critical components are susceptible to countless security threats that can severely compromise the network's functionality (Abdalla and Marojevic 2024). Vulnerabilities in the system can allow attackers to penetrate the O-DU and beyond through the O-RU or the fronthaul interface, utilizing unauthorized access points, especially within the open fronthaul Ethernet's L1 physical layer interfaces. Such breaches enable attackers to conduct numerous attacks, including MITM attacks over the M-plane, where the integrity and confidentiality of fronthaul communications are jeopardized. An unauthorized device connected to the Ethernet L1 interface has the potential to disrupt or degrade the performance of authorized network elements on the open fronthaul interface by flooding the L1 interface with excessive, unintended network traffic. Furthermore, the network's synchronization and control mechanisms, vital for coordinated network performance, face threats from DoS attacks against the master clock and impersonation attacks, where spoofing techniques are employed to mislead network components about the legitimate time source (Abdalla and Marojevic 2024).

Threats Against O-RU. A prominent threat involves the false base station attack, wherein an attacker sets up a rogue cell tower masquerading as a legitimate mobile network. This tactic facilitates a MITM attack between the subscriber's UE and the actual mobile network. Such false base stations, also known as SUPI/5G-GUTI catchers, coerce UEs to connect to them, thereby enabling attackers to intercept or disclose subscriber identities and conduct tracking attacks. This attack vector poses significant risks to subscriber privacy and network integrity.

Threats Against Near-RT RIC. Within the Near-RT RIC, malicious xApps present a formidable threat by exploiting UE identification data to track locations and alter UE priorities. Given their capabilities to influence the behavior of specific cells, UEs, or groups of UEs, malicious xApps can undermine network performance and privacy. For example, a malfunctioning xApp receiving commands via the A1 interface to control UEs can abuse this power to track VIPs or manipulate service priorities, revealing sensitive location information or degrading service quality. The security of Near-RT RIC and E2 nodes is compromised by such malicious xApps gaining unauthorized access or abusing radio network control capabilities. Furthermore, the vulnerability of Near-RT RIC APIs to nonauthentication or weak authentication extends the attack surface, especially when considering the risks associated with xApps handling user/password information (O-RAN Working Group 11 2024e).

Threats Against Non-RT RIC. The Non-RT RIC, accessible through the SMO, is susceptible to attacks aimed at disrupting its functionality or corrupting data. Attackers penetrating the Non-RT RIC can initiate DoS attacks to degrade its performance, directly impacting its ability to monitor network performance or enforce A1 policies effectively. Unauthorized access for purposes like UE tracking or data corruption/modification further exposes the network to significant security and operational risks (O-RAN Working Group 11 2024f).

Threats Against R1 Interface. The R1 interface, pivotal in facilitating the connection between rApps and the Non-RT RIC framework, acts as a means for the abstraction and exchange of services across the O-RAN architecture. Vulnerabilities within the R1 interface can be exploited through various attack vectors, including spoofing attacks aimed at gaining unauthorized access to R1 services. Attackers might also exploit password-based authentication mechanisms to request or intercept unauthorized data, leveraging weak password management practices. The critical nature of the R1 interface, especially in scenarios involving service management and exposure services, demands robust security measures. Implementing certificate-based mutual authentication using TLS and public key infrastructure (PKI) X.509 certificates is recommended to mitigate these threats, ensuring that only authorized Service Producers can produce and consume the services, thereby protecting the integrity and confidentiality of the data exchanged via the R1 interface.

Threats Against A1 Interface. The A1 interface serves a vital role in enabling policy-based guidance, ML model management, and the provision of enrichment information between the Non-RT RIC and Near-RT RIC, aiming to optimize RAN operations. However, this interface is susceptible to threats arising from weak mutual authentication mechanisms, which could allow malicious Non-RT RIC or Near-RT RIC entities to establish peering over the A1 interface improperly. Such vulnerabilities open the door to MITM attacks, where an internal threat actor can manipulate or inject malicious policy messages across the A1 interface. The consequences of such breaches include the Near-RT RIC receiving and acting on malicious policies, potentially leading to degraded network performance, unauthorized data access, or the disruption of critical RAN optimization processes. To counteract these threats, enhanced security protocols, focusing on strengthening mutual authentication and securing message integrity across the A1 interface, are crucial for maintaining the operational security and efficiency of the O-RAN system.

Threats Against xApps. xApps, crucial for enhancing RAN performance through the Near-RT RIC, can become avenues for significant security threats if sourced from untrusted or unmaintained entities. Attackers exploiting vulnerabilities in xApps can disrupt network service, potentially taking control over other xApps or the entire Near-RT RIC. Such malicious xApps can degrade NFs, execute DoS attacks, and even manipulate specific cell behaviors, UEs, or groups of UEs, raising concerns over unauthorized tracking or service manipulation. Malicious xApps may trigger Radio Resource Management decisions conflicting with the O-gNB's internal choices, creating conditions for service denial (O-RAN Working Group 11 2024e).

Threats Against rApps. rApps, operating within the Non-RT RIC, face risks from direct, indirect, and implicit conflicts among decisions from different rApps, resulting in network performance degradation or instability. Direct threats include the deployment of malicious rApps intentionally designed to disrupt network services, manipulate AI/ML model training, manage A1 policies erroneously, or sniff enrichment data such as UE locations and navigation information. These actions can lead to performance degradation, DoS conditions, and unauthorized access to sensitive information (O-RAN Working Group 11 2024g). Indirectly, rApps can suffer from conflicts among themselves—when different rApps attempt to change the same network parameter simultaneously or make changes that have opposing effects on the network's performance. Furthermore, even well-intentioned rApps from trusted sources might contain backdoors or vulnerabilities that attackers can exploit to take control over network operations or the Non-RT RIC itself. The lack of robust authentication and authorization mechanisms intensify these threats, allowing attackers to modify existing Near-RT RIC policies or introduce false policies, thereby undermining the network's security and operational efficiency.

These threats underscore the need for robust security measures, including secure authentication protocols, rigorous access controls, and vigilant monitoring of components and interfaces within the O-RAN architecture to protect against unauthorized access, data breaches, and service disruptions.

4.3.2.2 Threats Against O-Cloud

The O-Cloud infrastructure within the O-RAN system faces several security threats originating from the trust relationships among its various stakeholders. Key challenges include the decoupling of software from dedicated HW, shared HW among multiple organizations, involvement of third-party organizations in infrastructure management, and the utilization of open-source components (O-RAN Working Group 11 2024a). These challenges give rise to potential threats such as the compromise of VNF or containerized network function (CNF) images and embedded secrets, vulnerabilities in orchestrator configurations, access controls, and isolation mechanisms, as well as the misuse of VMs or containers to launch attacks on other components within the infrastructure. Additionally, threats include the risk of spoofing and eavesdropping on network traffic to gain unauthorized access to O-RAN network data processed within the system. The virtualization and containerization technologies deployed within the O-RAN architecture present the following threats concerns (O-RAN Working Group 11 2024a):

Generic Threats. Vulnerabilities in the host OS, hypervisor/container engine, and VNF/CNF software layers could be exploited by attackers, potentially enabling them to compromise the O-Cloud environment. Such compromises may lead to diverse malicious activities, including unauthorized access to underlying VMs or containers, exploitation of weak identity and access management policies, execution of adversary-controlled code, and elevation of privileges. Attackers may also exploit compromised credentials obtained through credential pharming or network compromise to gain initial access, persist within the environment, elevate privileges, or evade detection. Additionally, exposed services, such as Docker API, Kubernetes API server, etc., lacking authentication, particularly within containerized environments, pose risks of unauthorized access within the O-Cloud infrastructure. In multitenant and public cloud deployments, CNF applications running alongside unknown entities introduce further risks, with improperly authenticated interfaces potentially facilitating attacks compromising service availability and confidentiality. Furthermore, adversaries may exploit insecurely stored credentials within O-RAN NFs, virtual links, orchestration layers, or HW components, which could be dispersed across various locations within the O-Cloud platform, including plaintext files, OS repositories, or specialized files.

Threats Concerning VMs/Containers. VMs and containers in O-RAN environments are susceptible to various threats due to misconfigurations and lack of strong isolation. First, misconfigurations may result in VMs, or containers being granted excessive privileges, allowing attackers to perform unauthorized operations and gain elevated access to the host system. Additionally, unintended read/write access to directories on the host filesystem could enable attackers to manipulate sensitive data or install malicious software. Furthermore, the lack of strong isolation between VMs/containers and the host system poses risks of escape attacks, where a rogue VM/container breaches its confinement and affects other cohosted instances. This scenario could lead to severe consequences, such as compromising the confidentiality, integrity, and availability of resources within the O-RAN environment. Furthermore, insecure storage of sensitive data within VMs/containers or the deployment of malware on logical volumes allocated to VNFs/CNFs could facilitate unauthorized data access or manipulation by attackers. Additionally, attacks exploiting VM/container migration can target both the control plane and data plane, with techniques such as migration flooding and false resource advertising. Migration flooding involves an attacker moving all VMs/containers from a compromised server to an insufficiently resourced victim server, resulting in DoS for the VNFs/CNFs on the victim server due to resource exhaustion. This leads to performance degradation and potential crashes of the affected VMs/containers, impacting the overall availability and stability of the system. False resource advertising involves a hacked server falsely claiming to have ample free resources, enticing other servers to transfer some of their VMs/containers to it for workload distribution. Once the VMs/containers are moved, the attacker can exploit vulnerabilities to compromise them, as they are now hosted on a server under the attacker's control. This deceptive tactic allows the attacker to infiltrate and potentially manipulate the off-loaded VMs/containers, posing significant security risks to the O-Cloud infrastructure.

Threats Concerning VM/Container Images. Threats related to VM images in O-RAN involve various attack vectors that adversaries can exploit to compromise the integrity and security of the deployed infrastructure. One such threat involves attackers injecting malicious code or altering information within unprotected images during onboarding processes. Subsequently, when these tainted images are instantiated as VNF/CNF, the malicious code can trigger DoS attacks, data theft, or fraudulent activities. These attacks encompass various categories, including build machine attacks, where attackers manipulate the image creation process to insert malicious code. Supply chain attacks pose another risk, wherein attackers replace or modify images between the build and deployment stages, enabling the execution of arbitrary code. Additionally, vulnerabilities arise from connections to registries over insecure channels, increasing the likelihood of MITM attacks that compromise image integrity and confidentiality. Adversaries may also exploit vulnerabilities in container deployment processes, such as sending remote build requests to Docker APIs to create custom images containing malware. These threats underscore the importance of implementing robust security measures to safeguard VM and container images throughout their life cycle in O-RAN deployments.

Threats Concerning the Virtualization Layer (Host OS–Hypervisor/Container Engine). Threats concerning the virtualization layer in O-RAN highlight vulnerabilities within the host machines' components, including the hypervisor, container engine, and host OS, which must be protected against known vulnerabilities. Hyperjacking is a significant attack vector, wherein adversaries gain control over the server's host or install a malicious layer to exploit and manipulate VMs/containers running on top of the host. This grants the attackers control over all VMs/containers on the host, posing a severe security risk due to the single point of failure presented by the compromised hypervisor/container engine/host OS. Successful hyperjacking attacks typically involve injecting rogue layers beneath or on top of the original virtualization components, directly compromising the original layer, or running rogue layers on top of existing ones. Furthermore, attackers may tamper with the bootloader of the virtualization layer, further exacerbating security vulnerabilities. In an O-Cloud environment, failures during the boot sequence necessitate prompt handling by the network function orchestrator (NFO) or Fault and Configuration Management (FOCOM) systems to address scenarios like physical machine failures or safe mode initiation. Additionally, adversaries may target supporting services like DNS within the container network, disrupting service coordination and leading to potential DoS situations that impede O-Cloud's scalability and responsiveness to resource demands.

Threats Concerning O-Cloud Interfaces. The threats to O-cloud interfaces, specifically the O2 interface and O-Cloud API, encompass a range of potential attacks that could severely impact the integrity, availability, and confidentiality of the O-Cloud services. One primary concern is the tampering and alteration of requests and responses between the O-Cloud and the SMO, including FOCOM and NFO. This tampering can lead to incorrect resource provisioning, operational malfunctions, and even DoS attacks, which render the O-Cloud services unavailable to legitimate users. Furthermore, such unauthorized alterations could lead to critical service disruptions.

Another significant threat involves information leakage, where attackers gain access to sensitive data such as the state of virtualized resources or HW configurations. This not only compromises the security of the O-Cloud but also provides attackers with detailed knowledge to launch more targeted attacks. Resource assignment errors due to tampering can result in inefficient resource allocation, directly affecting service quality and performance. The compromise of infrastructure management systems (IMS can alter or delete HW state information, leading to unnoticed HW malfunctions and operational issues. Additionally, compromising the virtualization layer can lead to the theft of cryptographic keys or the manipulation of virtualized resources, severely undermining the security of VNFs or CNFs.

To mitigate these threats, it is essential to employ a multilayered security approach. Encrypting communication over the O2 interface, implementing robust access control measures, and utilizing monitoring and anomaly detection tools are crucial steps in protecting the O-Cloud environment. Regular updates and patches for the virtualization infrastructure, along with adopting a zero-trust architecture, further strengthen the security against potential attacks. These strategies are vital for maintaining the integrity, availability, and confidentiality of O-Cloud services, ensuring that they remain secure and resilient against a wide range of cybersecurity threats (O-RAN Working Group 11 2024h).

For a more detailed security analysis on the virtualization and containerization technologies, refer to 3GPP (2018), Chaudhuri (2015), and NIST SP 800-190 (n.d.).

4.3.2.3 Threats to Open-Source Code

The increasing reliance on open-source code within the O-RAN Software Community, a Linux Foundation project supported by O-RAN, introduces a list of security risks that require critical consideration. Open-source software, known for its accessibility and collaborative development model, brings with it an inherent vulnerability to cyber threats. One primary concern is the public availability of information about vulnerabilities in the National Vulnerability Database (NVD). While this database serves a crucial role in informing developers about existing vulnerabilities for prompt remediation, it ironically also aids hackers in identifying and exploiting these weaknesses. The culture of reusing open-source code can unfortunately facilitate the spread of vulnerabilities. This situation is exacerbated when developers incorporate libraries and dependencies that contain known vulnerabilities or originate from untrusted repositories, introducing backdoors for cyberattacks. Such vulnerabilities are not merely theoretical; there have been significant instances of security breaches attributed to compromised open-source libraries and dependencies, highlighting the risks within the supply chain. The lack of rigorous inventory management and vulnerability tracking among some O-RAN vendors and operators compounds these risks. Without comprehensive inventories of open-source dependencies or effective processes for managing vulnerability notifications and patch updates, these entities are ill-equipped to protect their applications against emerging threats. Furthermore, inconsistent supply chain traceability and security practices, coupled with a disregard for coding best practices, clash with the principles of Security by Design, further exposing systems to potential breaches.

The use of modules with known vulnerabilities and reliance on untrusted libraries represents a tangible threat. Developers, either due to oversight or under the pressure of deadlines, may fail to vet the security of these components adequately. Such lapses provide attackers with opportunities to exploit these vulnerabilities, executing backdoor attacks that compromise critical infrastructure components like hypervisors, OSs, VMs, or containers with malware. These exploits not only undermine the integrity and availability of the affected systems but also pose significant risks to the confidentiality of sensitive information.

4.3.2.4 Physical Threats

The physical security of O-RAN deployments is important, given their critical role in modern telecommunications infrastructure. These physical networks manage and process extremely sensitive data, including keys, user plane data, control plane data, and management data, often in cleartext, making them attractive targets for physical attacks. The vulnerabilities in physical components can range from inadequate protection of data centers and PNFs to insufficient protect against power outages, environmental disasters, and neglect in the maintenance and monitoring of HW parameters. Additionally, HW backdoors present a significant risk, potentially allowing unauthorized access to sensitive O-RAN components and data. The lack of proper physical security measures in O-RAN deployments can lead to several exploitable scenarios. Attackers might gain physical access to O-RAN components through unsecured management ports and consoles, such as JTAG and serial consoles, or via dedicated management ports left unprotected due to relaxed administrator credentials management. Such access could enable attackers to inject malware, manipulate software, extract sensitive data like private keys and certificates, and disable security features. These actions could have dire consequences, including eavesdropping, wiretapping on the communication and management planes, DoS, and other intrusive attacks, compromising the integrity, confidentiality, and availability of the O-RAN services. Furthermore, the disaggregation and the connection of the O-RAN components, such as the O-RU and O-DU, via cable networks for the fronthaul link, present additional vulnerabilities. Attackers could exploit these connections, particularly at cable landing sites or premises housing these units, to gain control over data traffic. Such breaches could facilitate unauthorized access to the network, allowing attackers to monitor, intercept, or manipulate data traffic between these critical O-RAN components.

4.3.2.5 Threats Against 5G Radio Networks

5G radio networks, as the latest advancement in wireless communication, come with many security vulnerabilities due to their reliance on airwaves for data transmission. These vulnerabilities are exposed through various forms of attacks, notably jamming and spoofing, which can significantly disrupt network operations. Jamming attacks are particularly concerning as they can interfere with essential signals and control channels within the 5G infrastructure. This interference affects everything from device synchronization and network access to the transmission of control information and the initiation of new connections. Furthermore, the Physical Broadcast Channel (PBCH), a crucial component for initial network access and synchronization, is susceptible to both jamming and spoofing attacks. Such vulnerabilities can lead to severe network disruptions, including DoS, where devices are either unable to connect to the network or deceived into connecting to malicious entities. Cognitive radio (CR) technology, designed to optimize spectrum usage, also faces significant threats from malicious users who aim to disrupt the system by jamming the spectrum or emulating primary user signals. This deceit forces legitimate secondary users to vacate the spectrum unnecessarily, leading to ineffective spectrum utilization and potential DoS scenarios. Furthermore, the CR system's reliance on users complying with a liability rule to ensure that fair spectrum access is undermined by selfish users prioritizing their own benefits over network regulations, thereby increasing the risk of interference with primary users.

4.3.2.6 Threats Against ML System

ML systems, while transforming numerous sectors through advanced data analysis and predictive capabilities, also introduce unique vulnerabilities that attackers can exploit. One of the primary threats to ML systems arises when an attacker gains unauthorized access to the training dataset. By tampering with this data, the attacker can significantly degrade the quality and integrity of the information used to train the ML model. This tampering can lead to the model being trained on incorrect or maliciously modified data, resulting in unreliable outputs. The consequences of such an attack are far-reaching, affecting the model's decisions, predictions, classifications, and detections, rendering them inappropriate for any practical application. Another significant threat emerges in scenarios where an ML model is designed to learn continuously from new data during its operational phase. In such cases, an attacker can introduce "bad" data into the model's learning process, corrupting its learning algorithms, and causing it to make incorrect decisions or predictions over time. This type of attack not only undermines the model's performance but also potentially retrains it to function in ways contrary to its intended purpose. Furthermore, attackers can target the ML models directly by altering their parameters or the algorithms that govern their behavior. Such modifications can skew the model's outputs, leading to incorrect predictions or decisions. This scenario is particularly alarming when ML models are employed to make critical business or operational decisions, as the ramifications of incorrect outputs could be catastrophic. Sensitive data extraction from ML models represents another vulnerability. Through advanced techniques, attackers can reverse-engineer ML models to extract the data on which they were trained. This is especially concerning when models are trained using confidential or proprietary information, posing significant privacy and security risks. The transfer learning attack vector is of concern when ML systems are developed by fine-tuning pretrained models available publicly.

Attackers can exploit these widely used models to introduce or amplify malicious behavior in the fine-tuned versions, leveraging the trust in the original model to propagate their attack.

4.3.2.7 Protocol Stack Threats

These threats exploit vulnerabilities in web technologies and protocols such as REST APIs and JSON, alongside traditional network protocols like Hypertext Transfer Protocol (HTTP) and Transmission Control Protocol (TCP). For instance, REST API vulnerabilities, including injection and cross-site scripting (XSS), pose significant risks, allowing attackers to execute unauthorized commands or disrupt services. Similarly, JSON, essential for data interchange within O-RAN, is susceptible to attacks like injection and deserialization, which can compromise data integrity and application security. Furthermore, JSON Web Token (JWT) vulnerabilities can lead to authentication bypasses, further exacerbating security challenges. Distributed denial-of-service (DDoS) attacks represent another critical threat vector, with methods such as HTTP GET floods, Garbage floods, and Reverse Bandwidth floods aiming to overload O-RAN systems and render them inoperative. These are complemented by traditional HTTP attacks, including further instances of XSS and SQL injection, which target application-level vulnerabilities to access or manipulate sensitive data. At the TCP level, attacks like SYN flood, ACK flood, and RST flood seek to disrupt network communication, hindering the O-RAN's ability to process legitimate requests and maintain stable connectivity.

4.3.3 O-RAN Threat Agents

Threat agents refer to individuals or entities with the potential to exploit vulnerabilities within the O-RAN architecture, pose significant risks to the integrity, availability, and confidentiality of the O-RAN system. These agents range from cybercriminals and state-sponsored hackers to insider threats, each armed with the motive and means to compromise the O-RAN system. Understanding the nature and capabilities of these threat agents is crucial for developing effective security measures and protecting the future of telecommunications infrastructure. These agents can be categorized as shown in Figure 4.5.

- **Cybercriminals.** Cybercriminals represent a significant threat agent in the O-RAN system, primarily motivated by financial gain, espionage, or disruption. With the open and interoperable nature of O-RAN architectures, cybercriminals find a fertile ground for launching complex attacks. They exploit vulnerabilities within the network's components, such as the RAN's open interfaces or the SDN elements, to conduct activities ranging from data theft and ransomware attacks to DoS. Their methods are diverse and evolving, including malware insertion, system infiltration, and exploiting protocol weaknesses. The open-source aspect of O-RAN can both aid in rapid development and potentially expose security gaps, making it imperative for network

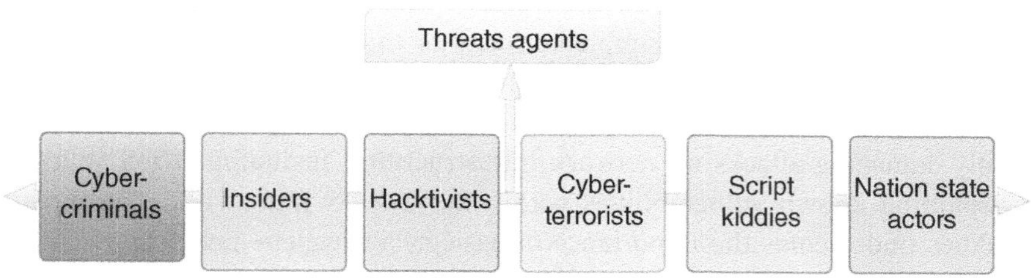

Figure 4.5 O-RAN Threat Agents.

operators to continuously monitor, update, and patch their systems against such malicious activities. Cybercriminals' deep technical expertise and often high level of organization enable them to orchestrate wide-reaching attacks with potentially severe implications for network reliability, user privacy, and national security.

- **Insiders.** Insiders pose a uniquely challenging threat to the O-RAN system, leveraging their access and knowledge of the network's operation to conduct malicious activities. Insiders, consisting of unhappy employees, contractors, or partners with privileged access, inherently bypass the initial security barriers facing external attackers. Unlike external cybercriminals, these insiders leverage their inside position to potentially compromise the network. Their motivations can range from personal grievances to financial incentives or ideological reasons. The damage an insider can inflict is profound, from sabotaging network operations and leaking sensitive information to installing backdoors for future attacks. Given their understanding of the network's architecture and its defensive measures, insiders can carefully navigate around security controls, making their detection particularly challenging.
- **Hacktivists.** Hacktivists pose a special category of threat to the O-RAN system, driven not by financial gain but by ideological motives. These individuals or groups leverage their hacking skills to promote political agendas, to drive social change, or to bring attention to specific issues. Their attacks on O-RAN networks could range from defacing web sites to launching DoS attacks, aimed at disrupting services or broadcasting their message. Unlike cybercriminals or insiders, hacktivists may seek public visibility for their actions, using the compromise as a platform to gain attention for their cause. The open and interconnected nature of O-RAN makes it a potential target for such actors, who might exploit its vulnerabilities to make a statement. Defending against hacktivist threats requires not only robust technical defenses but also an awareness of the sociopolitical contexts that might trigger such attacks.
- **Cyber-terrorists.** Cyber-terrorists represent a grave threat to the O-RAN system, operating with intentions to cause widespread fear, harm, or disruption for ideological, political, or religious motives. Unlike other threat agents, their primary objective is to undermine national security, cripple critical infrastructure, or instigate chaos, often targeting the communications backbone provided by O-RAN networks. Their methods could include sophisticated cyberattacks aimed at disabling network operations, causing physical damage to network components, or disrupting emergency services. The decentralized and software-driven nature of O-RAN potentially offers new avenues for such attackers to exploit. Defending against cyber-terrorist threats necessitates a comprehensive security that includes advanced threat detection, rigorous network monitoring, and close collaboration with law enforcement and intelligence agencies to mitigate the risk of catastrophic impacts on public safety and national security.
- **Script Kiddies.** Script kiddies constitute a threat to the O-RAN system, primarily due to their opportunistic nature and reliance on readily available hacking tools. Lacking the technical skills of professional hackers, these individuals exploit preexisting vulnerabilities using scripts, tools, and methods developed by others. Their motives can vary from seeking attention and recognition within certain communities to simple curiosity or the desire to cause disruption without a clear ideological or financial goal. While they may not possess the advanced technical knowledge of other threat agents, the accessibility of powerful hacking tools allows them to launch potentially damaging attacks on network infrastructures, including DDoS attacks, defacements, or exploiting known vulnerabilities within the O-RAN framework. The threat posed by script kiddies underscores the importance of basic cyber hygiene practices, such as regular software updates, vulnerability patches, and monitoring network traffic for unusual patterns, to protect against the broad range of attacks they might deploy.

- **Nation-State Actors.** Nation-state actors represent one of the most complicated and potentially damaging threats to the O-RAN system. Sponsored by governments, these actors engage in cyber espionage, sabotage, or warfare, aiming to advance national interests, gather intelligence, or disrupt the communications of adversaries. Their operations are characterized by elevated levels of complexity, significant resources, and advanced technical capabilities, allowing them to exploit vulnerabilities in O-RAN networks stealthily and persistently. Unlike other threat agents, nation-state actors may target critical infrastructure for long-term strategic goals rather than immediate financial gain or public recognition. Their activities can include the creation of backdoors, interception of sensitive communications, or even the establishment of sleeper functionalities for future attacks. The threat from nation-state actors necessitates a defense-in-depth strategy for O-RAN networks, incorporating advanced threat intelligence, robust access controls, and continuous monitoring to detect and mitigate state-sponsored cyber activities.

4.3.4 Potential Security Vulnerabilities of O-RAN

The O-RAN architecture, while pioneering in its approach to more open and flexible radio access networks, introduces specific vulnerabilities that could be exploited by various threat agents. Among these, unauthorized access to key O-RAN components, such as the O-DU, the O-CU-CP, the O-CU-UP, and the O-RU, stands out as a critical risk. Such access could lead to degraded RAN performance or enable broader network attacks, directly impacting the availability of services. Another area of concern is the unprotected synchronization and control plane traffic over the open fronthaul interface, where a lack of protection jeopardizes both the integrity and availability of network communications.

In terms of confidentiality, disabling over-the-air ciphers presents a significant vulnerability, allowing for potential eavesdropping on communications. The Near-RT RIC may conflict with the operations of the O-gNB (gNodeB), and similar conflicts could arise between xApps and rApps, further affecting network availability. Additionally, these applications' access to network and subscriber data could compromise confidentiality if not properly secured. The management interface, if left unprotected, exposes the network to risks across confidentiality, integrity, and availability, highlighting the need for robust security measures.

General vulnerabilities also pose threats to the O-RAN system. The decoupling of functions without a HW root of trust and a software trust chain can lead to integrity issues. The use of open-source code exposes the network to public exploits, threatening confidentiality, integrity, and availability. Furthermore, misconfiguration, poor isolation, or insufficient access management within the O-Cloud platform could lead to breaches.

4.4 Risk Assessment

After the identification of various assets, threats, and vulnerabilities within the O-RAN system, the next critical step is conducting a risk assessment. This process is pivotal for understanding the potential impacts these identified cybersecurity challenges could have on the O-RAN's operation and overall security. By evaluating the likelihood of threat exploitation against the severity of potential impacts, stakeholders can effectively prioritize security interventions. This strategic approach enables the development of a targeted action plan to mitigate risks, bolstering the defense mechanisms of O-RAN networks against unauthorized access, data breaches, and other cyber threats, ensuring that their integrity, availability, and confidentiality remain uncompromised.

4.4.1 Key Elements of Risk Assessment

Figure 4.6 offers a structured overview of the risk assessment process, which is critical for managing potential security issues within the O-RAN system.

At the core of this process is the identification of assets, which are valuable components such as infrastructure, data, or services that are crucial for the operation of the O-RAN. These assets are what the organization seeks to protect, as their compromise could lead to operational, financial, or reputational damage.

Threats to these assets in the O-RAN environment can take various forms, from deliberate attacks by cybercriminals or state actors to unintentional threats such as system failures or human errors. Vulnerabilities are the specific weaknesses within the system that threats might exploit. These could be outdated systems, misconfigured networks, or even lapses in physical security. When a threat successfully capitalizes on a vulnerability, an event occurs—this could be a data breach, service disruption, or any other form of security incident that negatively impacts the asset. The risk is then determined by evaluating the likelihood of such events occurring against the potential consequences they would entail. The likelihood is assessed based on factors like the O-RAN's exposure to threats, historical incidents, and the effectiveness of current security measures. Consequences are considered in terms of their impact on the organization's objectives and operations. The combination of these factors gives a measure of the risk associated with each identified threat. Finally, treatment refers to the strategies employed to manage this risk, ranging from applying security controls to mitigate it, transferring it through insurance, accepting it if it is within the organization's risk level, or avoiding it altogether by changing business practices. In the context of O-RAN, this might involve a mix of technical solutions, policy changes, and continuous monitoring to protect the network infrastructure from the identified risks.

When evaluating the severity of impact from threats that successfully exploit vulnerabilities in an O-RAN system, it is imperative to consider several angles: the type of impact, the scale of the impact, and the specific configurations or requirements that are affected.

The severity level is first distinguished by the type of impact on four key properties.

- **Privacy.** Privacy refers to unauthorized access to personal data. A low impact might involve limited, nonsensitive data exposure; a medium impact could be a more extensive compromise of personal data not resulting in direct harm; and a high impact would be a severe breach of sensitive personal data that could lead to significant harm to individuals.

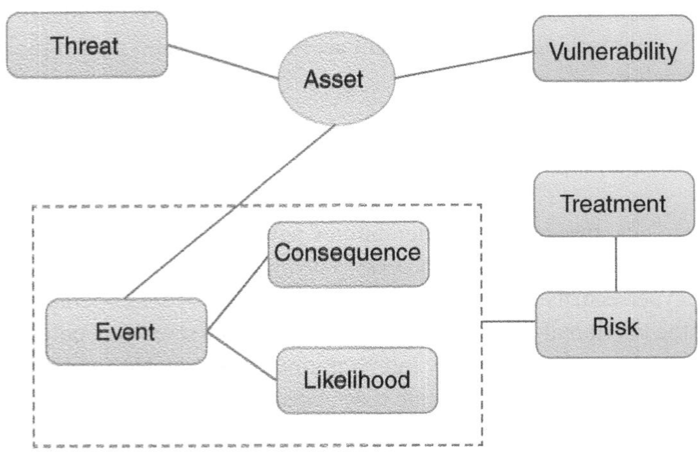

Figure 4.6 Concepts of Risk Assessment. Source: O-RAN Working Group 11 (2024a). Reproduced with permission.

- **Confidentiality.** Confidentiality involves the protection of information from unauthorized access and disclosure. A low impact here might mean minimal access to noncritical information, medium could indicate unauthorized access to more sensitive data without widespread exposure, and high severity would entail a major breach of highly confidential information with the potential for substantial damage.
- **Integrity.** Integrity relates to the accuracy and completeness of information and processing methods. A low impact might be minor inaccuracies without operational consequence, medium could affect system accuracy with some operational impacts, and high severity would imply a significant compromise of data integrity, undermining trust and causing major operational issues.
- **Availability.** Availability concerns the accessibility and usability of information and services. A low-level impact could result in short, inconsequential downtimes; a medium impact might cause moderate disruption to operations; and a high-level impact would be a prolonged and widespread disruption causing significant operational challenges.

The severity levels in an O-RAN system are also categorized based on the extent to which the network's units are affected by a given threat, aligning the impact level with the number of compromised components. A singular affected O-RU, or Radio Unit, constitutes a "low" severity level, as the impact is isolated. When a threat impacts an O-DU, along with its associated multiple O-RUs, the severity level escalates to "medium" due to the broader network involvement. The situation becomes even more critical with multiple O-DUs and their related O-RUs affected, which is classified as a "high" severity level, indicating a substantial threat to network operations. Operators must factor in the likelihood of such threats and the specific configuration of their deployed O-RAN system when conducting a comprehensive risk analysis.

The severity of impact is further influenced by the clock model and synchronization topology configurations within the O-RAN network. For example, in LLS-C1, where the O-DU directly synchronizes the O-RU, a DoS attack affecting only one O-RU or multiple O-RUs in a point-to-point connection results in a "low" severity level due to the limited and contained nature of the impact. In LLS-C2, the presence of Ethernet switches and the potential for a neighboring O-DU to serve as a backup master clock mitigates the severity to "low" as well, even if multiple O-RUs are affected. Contrastingly, in LLS-C3, where timing is distributed through the fronthaul network, a DoS attack on the master clock could disrupt multiple O-DUs and their related O-RUs, leading to a "high" severity level. LLS-C4, which involves local timing provision to the O-RU, would also result in a "low" severity level if attacked, as the impact is confined to a single O-RU.

The severity level also accounts for the presence or absence of normative O-RAN security requirements and controls. In the absence of such provisions, the severity level is considered "high," while the existence of agreed-upon security measures could justify a "low" or "medium" severity level. Thus, implementing O-RAN ALLIANCE security requirements and controls is a critical factor in mitigating risk and managing the severity of impacts from potential threats, reinforcing the importance of comprehensive security frameworks within O-RAN deployments.

The likelihood level in risk assessment quantifies how probable it is that a given threat will successfully exploit vulnerabilities in an O-RAN system. This evaluation is influenced by a set of factors, with the established O-RAN security requirements and controls being of primary importance. If an O-RAN system lacks these normative security measures, the likelihood of a threat successfully exploiting vulnerabilities is deemed high. Conversely, if robust security controls are in place and accounted for in the risk assessment assumptions, the likelihood level could be rated as low or medium.

Another factor in determining the likelihood level is threat event initiation, which considers the skill level of the potential attackers and the accessibility of potential entry points for launching

	3-High	3-Medium	6-High	9-High
Severity	2-Medium	2-Low	4-Medium	6-High
	1-Low	1-Low	2-Low	3-Medium
		1-Low	2-Medium	3-High
		Likelihood		

Figure 4.7 Risk Assessment Matrix. Source: O-RAN Working Group 11 (2024b). Reproduced with permission.

an attack. A high likelihood level is assigned if attackers with sufficient capabilities can initiate an attack from the internet or an untrusted network, where entry points are more accessible. External interfaces and services also influence the likelihood level, with more exposed elements increasing the chance of an attack.

Additionally, the adoption of a zero-trust architecture is factored into the likelihood assessment. In a zero-trust framework, it is recognized that internal threats exist alongside external ones, thus perimeter defenses alone are not deemed sufficient. This approach raises the baseline for likelihood scores, with a minimum of medium assigned. Specific types of attacks, such as reconnaissance actions, may be rated with a high likelihood given that they are less intrusive and harder to detect. In contrast, more damaging attacks, which can cause immediate and observable disruptions, might be assigned a medium likelihood, as threat actors often avoid actions that lead to rapid detection. Advanced persistent threats are particularly stealthy and aim to move laterally within a system to gather information without alerting the defense mechanisms, reflecting the need for comprehensive internal as well as external defenses in a zero-trust model.

The risk evaluation within the O-RAN system is structured and integrates findings from severity and likelihood assessments. To combine these insights into a coherent risk profile, a risk assessment matrix is utilized as shown in Figure 4.7. This matrix is a tool that cross-references the likelihood of a threat event occurring (probability) with the potential impact (severity) it would have if it did occur. In one axis of the matrix, you have the likelihood of the threat, which ranges from low to high, reflecting the probability of the threat successfully exploiting a vulnerability. On the other axis is the severity, which also ranges from low to high, indicating the level of impact or damage to the network's assets. At the intersection of these two axes are risk scores, which are determined by combining individual severity and likelihood assessments.

The process involves a simple multiplication of the severity score with the likelihood score to yield a final risk score. This scoring system allows for a straightforward yet effective way to estimate risk levels. For instance, even if a threat has a low likelihood of occurrence but the potential severity is high, the resulting risk could be considered medium. This means that even unlikely events cannot be ignored if their potential impact is substantial. The final risk score guides decision makers in prioritizing risks and determining the most appropriate risk treatment strategies, such as risk avoidance, mitigation, transfer, or acceptance.

4.4.2 O-RAN Risk Assessment Analysis: A Brief Overview

In assessing the risks to the O-RAN system, one must pay attention to threats with high-risk scores, as these represent the most severe and probable issues that need to be addressed urgently. These threats have been identified and will be explained shortly. For full risk analysis of all threats, refer to Table 7.2 in O-RAN WG11 Threat Modelling and Remediation (O-RAN Working Group 11 2024a).

Threat ID—T-O-RAN-01 (Near-RT RIC) illustrates a high-risk scenario where an attacker exploits insecure designs or the lack of adoption of comprehensive security controls in O-RAN components. The consequences of such an attack are diverse, resulting in loss of service, privacy breaches, and performance degradation across the O-RAN components. The impact is significant, as unauthorized access to an O-RAN component can not only compromise its performance and functionality but also pave the way for further attacks on other components within the system, leading to a cascade of failures. The potential for loss or tampering of sensitive data also poses a severe threat to the confidentiality, integrity, and availability of the network's information.

Several factors contribute to the high-risk level of this threat. The exposure factor is rated high due to the wide range of O-RAN components that could potentially be affected, such as A1, O1, E2 interfaces, and xApps/rApps. The possibility of adverse impacts is also high, underscoring the substantial negative consequences that could result from such a breach. The severity level is determined as high due to the profound implications of unauthorized access, while the likelihood is also high, reflecting the probability of such an event occurring, especially since industry requirements have not been clearly specified and it is uncertain whether existing industry practices adequately mitigate the identified threats.

The risk level remains high because the necessary security requirements and controls, particularly those that protect the hosting platforms of O-RAN functions against both physical and remote access, have not yet been fully articulated by the O-RAN ALLIANCE. Although the O-RAN ALLIANCE has addressed security for O-RAN interfaces at the transport layers, there is still a gap in securing the platforms themselves, necessitating prompt attention to refine and implement comprehensive security measures.

Threat ID—T-O-RAN-01 (Non-RT RIC + SMO) presents another high-risk concern within the O-RAN framework. This threat also involves the exploitation of potentially insecure designs or the insufficient implementation of security measures such as system hardening. The repercussions of such a security breach are wide ranging, potentially resulting in a loss of service, privacy violations, and a general decline in system performance within the O-RAN components. The impact of this threat extends to unauthenticated or unauthorized access to O-RAN components, leading to a degradation of system performance and service functionality. This vulnerability not only affects the directly compromised component but also poses a risk of lateral attacks spreading within the O-RAN system, affecting other interconnected components. Furthermore, there is a heightened risk of sensitive data being lost, stolen, or tampered with, endangering the confidentiality, integrity, and availability of information across the network.

Contributing to the high-risk score are several factors, including a high exposure level due to the many elements potentially affected, such as the A1, O1, O2 interfaces, and rApps. The assessment also notes high adverse impacts, reflecting the serious consequences of such threats. The overall risk score remains high because, while the O-RAN ALLIANCE has put forward security specifications for the transport layers of O-RAN interfaces, it has yet to define comprehensive security measures for the O-Cloud and the physical and remote protection of platforms that host O-RAN functions. The current state of industry practices in covering these threats remains uncertain, signaling a pressing need for the development and implementation of robust security hardening requirements to mitigate these high-severity and high-likelihood risks.

Threat ID—T-O-RAN-01 (O-CU) highlights a critical risk within the O-RAN system that revolves around the exploitation of vulnerabilities due to the lack of secured system designs or the insufficient implementation of security controls. The threat specifically targets the O-CU component of O-RAN, with an attacker causing disruptions that can lead to a loss of service, privacy infringements, and overall system performance issues. The impact of this threat is substantial, as unauthorized access to the O-CU could compromise its performance and lead to malfunctioning services. Such an incursion not only affects the O-CU but also carries the potential for further attacks, which could extend to other components within the O-RAN system. The threat further increases the risk of sensitive data being lost, stolen, or altered, thereby impacting the confidentiality, integrity, and availability of the network's information.

The risk score for this threat is deemed high, due to the significant exposure of critical interfaces like O1 and E2, and administration interfaces, which if compromised, could have widespread adverse effects. The likelihood of this threat is also rated high, especially given that specific O-Cloud security hardening requirements have yet to be specified, and there is uncertainty about whether current industry practices are adequately protective against such threats. Although the O-RAN ALLIANCE has established security protocols for O-RAN interfaces at the transport layers, which helps mitigate unauthorized access through these interfaces, there remains a gap in security provisions for the platforms hosting O-RAN functions against physical or remote breaches. The absence of such defined security measures in the hosting platforms raises the risk score, underlining the necessity for detailed and effective security requirements to address these vulnerabilities.

Threat ID—T-O-RAN-01 (O-DU) identifies a significant risk associated with the O-DU within the O-RAN architecture, arising from potentially insecure designs or the inadequate implementation of necessary security measures such as system hardening. This threat scenario consists of a wide range of consequences, including loss of service, privacy breaches, and a general decline in the system's performance across O-RAN components. The primary impact involves unauthorized or unauthenticated access to the O-DU component, which could severely compromise its functionality, leading to degraded performance or complete service disruption. This vulnerability does not only pose a direct risk to the O-DU but also opens the door for lateral attacks that could extend across the O-RAN system, targeting additional system components. Furthermore, the threat encompasses the potential for sensitive data to be lost, stolen, or tampered with, jeopardizing the network's confidentiality, integrity, and availability.

The high-risk score assigned to this threat stems from several factors, notably the high level of exposure of critical interfaces such as E2, O1, and the fronthaul. These interfaces, if compromised, could have widespread negative consequences. The severity and likelihood levels are both evaluated as high, given the current uncertainty surrounding the sufficiency of industry practices in addressing such vulnerabilities, compounded by the lack of specified O-Cloud security hardening requirements. Although the O-RAN ALLIANCE has implemented security measures for O-RAN interfaces at the transport layers, effectively mitigating unauthorized access through these channels, there remains a notable gap in security for the physical and remote protection of the platforms hosting O-RAN functions.

Threat ID—T-O-RAN-02 delves into the risk posed by misconfigured or poorly configured O-RAN components, a scenario that can lead to significant disruptions and vulnerabilities within the O-RAN system. This situation facilitates a myriad of issues, including loss of service, breaches in privacy, and a decline in overall system performance. The essence of the threat lies in the potential for unauthorized or unauthenticated access to O-RAN components, which not only degrades the performance and functionality of the affected components but also opens avenues for lateral attacks on other parts of the O-RAN system. The risk extends to the loss, theft, or tampering of sensitive data, threatening the confidentiality, integrity, and availability of the network's information assets. The severity and likelihood levels for this threat are both considered high. This assessment is driven by the high potential for threat event initiation—given the accessibility and attractiveness of misconfigured components to attackers—and the significant adverse impacts that could result from such a breach. The high-risk score reflects the serious implications of compromised O-RAN components.

The rationale behind the high-risk score stems from the current ambiguity regarding the adequacy of industry practices in preventing such threats, compounded by the lack of specific O-Cloud security hardening requirements. While the O-RAN ALLIANCE has made significant efforts in securing O-RAN interfaces at the transport layer, the broader issue of ensuring the secure configuration and protection of the platforms hosting O-RAN functions from both physical and remote access remains unaddressed.

Threat—ID T-O-RAN-03 addresses the vulnerabilities within the O-RAN network that stem from attacks originating from the internet, capitalizing on weak authentication and access control mechanisms. These vulnerabilities make it possible for attackers to breach the O-RAN network boundary, leading to a range of detrimental outcomes such as network flooding, resulting in service disruptions and performance degradation, along with unauthorized access to O-RAN components. The impacts of such attacks are numerous. They can cause DoS, preventing legitimate access to components and the services they offer. Unauthorized access due to compromised authentication and access control can lead to malfunction of O-RAN components, enable several attacks to spread to other components within the system, and result in the loss, theft, or tampering of sensitive data. The broad scope of these impacts affects the confidentiality, integrity, and availability of the O-RAN network.

Despite the severity of the impacts being high, the overall likelihood level is assessed as medium, influenced by the high threat event initiation factor but moderated by the medium level of adverse impacts. This adjustment reflects the recognition that, while the initiation of such attacks is probable given the attractiveness of the target and the potential weaknesses in security measures, the full realization of adverse impacts might be mitigated by existing but unspecified layers of defense within the network. The high-risk score is justified by the absence of specific security measures for xApps, especially regarding their onboarding onto the SMO and deployment onto the Near-RT RIC, including aspects such as authentication, access control, and secure configuration. The lack of a testing framework for xApps further exacerbates this vulnerability. This risk scenario assumes that O-RAN network elements are adequately shielded from direct exposure to untrusted internet endpoints, underscoring the necessity for robust network access control to prevent such breaches.

Threat ID—T-O-RAN-04 focuses on the risk associated with an attacker leveraging IoT devices to flood the airlink signal, disrupting legitimate communications within the O-RAN network. This type of attack primarily targets the network's availability, aiming to overload the system with excessive traffic and thereby cause a loss of service. The impact of such an attack is a DoS, rendering components inaccessible and disrupting the functions or services they provide. Given that this threat specifically compromises the availability aspect of the network, its severity level is determined to be medium. This reflects the significant, although not catastrophic, potential disruption to network operations and service delivery.

Despite the high threat event initiation factor, indicating a high probability of such attacks being attempted due to the increasing proliferation and connectivity of IoT devices, the overall likelihood level for this threat is assessed as medium. This is due to considerations of existing network defenses and the capacity to mitigate the impact of such attacks. However, the risk score is classified as high, driven by the observation that the O-RAN ALLIANCE has not yet specified targeted security measures to address flooding attacks through IoT devices specifically. This gap in the security framework highlights the need for the development and implementation of comprehensive measures to protect against such vulnerabilities, particularly as IoT devices continue to play a more significant role in network systems.

Threat ID—T-O-RAN-05 outlines a significant risk where an attacker gains unauthorized entry into the O-RAN system through its various open interfaces, including the fronthaul, O1, O2, A1, and E2. This breach could result in a variety of detrimental outcomes, such as data tampering, information disclosure, and the disruption of service due to DoS attacks targeting specific components within the system. Furthermore, unauthenticated or unauthorized access to O-RAN components can severely affect the system's performance and functionality. It also opens the possibility for other attacks, spreading the breach to other components within the network, and leading to the loss, theft, or alteration of sensitive data. The impacts of this threat encompass all three pillars of cybersecurity: confidentiality, integrity, and availability, each of which is crucial for the secure and efficient operation of the O-RAN system. Despite the severity of these potential impacts being rated as high, the overall likelihood of this threat occurring is assessed as medium. This assessment is influenced by factors such as the medium level of threat event initiation—indicating that while possible, such attacks may require specific conditions or knowledge to be successfully executed—and the existence of moderate (but not entirely sufficient) existing controls within the system. Additionally, for a malicious user to exploit these vulnerabilities, direct physical access to the target system is often necessary, which may not always be feasible.

Given these considerations, the risk score for Threat ID T-O-RAN-05 remains high, underscoring the serious nature of the threat despite the presence of some mitigating factors and controls. This high-risk designation highlights the critical need for continuous evaluation and enhancement of security measures within the O-RAN system. Protecting against such vulnerabilities requires not only maintaining and improving existing controls but also developing new strategies to protect the system's open interfaces against unauthorized access and exploitation.

4.5 Security Principles, Security Requirements, and Security Controls in O-RAN

Security principles lay the foundational guidelines that govern the design and operation of secure O-RAN systems, emphasizing the necessity for confidentiality, integrity, and availability in network communications and operations. Building upon these principles, security requirements describe specific criteria and standards that O-RAN components, and systems must meet to ensure that they are protected against a wide range of cyber threats. These requirements address vulnerabilities across the network, from the physical layer to the application layer, and include mandates on encryption, access control, and threat detection mechanisms. Finally, security controls are the practical implementations and measures taken to fulfill the security requirements, effectively mitigating potential risks. These controls range from technical solutions, such as intrusion detection systems, to procedural and administrative strategies, like security policies and incident response protocols. Together, security principles, security requirements, and security controls form a comprehensive security framework essential for protecting the O-RAN system against cyber threats, ensuring the resilience and reliability of next-generation telecommunications networks.

4.5.1 O-RAN Security Principles

Before diving into the security principles of O-RAN, it is important to acknowledge that this area of focus is a work in progress within the O-RAN community. These principles are foundational to developing a comprehensive and resilient defense strategy against the broad spectrum of cyber threats targeting O-RAN systems. As the O-RAN system continues to evolve, these security principles are being refined and will be formally described and specified in detail in future O-RAN specifications. This ongoing work highlights the commitment of the O-RAN community to not only advance the capabilities and performance of radio access networks but also to ensure that these innovations are matched with equally robust security measures. By setting a clear framework of security principles, O-RAN aims to provide a solid foundation upon which all other security requirements and controls can be developed, ensuring that the entire network architecture is built with security at its core. This process reflects the dynamic nature of cybersecurity, where continuous improvement and adaptation are necessary to address emerging threats and vulnerabilities.

SP-AUTH Mutual Authentication. The mutual authentication security principle within the O-RAN framework emphasizes the critical importance of bidirectional verification between the system and its components, including base stations, users, and administrators. This principle aims to improve network security by ensuring that each party in a communication exchange is authenticated, thereby effectively identifying and preventing threats from fake base stations, unauthorized components, malicious applications, and ill-intentioned users or administrators. To facilitate this, it mandates that each component involved in the communication is assigned a unique identifier and maintains one or more securely protected credentials. This measure is foundational in protecting the integrity and confidentiality of the O-RAN system, ensuring that access and communication privileges are extended solely to verified and legitimate entities.

SP-ACC Access Control. The access control principle underlines the imperative need to restrict access to O-RAN resources and services, ensuring that only authorized administrators or components can interact with or modify the system. This principle is foundational in preventing unauthorized or unintended access that could compromise the network's security or operational integrity. It encompasses a range of control measures including network access controls, which are essential for filtering out unauthorized or unexpected traffic across the O-RAN components through their interfaces, protecting the network from potentially malicious activities. Additionally, access controls specifically designed to limit interaction with component configurations are crucial for maintaining system stability and security. Furthermore, HW access controls play a pivotal role in sustaining the trust chain, ensuring that the HW components and the underlying infrastructure remain secure against tampering or unauthorized access. Collectively, these access control measures are critical for protecting the O-RAN infrastructure from potential security breaches and ensuring the network's resilience against cyber threats.

SP-CRYPTO Secure Cryptographic, Key Management, and PKI. The SP-CRYPTO principle advocates for the utilization of well-known and unbroken cryptographic schemes that have been standardized, steering clear of proprietary schemes and protocols due to their potential security vulnerabilities and lack of widespread scrutiny. Key management is highlighted as a vital component, necessitating a secure process for managing the life cycle of O-RAN keys, such as KgNB, KRRC-enc, KRRC-int, KUP-enc, and ksn. This includes ensuring keys are generated with sufficient entropy from reliable sources, securely stored, and appropriately rotated or revoked, when necessary, alongside secure key destruction practices. Furthermore, the principle underscores the importance of a trustworthy PKI for authentication and data encryption, recommending support for public certificate authorities (CAs) and insisting that certificates should be issued by trusted or rooted CAs. These CAs should follow a clear certificate policy outlining the rules and policies governing certificate issuance. Additionally, the configuration of O-RUs with prioritized lists of allowed integrity and ciphering algorithms, as decided by the operator, is crucial for maintaining the integrity and confidentiality of the communications within the O-RAN network.

This comprehensive approach to cryptographic security, key management, and PKI is foundational in building a secure and trustworthy O-RAN framework.

SP-TCOMM Trusted Communication. The SP-TCOMM principle, or trusted communication, within the O-RAN framework, underscores the necessity of protecting communication channels to maintain the integrity, confidentiality, availability, authenticity, and replay protection of resources as they traverse the O-RAN interfaces. This principle is pivotal in ensuring that data and information, identified as critical assets, are protected from unauthorized access, alteration, or disruption during transmission across the network. By adhering to industry best practices, the O-RAN system aims to implement robust security measures that encompass encryption to maintain confidentiality, mechanisms to verify data integrity, protocols to ensure system availability, authentication to confirm the identity of communicating parties, and strategies to protect against replay attacks.

SP-SS Secure Storage. The secure storage principle, in the O-RAN framework, emphasizes the importance of protecting resources, especially those classified as "critical assets," when they are stored or at rest. A particular focus is placed on the secure storage of highly sensitive elements such as private keys, certificates, and hash values, recognizing their critical role in the network's overall security. By ensuring that these assets are securely stored, the O-RAN system mitigates risks associated with data breaches, unauthorized access, and potential compromises that could impact the network's integrity and confidentiality. Implementing strong encryption for stored data, access controls to limit who can retrieve or alter this information, and regular audits to assess the security of stored data are essential components of this principle. Through dedicated attention to secure storage practices, the O-RAN framework aims to establish a resilient and trustworthy infrastructure capable of protecting vital assets against evolving cybersecurity threats.

SP-SB Secure Boot and Self-configuration. This principle is designed to ensure the integrity and trustworthiness of the O-RAN components by securing their firmware and configurations. By establishing a verified platform firmware, the system creates a foundation of trust that can extend upward through the software stack. This means that the firmware can authenticate the OS bootloader, which in turn can verify subsequent software components, up to and including the OS, hypervisor, or container runtime layers, and the O-RAN components themselves. The essence of this principle lies in the concept of a chain of trust (CoT), a security technique where each stage of the boot process is responsible for verifying the integrity and authenticity of the next stage before allowing it to execute. This method ensures that only verified and authenticated code can run on the system, significantly reducing the risk of malicious software injection and execution. To effectively implement this principle, secure boot processes, signature verification, and self-configuration mechanisms must be securely and accurately executed across all O-RAN components. This approach ensures that each component is authenticated before it is loaded, thereby maintaining the system's integrity from the outset and throughout its operation. The secure boot and self-configuration principle is fundamental in building a secure and resilient O-RAN system, capable of resisting unauthorized access and tampering at every layer of the network.

SP-UPDT Secure Update. The SP-UPDT principle, secure update, is important for maintaining the integrity and security of the O-RAN system. It emphasizes the need for a secure update management process that facilitates the introduction of new components or software changes. This process is essential for regular enhancement of the system responding to emerging security challenges and updating cryptographic algorithms as needed. O-RAN vendors are expected to provide continuous security monitoring and timely patches for all components, ensuring that vulnerabilities are addressed promptly to maintain system integrity and protect against potential threats.

SP-RECO Recoverability and Backup. The SP-RECO principle outlines the strategies for ensuring the O-RAN system's ability to recover from disruptions, particularly those caused by DoS attacks. It mandates the implementation of mechanisms for detecting and mitigating DoS attacks, alongside defining a comprehensive recovery plan. This plan should be capable of resetting O-RAN components to a trustworthy state after any malfunction or security breach, thereby safeguarding the network's operational continuity. These systems should guarantee an adequate level of availability and reliability, ensuring that the network can swiftly recover from incidents without significant loss of data or functionality.

SP-OPNS Security Management of Risks in Open-Source Components. The SP-OPNS principle focuses on the security management of risks in open-source components within the O-RAN system. This principle underscores the necessity for vendors utilizing open-source code to significantly enhance security by adhering to industry coding best practices. Given the inherent exposure to public exploits that open-source components may bring, a higher level of diligence is recommended to mitigate potential risks effectively. Key to managing these risks is the maintenance of a Software Bill of Materials (SBOM), which provides a comprehensive inventory of all open-source components utilized within the system. This inventory allows for a clearer understanding of where and how open-source code is integrated into the O-RAN infrastructure, facilitating targeted security measures. Furthermore, conducting regular security analyses—including audits and vulnerability scans—is crucial to identifying and addressing any security vulnerabilities within these components promptly.

SP-ASSU Security Assurance. The SP-ASSU principle addresses the importance of establishing and maintaining trust in the security of mobile networks, which are recognized as critical infrastructure. First, vendors are required to ensure their software or HW complies with the 3GPP Security Assurance Specifications (SCAS), offering proof of adherence to these specifications. This step is crucial for aligning with globally recognized security benchmarks in mobile communications. Additionally, vendors must verify that their offerings meet the security tests, requirements, and recommendations set forth by the O-RAN ALLIANCE, further reinforcing the security of O-RAN components. Beyond adherence to specific technical standards, vendors are also tasked with ensuring that their products align with various national and international cybersecurity regulations, such as the Cybersecurity Act (Cyber Act), General Data Protection Regulation (GDPR), and others. This ensures a holistic approach to security that not only protects network integrity but also addresses privacy and data protection concerns. To solidify this commitment to security assurance, vendors should actively engage in comprehensive risk assessments, secure code reviews, penetration testing, vulnerability analysis, and provide hardening guidelines for their O-RAN components.

SP-PRV Privacy. Recognizing that communication services inherently collect data and personal information from end users, it is crucial to protect the privacy of this information, which includes data privacy, identity privacy, and personal information privacy. The principle highlights the risk of adversaries exploiting this data to extract sensitive personal details about end users, such as their priorities, location, movements, and preferences. Therefore, implementing measures to protect end user privacy against such threats is a fundamental aspect of O-RAN's security strategy.

SP-SLC Continuous Security Development, Testing, Logging, Monitoring, and Vulnerability Handling. The SP-SLC principle emphasizes the need for continuous security development, testing, logging, monitoring, and vulnerability handling. This includes adopting continuous development and continuous integration (CD/CI) practices complemented by continuous regression testing and software security auditing. It stresses the importance of logging relevant activities and analyzing logs in real time to detect potential security attacks and for security auditing purposes. Continuous monitoring ensures that the security state is maintained across the life cycle of O-RAN components, while a robust vulnerability management system is necessary for tracking, identifying, and remediating vulnerabilities efficiently.

SP-ISO Robust Isolation. The SP-ISO principle, robust isolation, addresses the need for stringent isolation measures in a multivendor environment. It mandates intradomain host isolation and the security isolation of resources within the same host, including VMs, container networks (CNs), and the virtualization/container layer, as well as CPU, storage, and network resources. This is achieved through system security orchestration, segmentation, life cycle management, time scheduling, and comprehensive monitoring and auditing across the management, signaling, control, and data planes, as well as the execution of virtualized O-RAN components.

SP-PHY Physical Security. The SP-PHY principle advocates for situating the O-RAN system within a physically secure environment to reduce the risks associated with resource theft and destruction. Ensuring a secure location is just the beginning; the principle extends to the support for secure storage of sensitive data, such as cryptographic keys and configuration data. It also includes the secure execution of critical functions, including encryption/decryption and authentication processes, and the secure execution of boot and update processes.

Furthermore, the SP-PHY principle draws special attention to the threats of site intrusion and unauthorized physical access, particularly at O-RU sites. To mitigate these risks, the principle recommends disabling all unnecessary physical and logical ports, protocols, and interfaces on O-RU equipment. This action is aimed at minimizing potential entry points for unauthorized access. Additionally, the establishment of secure physical connections for operation and maintenance (O&M) operations is highlighted. An example of this would be the use of a secure laptop equipped with secure credentials for carrying out O&M tasks. Implementing these physical security measures ensures that the O-RAN infrastructure is not only protected against digital threats but also against physical tampering and unauthorized access, thereby enhancing the overall security of the O-RAN system.

SP-CLD Secure Cloud Computing and Virtualization. The SP-CLD principle on secure cloud computing and virtualization within the O-RAN framework includes the implementation of various virtual machine-based security measures such as intrusion detection, isolation, kernel protection, access control, and trusted computing. To achieve a fully trusted virtualization layer, there is a need for trust across all layers in the technology stack—from the HW and firmware up to the virtualized software. This comprehensive trust is essential because the security of virtual machines or containers is inherently dependent on the trustworthiness of the host system. Ensuring this trust across all layers helps safeguard against vulnerabilities that could compromise the virtualized components of the O-RAN system.

SP-ROB Robustness. The O-RAN system is required to maintain not just the physical and software robustness but also the robustness of the CR channel. This is crucial for meeting the quality of service (QoS) expectations for communication services as demanded by users. Particularly in scenarios where spectrum sensing could be compromised—such as when sensing nodes like O-RUs malfunction—enhancing robustness becomes even more critical. Robustness is a key factor in defending against various security threats, including jamming and DoS or DDoS attacks, ensuring that the system can resist and recover from such adversarial actions without significant degradation in performance or service delivery.

Each of these principles plays a crucial role in fortifying the O-RAN system against a myriad of security challenges, ensuring the integrity, reliability, and privacy of the network. In addressing the various threats to the O-RAN system, these threats have been systematically associated with applicable security principles as outlined in O-RAN WG11 Threat Modelling and Remediation Documentation. This approach ensures a comprehensive understanding and mapping of how each security principle addresses specific threats, providing a structured framework for implementing targeted security measures.

4.5.2 Security Requirements for O-RAN

Establishing robust security requirements is fundamental to protecting the O-RAN network and its users from a myriad of cybersecurity threats. These requirements serve as a detailed blueprint for implementing the security principles within the O-RAN system, addressing specific vulnerabilities and threat vectors identified through comprehensive risk assessment. Covering a wide range of areas, from access control and data protection to secure communication and physical security, the security requirements in O-RAN are designed to ensure the confidentiality, integrity, and availability of the network's resources and services. By meticulously defining these requirements, O-RAN aims to provide a secure and resilient infrastructure capable of supporting the next generation of mobile communications while protecting against the constantly evolving landscape of cyber threats.

The SMO framework within the O-RAN system is guarded by a comprehensive set of security requirements designed to ensure the robustness of network operations against any form of cyber threats. These requirements emphasize the critical importance of stringent authentication and authorization mechanisms. Specifically, the SMO is mandated to authenticate both its internal functions and external systems, ensuring that every entity interacting with the network is verified and authorized. This dual-layered approach to authentication and authorization is pivotal in maintaining a secure operational environment, effectively protecting against unauthorized access and potential security breaches. Furthermore, the resilience of the SMO against DDoS attacks is a cornerstone of its security requirements. The framework is expected to withstand volumetric DDoS attacks targeting the O2 interface, external interfaces, and internal communications among SMO functions. This capability to recover without catastrophic failure from such attacks is essential for maintaining service continuity and network availability. In addition to these protective measures, the SMO security requirements also cover secure communications and comprehensive logging and event management. Internal communications within the SMO must be encrypted and protected to preserve the confidentiality, integrity, and availability of data. Mutual authentication between SMO functions further enhances this security layer. The management of event logs, including their secure storage, integrity protection, and authorized access, forms a critical aspect of the network's operational integrity.

Non-RT RIC and rApps within the O-RAN architecture require a robust level of security in their operations and interactions. The security requirements for the Non-RT RIC and rApps focus on authorization, resilience to attacks, and secure communication protocols. Authorization plays a pivotal role in the interaction between the Non-RT RIC and rApps. The Non-RT RIC is required to support authorization as a resource owner/server and client, establishing a secure environment for accessing and managing resources. Additionally, the Non-RT RIC framework must authorize requests from rApps, which act as clients, ensuring that only authorized applications can communicate with and utilize the resources of the Non-RT RIC. Conversely, rApps are tasked with providing appropriate client authorization requests to the Non-RT RIC framework, further reinforcing the mutual authentication and authorization process.

Resilience against DDoS attacks is another critical aspect of these requirements. Both the Non-RT RIC and rApps are expected to recover without catastrophic failure from volumetric DDoS attacks targeting the A1 and R1 interfaces. Secure communication protocols are also emphasized, with the SMO/Non-RT RIC framework required to authenticate both the API producer and API consumer using a Kafka-based protocol for data streaming across the R1 interface. This ensures that data streaming activities are securely managed and that only authenticated entities can produce or consume APIs. Additionally, an authorization mechanism for the Kafka-based protocol is mandated to provide access for data streaming, further securing the communication channel and safeguarding against unauthorized data access.

The security requirements for the **Near-RT RIC and xApps** within the O-RAN framework emphasize the need for rigid security measures to protect data access, ensure secure communication, and provide a resilient architecture against cyber threats. Authentication and authorization are central to these requirements. The Near-RT RIC must authenticate xApp access to its databases during the Service Data Layer (SDL) registration process, ensuring that only authorized xApps can access critical data stored in the Near-RT RIC databases. This process extends to providing authorized access to these databases, further securing sensitive information. Communication between xApps and the Near-RT RIC platform APIs is required to be mutually authenticated, establishing a secure channel for data exchange and interactions between these components.

An authorization framework within the Near-RT RIC architecture is mandated to manage the consumption of services exposed by platform APIs by xApps, taking operator policies into account. This framework ensures that access to platform services by xApps aligns with the specific policies and security requirements set by operators. Additionally, the Near-RT RIC is expected to support authorization both as a resource owner/server and as a client, highlighting the dual role it plays in the O-RAN security architecture.

Resilience against DDoS attacks and content-related attacks, such as injection attacks and buffer overflows, is crucial for maintaining the integrity and availability of the Near-RT RIC. The system should be capable of recovering from volumetric DDoS attacks across the A1 interface without catastrophic failure and possess mechanisms to detect and defend against content-related attacks, ensuring the continuous and secure operation of the Near-RT RIC and its communication with xApps.

The security requirements for the O-DU, O-CU, and O-RU components are defined in 3GPP (2017). This document provides a comprehensive framework for the security architecture and procedures applicable to 5G systems, detailing the measures and protocols necessary to ensure the confidentiality, integrity, and availability of the network and its services. By adhering to these established security requirements, the O-DU, O-CU, and O-RU components of O-RAN are aligned with global standards for 5G security, ensuring robust protection against a wide range of cyber threats and vulnerabilities.

The security requirements for the A1 interface within the O-RAN architecture are designed to address the need for secure communications between the Non-RT RIC and one or more Near-RT RICs. Key to these requirements is the implementation of confidentiality, integrity, and replay protection mechanisms, which collectively ensure that data is not only secure from unauthorized access and alteration but also safeguarded against attempts to reuse captured messages maliciously.

Furthermore, the A1 interface mandates mutual authentication and authorization, emphasizing a security approach where both communicating parties verify each other's identity. This mutual authentication process is crucial for establishing a trusted communication channel, thereby significantly mitigating the risk of impersonation or unauthorized access. The implementation of TLS is specified for encrypting data in transit, ensuring that all communications over the A1 interface are secure and protected from potential eavesdropping or interception. Additionally, mutual TLS (mTLS) is recommended for enhancing the mutual authentication process, requiring both the client and server to authenticate themselves to each other, thus offering a higher level of security. Authorization via OAuth 2.0 is also supported, providing a secure and efficient mechanism for granting access to resources without exposing user credentials. These security measures, as specified in the O-RAN Security Protocols Specifications, are fundamental to fortifying the A1 interface against cyber threats, ensuring that the O-RAN system remains robust, secure, and resilient.

The open fronthaul interface, particularly within its C-plane and S-plane, plays a critical role in the operation and performance of the O-RAN architecture. For the C-plane, the integrity and timeliness of messages between the O-DU and O-RU are essential for the successful transmission and reception of RF signals. Adversaries exploiting vulnerabilities to inject spoofed C-plane messages could severely impact network performance by causing DoS and degraded cell performance. To mitigate such risks, the support for authentication and authorization of O-DUs exchanging C-plane messages with O-RUs is required, which is aimed at authenticating legitimate communications and preventing unauthorized message injection.

The S-plane, which encompasses the Precision Time Protocol (PTP) for clock synchronization within the network, is susceptible to attacks that could compromise timing accuracy and network coordination (O-RAN Working Group 11 2024j). An attacker masquerading as a grandmaster (GM) clock or manipulating the PTP could disrupt synchronization across the network. Addressing this threat, the S-plane is required to support the authentication and authorization of PTP nodes to ensure that only legitimate nodes participate in time synchronization. Furthermore, the S-plane is mandated to prevent the spoofing of master clocks and protect against MITM attacks that could degrade clock accuracy, particularly in scenarios where the O-DU is in a data center.

4.5.3 Security Controls for O-RAN

Security requirements within a network or system architecture, such as O-RAN, establish the objectives and standards necessary to protect against cyber threats, unauthorized access, and data breaches. Meeting these requirements effectively involves implementing security controls designed to address identified vulnerabilities and mitigate potential risks. For instance, if a security requirement aims to ensure the confidentiality and integrity of data in transit, deploying TLS protocols serve as a direct control to encrypt data and authenticate communication channels. Similarly, requirements around access control can be met through the use of mechanisms like IEEE 802.1X for network access control, which authenticates and authorizes devices before they can access network resources. Security controls act as practical tools and procedures that operationalize security requirements, transforming them from conceptual goals into tangible protections that safeguard the system's operations and assets. By carefully selecting and applying these controls, organizations can fulfill their security requirements, ensuring a robust defense against diverse cybersecurity threats.

The security controls for the SMO, Non-RT RIC, rApps, Near-RT RIC, and xApps within the O-RAN architecture are comprehensive, addressing various aspects of security including authentication, authorization, confidentiality, integrity, and resilience to attacks. These controls are specified in the O-RAN Security Protocols Specifications and are designed to safeguard the O-RAN system against potential cybersecurity threats.

For the SMO, security controls include the implementation of OAuth 2.0 for authorization purposes, allowing for secure token-based access to resources. The SMO also supports mutual authentication of SMO functions using mTLS with PKI X.509v3 certificates, and it may utilize TLS with a pre-shared key for authenticating other SMO functions. These measures ensure that only authorized entities can access and interact with SMO functions, maintaining the security of the system.

In the Non-RT RIC and rApps, security controls focus on supporting OAuth 2.0 for both resource owner/server and client roles, facilitating secure authorization for service requests. For communications across the A1 and R1 interfaces, the framework may support TLS for secure data transmission and OAuth 2.0 for authorization, ensuring that interactions between the Non-RT RIC, rApps, and the SMO are protected against unauthorized access and data breaches (O-RAN Working Group 11 2024f).

Near-RT RIC and xApps security controls emphasize API security, with transactional APIs supporting mTLS authentication via X.509v3 certificates and OAuth 2.0 for authorization, considering operator policies in API consumption. For time-critical APIs, IPsec with IKEv2 certificated-based authentication provides additional security. These controls extend to ensuring the confidentiality and integrity of API messages, with TLS and IPsec used to protect data in transit. Additionally, Near-RT RIC is equipped to verify policies received through the A1 interface against predefined schemas and values, enhancing the integrity and validity of policy enforcement (O-RAN Working Group 11 2024e). Table 4.1 depicts an overview of security controls for Near-RT RIC APIs (O-RAN Working Group 11 2024b).

The security control for the C-plane focuses on the authentication and authorization of network elements that support the C-plane operations, addressing the requirements outlined in Section 5.2. This is achieved through the implementation of IEEE 802.1X-2020 Port-based Network Access Control, a standard that provides a mechanism for authenticating and authorizing devices on a network, ensuring that only trusted PTP nodes can participate in the network's synchronization activities. For the S-plane, particularly in relation to synchronization architecture redundancy, the architecture is designed to enhance robustness against potential security breaches through redundancy mechanisms. Security controls recommend the support for simultaneous GMs and the assignment of GMs to physically separated PTP ports. This approach provides topology resilience by allowing multiple masters to be connected, thereby enhancing the redundancy and resiliency of the open fronthaul synchronization architecture. These recommendations are further elaborated in the O-RAN synchronization architecture and solution specification, highlighting the importance of redundancy and resiliency in maintaining secure and uninterrupted synchronization across the network. Additionally, the authentication and authorization of PTP nodes within the S-plane are crucial for protecting the synchronization processes. Adhering to the security requirements of the S-plane, the implementation of IEEE 802.1X-2020 Port-based Network Access Control for PTP nodes ensures that synchronization activities are restricted to authenticated and authorized devices. This not only secures the network against unauthorized access but also strengthens the network's defense against synchronization-related attacks.

Table 4.1 Security Controls for Near-RT RIC APIs.

API Protocol	Authentication Method	Authorization Method	Confidentiality Method	Integrity Method
gRPC	mTLS	OAuth2	mTLS	mTLS
SCTP	IKEv2	—	IPsec	IPsec
REST/HTTP	mTLS	OAuth2	mTLS	mTLS

Collectively, these security controls form an integral part of the O-RAN security architecture, offering a multilayered defense strategy that spans from the transport layer to application layer interactions. By adhering to these specified protocols and standards, O-RAN ensures a secure and resilient network capable of withstanding various cyber threats, protecting the system's operations, and protecting sensitive data.

4.6 O-RAN Security Protocols

4.6.1 Security Protocols in a Nutshell

In the O-RAN architecture, a suite of key security protocols plays a pivotal role in protecting network operations and communications. SSH establishes a secure channel over an unsecured network, enabling secure data communication between devices. TLS ensures privacy and security between communicating applications over the internet, protecting against eavesdropping and data tampering. NETCONF over secure transport is utilized for secure configuration management of network devices, maintaining the integrity of network settings. Datagram Transport Layer Security (DTLS) offers security for datagram-based applications, ensuring the confidentiality and integrity of data in motion. IPsec secures IP communications by authenticating and encrypting each IP packet in a data stream, providing robust network security. Certificate Management Protocol version 2 (CMPv2) facilitates the management of certificates within a PKI, streamlining the distribution and authentication of digital certificates. Lastly, OAuth 2.0, an authorization framework, allows for controlled access to resources, enabling secure and limited access to user accounts on HTTP services. Together, these protocols form the backbone of O-RAN's security framework, ensuring comprehensive protection across all aspects of the network's operations (O-RAN Working Group 11 2024c).

4.6.2 SSH

SSH, is a protocol used to securely access and manage applications over an unsecured network (IETF 2006). Imagine SSH as a secure pipeline between your computer and another computer you want to control or access data from. It is like sending a letter in a locked box through regular mail, where only you and the recipient have the key to open it. This way, even if someone intercepts the box, they cannot see what is inside. Figure 4.8 presents a high-level overview of SSH.

SSH works by using a pair of keys to establish a secure connection:

- **Public Key.** This is like a padlock. You can share it with anyone, and they can use it to lock (encrypt) information they send to you.
- **Private Key.** This is like the key to the padlock. Only you have it, and you can use it to unlock (decrypt) information sent to you with your public key.

Figure 4.8 SSH Overview.

When you want to connect to another computer using SSH, the process typically goes like this:

- **Authentication.** The two computers use the keys to prove their identities to each other. You prove you have the right key to access the other computer, and it proves it is the computer you are expecting to connect to.
- **Encryption.** Once both sides are authenticated, all data sent back and forth is encrypted. This means that even if someone intercepts the data, they cannot understand it because it is all scrambled.

SSH is widely used by system administrators for controlling web servers, accessing computers remotely, securely transferring files, and more because it is a secure and versatile tool.

Within the O-RAN architecture, SSH is instrumental in securing communications across various critical components and interfaces, thereby maintaining the network's integrity and confidentiality. It is extensively employed for the secure management and configuration of key elements such as the O-DU, O-CU, and O-RU, ensuring that all operational commands and configuration changes are securely transmitted. The SMO component, along with the Near-RT RIC and Non-RT RIC, also rely on SSH for secure access and management. By facilitating encrypted communications, SSH plays a pivotal role in protecting these components from unauthorized access and cyber threats, significantly enhancing the security of the O-RAN system.

4.6.3 TLS

TLS is like a protective envelope that ensures any letter (data) you send through the internet reaches its destination safely and privately (IETF 2009). Imagine you are sending a confidential letter to a friend; TLS works by sealing this letter in an envelope that only your friend can open, ensuring no one else can peek inside while it is being delivered. This envelope is sealed through a process called encryption, which scrambles the contents of the letter so that it is unreadable to anyone except the intended recipient. When your friend receives the letter, they use a special key to unscramble (decrypt) the message and read it.

TLS plays a critical role in securing communications within the O-RAN architecture by encrypting data transmitted between different network components. This encryption protects the data from being intercepted, tampered with, or read by unauthorized parties as it moves across the network. TLS ensures that sensitive information, such as configuration details, user data, and operational commands, remains confidential and intact. Within the O-RAN system, the utilization of TLS across key interfaces plays a pivotal role in ensuring the confidentiality, integrity, and availability of communications. The O1 interface benefits from TLS by securing management and orchestration messages between the SMO framework and NFs like the O-CU and O-DU. Similarly, the A1 interface leverages TLS to protect communications between the Non-RT RIC and Near-RT RIC, ensuring a secure exchange of policy and control information. Furthermore, the E2 interface employs TLS to safeguard the transmission of control, management, and user plane data between the Near-RT RIC and network elements such as the O-CU and O-DU.

4.6.4 DTLS

DTLS is the sibling of TLS, but it is specifically designed for securing datagram-based communications instead of stream-based (IETF 2015b). Imagine sending a series of post-cards (datagrams) instead of a continuous roll of paper (stream). Unlike a stream, where you can ensure the order and reliability of the message delivery (like a telephone conversation), datagrams are more like individual, separate postcards sent through the mail. Each postcard might take a different route to the destination and arrive in any order. DTLS ensures that each of these postcards is sealed and stamped correctly, making sure that each one arrives intact, has not been tampered with, and can be verified as genuine by the recipient. DTLS is crucial for scenarios where the underlying transport protocol does not guarantee in-order delivery of packets (like UDP), making it ideal for time-sensitive or real-time applications, such as video streaming, VoIP, or gaming. It provides the same level of security as TLS, including encryption, integrity checks, and authentication, but is adapted for the unpredictable nature of datagram services.

In the O-RAN architecture, DTLS plays a crucial role in enhancing the security of datagram-based communications across various interfaces, ensuring the integrity, confidentiality, and authentication of the transmitted data. The E2 interface benefits significantly from DTLS, facilitating secure and timely communication between the Near-RT RIC and network elements such as the O-CU and the O-DU, which is essential for the transmission of critical control signals. For the F1/U interface, which connects the O-DU and the O-CU, DTLS is particularly valuable in scenarios that leverage datagram transport for the efficient exchange of control and user plane communications, offering a secure channel that does not compromise on speed and efficiency. Although the O1 interface primarily relies on secure HTTP-based protocols for management purposes, DTLS finds its application in specific cases where datagram transport is utilized for conveying telemetry data or providing real-time status updates, thereby broadening the scope of secure communications within the O-RAN system.

4.6.5 NETCONF over Secure Transport

NETCONF (Network Configuration Protocol) over Secure Transport is a network management protocol that provides mechanisms to install, manipulate, and delete the configuration of network devices. Unlike traditional management protocols that operate over insecure channels, NETCONF is designed to run over secure transport layers, ensuring confidentiality, integrity, and authentication of the management sessions. The fundamentals of NETCONF involve structured data (using XML or JSON) and a set of operations that allow for precise control over the configuration data stored in network devices. It operates in a client–server model where the client sends requests to manipulate configuration data, and the server responds with the outcomes of those operations. The "secure transport" aspect typically involves using SSH or TLS protocols, which wrap the NETCONF communication in an encrypted tunnel, protecting it from eavesdropping or tampering (IETF 2015a).

It is extensively utilized across O-RAN for tasks such as the configuration management of key components including the O-DU, the O-CU, and the O-RU, ensuring that all modifications to the network's configuration are securely transmitted and applied. Furthermore, NETCONF plays a significant role in the secure deployment of software updates and patches across O-RAN components, protecting the network against vulnerabilities. Additionally, it facilitates the secure collection of operational data and telemetry from these components, enabling reliable network monitoring and performance assessment.

4.6.6 IPsec

IPsec is a suite of protocols designed to secure IP communications by authenticating and encrypting each IP packet of a communication session. IPsec provides a robust layer of security for data transmitted across IP networks, ensuring the confidentiality, integrity, and authenticity of the data exchanged between devices. At its core, IPsec operates in two main modes (IETF 2005):

- **Transport Mode.** In this mode, IPsec encrypts only the payload of an IP packet, leaving the header untouched. This mode is typically used for end-to-end communication between two devices, such as two computers or a computer and a server.
- **Tunnel Mode.** Unlike transport mode, tunnel mode encrypts both the payload and the header of an IP packet. This encapsulated data is then wrapped in a new IP packet with a new header. Tunnel mode is often used for site-to-site virtual private networks (VPNs), where entire networks communicate securely over the internet.

IPsec employs two main protocols to provide security services.

- **Authentication Header (AH).** AH provides data integrity, data origin authentication, and an optional antireplay service. It authenticates the IP packet data and ensures that the data has not been tampered with during transmission.
- **Encapsulating Security Payload (ESP).** ESP provides confidentiality (encryption), data integrity, data origin authentication, and an antireplay service. It encrypts the data being sent to protect it from eavesdropping and ensures the data integrity and authenticity.

IPsec also uses the Internet Key Exchange (IKE) protocol to establish shared security attributes and keys between communicating parties, facilitating a secure and authenticated channel for data exchange. IKE typically operates in two phases: Phase 1 establishes a secure channel between the parties for communication, and Phase 2 uses this secure channel to negotiate the security associations for encrypting and authenticating the actual IP data flows.

In the context of O-RAN and other IP-based networks, IPsec serves as a foundational security measure for protecting data communications over potentially unsecured networks, such as the internet, ensuring that sensitive information remains confidential and secure from unauthorized access and attacks.

4.6.7 CMPv2

CMPv2 is a protocol used in the management of digital certificates within a PKI. PKI is a framework that provides digital certificates to verify the identities of parties involved in electronic transactions. CMPv2 facilitates various certificate-related operations, making it easier for entities within a PKI to securely manage and exchange digital certificates. CMPv2 has the following core functionalities (IETF 4210 2005).

- **Certificate Enrollment and Renewal.** CMPv2 enables devices and users to request new certificates and renew existing ones. This is crucial for maintaining an up-to-date and secure communication environment, as certificates validate the identity of entities and encrypt communications.
- **Certificate Revocation.** If a certificate is compromised or no longer needed, CMPv2 allows for its revocation, ensuring that it cannot be used maliciously or mistakenly.
- **Key Update and Recovery.** CMPv2 supports updating cryptographic keys associated with certificates and, if necessary, recovering lost keys. This ensures that entities can maintain secure communication even when keys need to be changed.
- **Cross-certification.** CMPv2 facilitates the creation of trust relationships between different PKIs by allowing them to issue certificates that are recognized across these infrastructures.

CMPv2 operates over secure channels, often using protocols like TLS, to protect the exchange of certificate management messages. It defines a set of messages and procedures that enable secure and automated certificate issuance, renewal, revocation, and recovery. In the O-RAN system,

CMPv2 can play a significant role in securing communications between various components, such as the SMO framework, Near-RT RICs, and other network elements. By managing digital certificates and cryptographic keys efficiently, CMPv2 ensures that the identity of these components can be authenticated, and their communications can be encrypted, enhancing the overall security stance of the O-RAN system.

4.6.8 OAuth 2.0

OAuth 2.0 is a widely adopted authorization framework that enables applications to obtain limited access to user accounts on other services, such as Facebook, Google, or Twitter, without exposing users' passwords. Think of OAuth 2.0 as a way of giving a valet key to an application, allowing it to perform specific actions on behalf of a user without giving full access to the user's account. The OAuth 2.0 framework specifies several roles as mentioned below (IETF 2012).

- **Resource Owner.** Typically, the user who can grant access to their resources (data) on one service to another service.
- **Resource Server.** The server hosting the protected resources, capable of accepting and responding to protected resource requests using access tokens.
- **Client.** The application requesting access to the resource server on behalf of the resource owner, using an access token.
- **Authorization Server.** The server that issues access tokens to the client after successfully authenticating the resource owner and obtaining authorization.

OAuth 2.0 defines four main authorization grant types for different use cases (IETF 2012).

- **Authorization Code.** Used by web and mobile applications, where the client redirects the user to an authorization server, obtains an authorization code, and exchanges it for an access token.
- **Implicit.** Designed for clients implemented in a browser using a scripting language, simplifying the flow by obtaining an access token directly without an intermediate authorization code.
- **Resource Owner Password Credentials.** Allows the client to directly request and receive an access token by presenting the resource owner's credentials (username and password).
- **Client Credentials.** Used when the client is acting on its own behalf, not on behalf of a specific user, and can authenticate itself with its own credentials.

OAuth 2.0 also supports the use of refresh tokens, which allow clients to obtain new access tokens without requiring the resource owner to be present or reauthenticate, thus providing long-term access to resources. OAuth 2.0 can be used to secure communications in an O-RAN system between various components, such as the SMO, Near-RT RIC, and Non-RT RIC, by ensuring that only authorized clients can access specific resources or perform actions.

4.7 O-RAN Security Testing

O-RAN security testing is an essential aspect of ensuring the robustness and resilience of O-RAN architectures against cyber threats and vulnerabilities. This comprehensive testing process involves evaluating various components, interfaces, and protocols within the O-RAN system to identify and mitigate potential security weaknesses. By systematically examining the security measures implemented across the O-RAN system, including authentication, encryption, and access control mechanisms, security testing aims to safeguard the network's integrity, confidentiality, and availability.

Through a combination of penetration testing, vulnerability assessments, and compliance checks against established security standards and best practices, O-RAN security testing ensures that the network is capable of withstanding malicious attacks, protecting sensitive data, and maintaining reliable and secure operations for next-generation mobile communications.

This section is further divided into subsections for a clear understanding of the O-RAN security testing. Section 4.7.1 discusses the tools for security testing, outlining the necessary software and HW tools required for effective security assessment. Section 4.7.2 introduces the test report template, providing a standardized format for documenting test results and ensuring consistency in reporting. Lastly, Section 4.7.3 explains the security testing framework, offering a structured approach to executing and managing security tests within the O-RAN environment.

4.7.1 Tools for O-RAN Security Testing

Table 4.2 enumerates the testing and measurement tools necessary for conducting the security assessments outlined in O-RAN Working Group 11 (2024d).

4.7.2 Test Report Template

The test report aims to provide a comprehensive and detailed account of the tests conducted, ensuring that the descriptions are thorough enough to allow for reproducibility by different parties and to facilitate comparison. Utilizing the structured template outlined below, the report will include not only textual descriptions but also photos and screenshots to vividly illustrate the test environment. This approach enhances the clarity and understanding of the testing procedures and outcomes. Furthermore, the report will specify additional parameters for each test, as detailed in the subsequent sections, to provide a clear, complete, and accessible record of the testing process and its findings.

A security test template is structured to provide a comprehensive overview and detailed documentation of security testing activities. Here are the key elements of a test report.

1) **General Information.** This section captures the foundational details of the test campaign, including its name, the report version, testing dates, the contact person (usually the tester), the test location, and a brief description of the test campaign. It also includes a summary of test results and overall conclusions, providing a high-level overview of the testing outcomes and implications.
2) **Test and Measurement Equipment and Tools.** Here, all tools and equipment utilized during the testing process are listed with specifics such as the tool or equipment type, manufacturer, version, and any relevant notes that may impact the interpretation of the test results or the replication of the test setup.
3) **System Under Test (SUT).** This portion details the device(s) under test (DUTs) included in the SUT, the deployment architecture, and a thorough description of the SUT. This gives context to the test environment and clarifies the scope of the testing activities.
4) **Test Configuration.** It outlines the specific configurations used during testing, including function, service, and network settings. This ensures that the test conditions are clearly defined and can be replicated or referenced in future testing scenarios.
5) **Test Results.** This critical section documents each test executed, including the test number, name, execution date, reference to the test specification, and detailed test setup, including connection and block diagrams. It also covers the test procedure followed, observations made during testing, and conclusions drawn from the results. This comprehensive documentation is vital for understanding the test process, evaluating the outcomes, and making informed decisions based on the results.

Table 4.2 Testing and Measurement Tools.

Test Tool	Purpose
Commercial UE and/or UE emulator	Commercial UE or UE emulator establishes a stateful end-to-end connection for data traffic analysis. These UEs, designed with specific test and diagnostic capabilities, utilize (emulated) SIM cards with preprovisioned subscriber profiles. For tests requiring multiple UE sessions, either multiple commercial UEs or a UE emulator are used. Connections to the System Under Test (SUT) are made via RF cables or over-the-air (OTA), with UEs placed in RF-shielded environments to prevent external interference. Additionally, a logging tool is connected to the UE for capturing measurements and KPI logs essential for test validation and reporting.
4G/5G core or core emulator	A 4G/5G core or core emulator is used for terminating 4G/5G NAS sessions and facilitating core network procedures for RAN (SUT) testing. It must support end-to-end connections and data transfers between the application server and commercial UE/UE emulator.
Application (traffic) server	An application (traffic) server is used as an endpoint for initiating and/or terminating data traffic streams with commercial UEs/UE emulators. It is designed to generate data traffic relevant to the services being tested.
Network impairment emulator	For tests requiring the insertion of network impairments, such as packet delay and/or jitter, at interfaces like Open FH, a network impairment emulator is utilized.
Packet generation tool/DoS emulator	For generating DoS traffic in security tests, a packet generation tool or DoS emulator will be employed. This tool is designed to support crafting network traffic across a variety of protocols, including Ethernet, IP, UDP, TCP, PTP, eCPRI, TLS, and HTTP/HTTPS.
Packet capture tool	A packet capture tool is used for capturing data traffic samples for validation, analysis, and troubleshooting. It can also capture samples of legitimate traffic to serve as templates for fuzzing attacks. This tool is equipped to support capturing network traffic across various protocols, including Ethernet, IP, UDP, TCP, PTP, eCPRI, TLS, QUIC, and HTTP/HTTPS.
Network tap	A network tap is a device, either hardware or software, which allows you to see and access the data moving through a computer network.
Port scanner	A protocol scanner is a tool designed for examining network protocols and services, capable of identifying open ports and determining which services are actively exposed on those ports. Typically equipped with a built-in database of services, it utilizes a range of probes to query different services. In practice, protocol scanners are frequently utilized for service detection.
Fuzzing tool	A protocol fuzzing tool will be employed for generating unexpected input in security tests, designed to mutate and replay captured network traffic across protocols such as Ethernet, IP, UDP, TCP, PTP, eCPRI, TLS, and HTTP/HTTPS.
Vulnerability scanning tool	A vulnerability scanning tool will be utilized to identify and exploit known vulnerabilities during security tests, using a regularly updated database based on Common Vulnerabilities and Exposures (CVE). It is designed to scan network services operating on the TCP/IP protocol stack.

(Continued)

Table 4.2 (Continued)

Test Tool	Purpose
NFV benchmarking and resource exhaustion tool	An NFV tool will be employed to measure performance and generate resource exhaustion type of DoS attacks in O-Cloud systems. This tool is adaptable for use in any O-Cloud environment, whether public or private, and can test VNFs and CNFs.
SSH audit tool	An SSH audit tool will be utilized to check for specific properties including the protocol version, cipher suites, and any known vulnerabilities in both server and client SSH software.
TLS scanning tool	A TLS scanning tool will be employed to assess key properties such as the protocol version, cipher suites, and any known vulnerabilities within server TLS software.
DTLS scanning tool	A DTLS scanning tool shall be used to verify the following properties: version of protocol, cipher suites, and known vulnerabilities in server DTLS software.
IKE scanning tool	An IKE scanning tool will be utilized to evaluate essential attributes like the protocol version, cipher suites, and any identified vulnerabilities in server IPsec software.
Software image signing tool	A software image signing tool will be used for digitally signing and verifying software images, such as xApps or O-RAN components, provided by software producers/providers.

Together, these elements form a structured approach to documenting security tests, ensuring clarity, consistency, and thoroughness in capturing and communicating the test activities, findings, and implications for the security of the system under examination.

4.7.3 Security Testing Framework

Figure 4.9 lists key aspects of the overall O-RAN security testing framework.

4.7.3.1 Security Protocol Validation

SSH Protocol Test The SSH Protocol Test is designed to evaluate the implementation of the SSH protocol within the O-DU and O-RU components of the O-RAN architecture, ensuring it aligns with O-RAN specifications. The primary goal of this test is to confirm the use of SSH version 2, which is recognized for its enhanced security features, including the use of strong cryptographic algorithms for host key signatures, symmetric encryption, key exchanges, and message authentication codes (MACs). Additionally, this test aims to identify and eliminate any known vulnerabilities within the SSH implementations, thereby bolstering the security of the network elements.

The approach to testing involves two phases, focusing on both server and client aspects of SSH communication. On the server side, an SSH audit tool is employed to scrutinize the SSH server configurations, comparing them against a predefined set of approved protocols and algorithms to ensure compliance. This includes a thorough check for any vulnerabilities that have been reported in the context of the SSH server's deployment. Similarly, on the client side, the same audit tool is utilized in its client audit mode to examine SSH client implementations, ensuring that they too adhere to the required security standards and are free from recognized vulnerabilities.

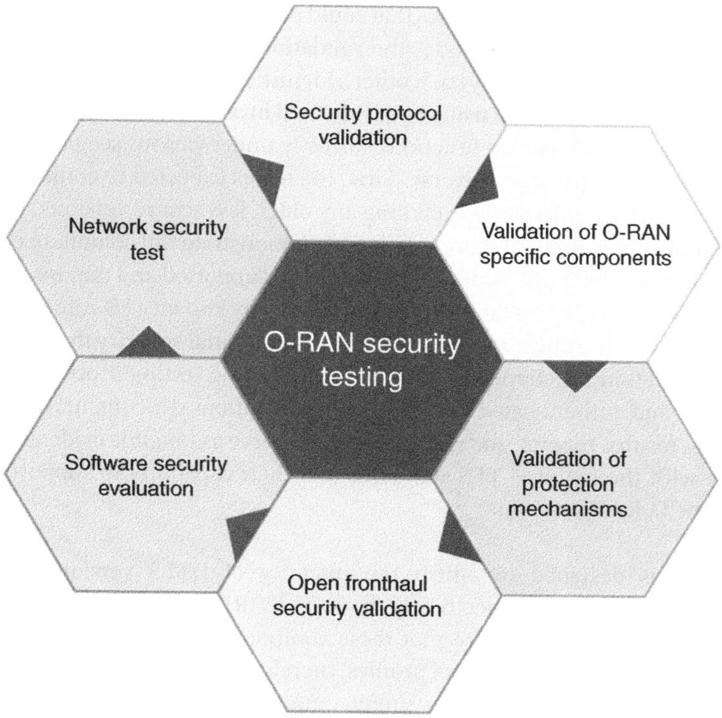

Figure 4.9 Key Aspects of O-RAN Security Testing.

The expected outcomes of this testing process include the confirmation that only SSH version 2 is supported and operational, excluding older, less secure versions. Furthermore, it is expected that all cryptographic algorithms used for SSH communication consisting of host key signatures, symmetric encryption, key exchanges, and MACs are strictly approved by the O-RAN security protocols. Importantly, the test aims to verify the absence of any well-known vulnerabilities within both server and client implementations of SSH, enhancing the security framework of O-RAN deployments. Successful completion of this test is documented through detailed findings, including logs and screenshots, which serve as evidence of the SSH implementations' compliance with O-RAN security specifications.

The TLS Protocol Test This test is aimed at ensuring the secure implementation of the TLS protocol across critical components of the O-RAN system, including the SMO, RICs, O-CU, O-DU, and O-Cloud. This test focuses on verifying that these components support the latest and more secure versions of TLS, specifically TLS v1.2 or v1.3, and that they implement mutual authentication mechanisms effectively. The requirement extends to ensuring adherence to specified TLS profiles and conducting thorough assessments to identify and mitigate any known TLS vulnerabilities.

The methodology adopted for the TLS Protocol Test involves a series of steps designed to thoroughly evaluate TLS implementations. First, TLS scanning tools are employed to meticulously identify the TLS versions in use, the cipher suites configured, and the overall TLS configurations of the components under test. This step is crucial for verifying that only the more secure TLS versions are enabled and the configurations align with best security practices. Following this, vulnerability scans are conducted to pinpoint common TLS vulnerabilities that could potentially compromise the security of communications. Additionally, the test includes the validation of mutual authentication, a process whereby both the client and server verify each other's identities before establishing a secure connection. This aspect of the test is evaluated using both valid and invalid client certificates to ensure that the mutual authentication mechanism functions correctly under various scenarios.

The expected outputs of the TLS Protocol Test are diverse. First, the test is expected to confirm the exclusive support for TLS v1.2 and v1.3, effectively excluding any older, less secure versions of the protocol. Furthermore, the components under test are expected to demonstrate full compliance with the specified TLS profiles, ensuring that only secure cipher suites are supported and that insecure ciphers are not enabled. Another critical expectation is the absence of any known TLS vulnerabilities within the tested components, indicating a robust stance against potential security threats. Finally, successful mutual authentication functionality, validated through the testing process, is essential for confirming the secure and authenticated nature of communications. Documentation of the test findings, including logs, results, reports, and screenshots, will serve as tangible evidence of the components' compliance with the required TLS standards and their capability to support secure communications within the O-RAN system.

The DTLS Protocol Test This test is designed to affirm the integrity of DTLS version 1.2 implementations within key O-RAN components, specifically the Near-RT RIC and the O-CU-CP, the O-CU-UP, and the O-DU. This test aims to ensure that these components not only support DTLS v1.2 but also adhere strictly to the specified protocol profiles, thereby enhancing the security of datagram-based communications within the O-RAN architecture. Furthermore, a critical aspect of this testing is to ascertain that the DTLS implementations are free from well-known vulnerabilities that could compromise the network's security.

DTLS scanning tools are deployed to methodically assess the DTLS versions supported by the components under test, along with the cipher suites and security configurations in use. This step is vital for verifying the implementation's alignment with the DTLS v1.2 standard and for ensuring that the security configurations are robust and in line with best practices. Following this, the findings from the DTLS scanning are meticulously compared against the approved DTLS versions and profiles, ensuring strict compliance and identifying any deviations.

Primarily, the test is expected to confirm the exclusive support for DTLS v1.2 in the Near-RT RIC and the O-CU, excluding support for any older, potentially less secure versions of the protocol. Additionally, these components are expected to demonstrate complete compliance with the DTLS protocol profiles, ensuring that only secure and approved configurations are utilized. Furthermore, a thorough examination of well-known DTLS vulnerabilities is expected to yield no identified vulnerabilities, affirming the security of the DTLS implementations.

The IPsec Protocol Test This test is designed to evaluate the implementation of IPsec within the Near-RT RIC, the O-CU-CP, the O-CU-UP, and the O-DU, with a focus on their capabilities to operate in tunnel mode. This mode of IPsec operation encapsulates IP packets entirely, providing a secure channel for data transmission by employing the ESP protocol in conjunction with the IKE version 2 (IKEv2) protocol. The primary goal is to ensure that these critical O-RAN components can securely communicate over potentially insecure networks, such as the internet, by establishing secure VPN tunnels. The specific requirement for this test mandates the activation of ESP in tunnel mode, supplemented by the support for IKEv2, which is essential for negotiating the security associations that enable the secure communication channels established by IPsec. Additionally, the test seeks to verify that the implementations are in line with specified IPsec capabilities, including the use of approved ESP encryption and authentication transforms, as well as supported Diffie-Hellman groups (DH group 19—256-bit ECP group) for key exchange, ensuring a robust level of security in the IPsec tunnels.

To conduct the IPsec Protocol Test, IKE scanning tools are utilized to identify the ESP encryption and authentication transforms being used, as well as the DH groups supported by the Near-RT RIC and the O-CU.

The expected outcomes of the test are critical for the security of the O-RAN architecture. First, the test aims to confirm that only IKEv2 is supported for establishing security associations, given its enhancements over previous versions in terms of security and efficiency. Second, the Near-RT RIC and the other relevant components should demonstrate compliance with the specified ESP transforms and DH groups, aligning with the security requirements outlined in the O-RAN specifications.

The OAuth 2.0 Test This test is focused on validating the implementation of OAuth 2.0-based authorization mechanisms across O-RAN components, particularly in managing API service requests. This test is crucial for ensuring that the O-RAN architecture can securely control access to its resources, making it a key aspect of the system's overall security framework. OAuth 2.0, as a standard for authorization, enables O-RAN components to grant service access through token-based authentication, coupled with mutual TLS authentication to enhance security further. The testing requirement involves the comprehensive implementation of OAuth 2.0 authorization processes. This encompasses the registration of resources and the facilitation of token-based access to these resources. Mutual TLS authentication serves as an additional layer of security, ensuring that both the client and the server can authenticate each other, thus preventing unauthorized access attempts. The approach to conducting the OAuth 2.0 test includes the following key steps.

- **Application Access Token Request Validation.** This step involves testing the process for requesting access tokens, using both valid and invalid client certificates. This process helps to ascertain that only clients with valid certificates can obtain access tokens, which are crucial for accessing protected resources.
- **Token-Based Service Request Validation.** This phase tests the handling of API service requests that utilize the previously obtained access tokens. The validation process is carried out with both correct and incorrect tokens to ensure that the system can correctly grant or deny access based on token validity.

Here are the expected outcomes of the OAuth 2.0 test (O-RAN Working Group 11 2024i).

- **Successful Authorization for Valid Scenarios.** For scenarios involving valid client certificates and correct access tokens, the system should successfully authorize the API service requests, demonstrating the effective implementation of OAuth 2.0 authorization within the O-RAN components.
- **Correct Handling of Invalid Requests.** When faced with invalid client certificates or incorrect access tokens, the system should appropriately deny access, providing detailed error responses that conform to OAuth 2.0 standards. This outcome is critical for ensuring the system's resilience against unauthorized access attempts.

4.7.3.2 Common Network Security Tests and System Security Tests for O-RAN Components

Network Protocol and Service Enumeration The Network Protocol and Service Enumeration test is designed to ensure that the active network protocols and services on O-RAN components match the vendor's documented list, focusing on minimizing the attack surface by disabling unused protocols. This verification is important in the O-RAN environment for components like the SMO, the

RICs, the O-CU, and the O-DU, given its emphasis on security and efficient network utilization. Through a meticulous process often referred to as network or port scanning, the test probes for active services across TCP, SCTP, and UDP ports, aiming to compile a comprehensive list of operational protocols and services. This list, detailing each service's protocol, port number, and name, is then cross-referenced with the vendor's documentation to ensure full alignment and the absence of unauthorized services.

The methodology for conducting this test involves utilizing a port scanner to systematically identify open ports and subsequently detect services on those ports, employing a combination of vendor-provided information and the IANA Service Name and Transport Protocol Port Number Registry for service identification. The expected result from this process is the exact match between the services found by the port scanner and those listed by the vendor, thereby affirming the accuracy of the vendor's documentation and the secure configuration of the O-RAN component. A successful test outcome is characterized by an exact match or a scenario where the scanner identifies fewer services than listed but remains in alignment with the vendor-provided list. Conversely, the discovery of undocumented services signifies a test failure, pointing toward potential security vulnerabilities and the existence of an undocumented attack surface, underscoring the necessity for accurate service documentation and stringent security measures in the O-RAN system.

Password-Based Authentication The O-RAN security tests focusing on password-based authentication encompass three main aspects: password guessing, unauthorized password reset, and password policy enforcement, addressing different vulnerabilities related to password security within the O-RAN components like the SMO, the RICs, the O-CU, and the O-DU.

The **Password Guessing** test ensures that O-RAN components are equipped with mechanisms to prevent password guessing attacks, such as brute-force or dictionary attacks. Techniques like increasing delay after failed attempts, implementing challenge–response mechanisms like CAPTCHA, or temporary lockouts after reaching a threshold of consecutive failed attempts are evaluated. This involves probing all services using password-based authentication across the O-RAN component to detect any active protection mechanism against such attacks.

For **Unauthorized Password Reset**, the goal is to verify that the password reset mechanisms cannot be bypassed or misused, ensuring that any out-of-band password recovery mechanisms are absent or deactivated. This is crucial for preventing unauthorized access to the O-RAN components, their configuration, and data. The test covers services using password-based authentication and evaluates physical DUTs for the presence of HW factory reset switches that could allow password resets without a full factory reset.

Last, the **Password Policy Enforcement** test verifies that O-RAN components enforce a secure password policy for services using password-based authentication. This test checks if the DUT rejects password changes that do not comply with the applied password policy, which may include rules on password length, character types, complexity, and deny-lists of unacceptable passwords. This ensures that only strong, policy-compliant passwords are used, enhancing the security of the O-RAN components.

Network Protocol Fuzzing Network protocol fuzzing is an automated testing technique aimed at identifying vulnerabilities within an SUT by bombarding it with invalid or random inputs to provoke malfunctions or crashes. This method is notably effective in exposing vulnerabilities within extensive input fields of modern programs, which often have limited test coverage. However, for fuzzing to be truly beneficial, especially in a multivendor environment like O-RAN, it must be clearly defined and incorporated throughout the system development life cycle.

Fuzzing in the context of telecommunication network protocols presents unique challenges compared to traditional software fuzzing. It involves sending malformed data through network ports rather than command lines or input files, complicated by the layered structure of network protocols within the SUT. Given that many O-RAN components rely on common software frameworks for general communication, assessing the risk profile of general transport protocols through extensive fuzzing in other activities is vital. This contrasts with O-RAN-specific protocols, which might have undergone less industry-wide testing, presenting potentially higher-risk profiles.

O-RAN-specific protocols, often based on state machines serving multiple endpoints simultaneously, require scale testing to uncover vulnerabilities like memory leaks or buffer overflows, which could lead to unauthorized code execution or software crashes. Fuzzing the M-plane protocol within the O-RAN fronthaul configuration is particularly critical due to its integration of multiple technologies. Effective fuzzing should be protocol and state-machine-aware, allowing for focused testing on high-risk aspects of the SUT and intelligent responses to SUT behaviors. Furthermore, fuzzing should encompass both authenticated and nonauthenticated sessions, as many vulnerabilities lie within the management protocols' authenticated sessions, posing risks for privilege escalation.

For comprehensive risk assessment, robust logging across HW, software, and O-RAN functions is essential to detect potential memory leaks, DoS attacks, and other vulnerabilities early on without impairing performance. Documentation of fuzzing tests should include a detailed list of externally reachable SUT protocols, tools used, their versions, user settings, and outcomes, emphasizing any input that caused unusual behavior. Since fuzzing test cases are inherently nonexhaustive and challenging to replicate, producing varied results across different vendors, continued investment in fuzzing activities is necessary to refine approaches and meet the risk-based models set by vendors and operators.

Denial of Service/Message Flooding The DoS or message flooding test is designed to assess the resilience of O-RAN components with external network interfaces against volumetric DDoS attacks, ensuring that such attacks do not cause system crashes and that the system can return to its normal service level post-attack. This requirement is vital for maintaining service continuity and availability in the face of malicious attempts to overwhelm the network or inadvertent network/operator errors. The types of DDoS attacks tested include protocol layer attacks (such as SYN, UDP, and TCP floods), volume-based attacks (like ICMP floods and Smurf DDoS), and application layer attacks (including GET/POST floods and low-and-slow attacks targeting specific applications or OS network services).

For testing, each O-RAN component or system interface is subjected to scenarios mimicking large volumes of requests typical of DoS/DDoS attacks. The test setup necessitates network access to the DUT and a vendor-provided list of the network protocols and services supported by the DUT. The test procedure differentiates between call flows requiring authentication and those that do not. For authenticated call flows, the test simulates repeated requests post-authentication, increasing the rate over time to identify the failure point. The process is repeated for pre-authentication request flows. For call flows not requiring authentication, a similar strategy is employed to mark the failure point based on the DUT's response to the escalating request rate.

The expected outcome of this test is for the O-RAN component to exhibit a potential failure in serving requests under attack scenarios but crucially, to recover and return to a functional state once the attack ceases. The DUT should be capable of responding to service requests normally after recovery. Failure of the DUT to return to a functional state posttest constitutes a test failure, highlighting vulnerabilities within the system's resilience against volumetric DDoS attacks. This testing ensures that O-RAN components are adequately prepared to handle and recover from such attacks, thereby maintaining network integrity and service availability.

4.7.3.3 Software Security Evaluation

Software Bill of Materials (SBOM) The O-RAN security specifications mandate a comprehensive approach to the SBOM, covering aspects such as SBOM signature, data fields, format, and depth to enhance the security and transparency of software components used within the O-RAN system.

- **SBOM Signature.** This requirement ensures that every SBOM provided by O-RAN software producers, including those for patches, is digitally signed, establishing the authenticity and integrity of the SBOM. Various formats like SPDX, CycloneDX, and SWID have specified methods for including and verifying digital signatures, such as separate signature files for SPDX and XML/JSON signature standards for CycloneDX and SWID formats. The validation of these signatures is crucial for ensuring that the SBOMs have not been tampered with and are from a trusted source.
- **SBOM Data Fields.** According to NTIA guidance, the SBOM must contain a minimum set of data fields that describe the software components in detail, including the supplier's name, component name, version, unique identifiers, dependency relationships, the author of SBOM data, and a timestamp. This test verifies that the SBOM includes all necessary information for effective software component management and risk assessment.
- **SBOM Format.** O-RAN software deliveries must provide the SBOM in one of the accepted formats—SPDX, CycloneDX, or SWID—to ensure standardized documentation that facilitates easier integration and analysis across different tools and platforms. The choice of format affects how the SBOM's digital signature is included and verified.
- **SBOM Depth.** The depth of the SBOM must be sufficient to cover top-level dependencies for every software delivery and to a second level for any software from the O-RAN Software Community or open-source software. This granularity helps network operators to understand the software composition fully and assess potential security risks more accurately.

Software Image Signing and Verification In the O-RAN system, ensuring the security and integrity of software images and application packages is crucial. This is addressed through requirements for digital signing and verification processes outlined in the O-RAN Security Requirements Specifications.

Software Image/Application Package Signing. This procedure mandates that any software images or applications of O-RAN components must be digitally signed by their respective providers before distribution. Furthermore, the service provider must also apply a digital signature to these software packages before publishing them within its catalog accessible to the SMO. The signing process involves generating a key pair, optionally requesting a signing certificate (preferably a short-lived one), hashing and signing the image or package with a strong algorithm like SHA256, and finally uploading the signed package for distribution or publication. This process ensures that the software's origin is authenticated and its integrity is intact from the producer to the operator.

Software Signature Verification. On the operator's side, there is an equally important requirement to verify the digital signatures of software images or application packages during the onboarding and instantiation processes. This verification ensures that the software has not been tampered with and is from a trusted source. The process involves using either manual methods or a software signing service to validate the signatures against shared digital certificates or public keys. The procedure specifies that the service provider's signature should be verified first during the instantiation, followed by the provider's signature.

The expected outcome of these processes is to maintain a high level of trust and security within the O-RAN network, ensuring that all software components and applications are verified for authenticity and integrity.

4.7.3.4 Open Fronthaul Security Validation

In ensuring secure network access, particularly within the open fronthaul network. The IEEE 802.1X-2020 standard plays a critical role by providing port-based network access control, incorporating roles such as the supplicant, authenticator, and authentication server. This framework is essential for validating entities attempting to access the network, leveraging mechanisms like EAP TLS authentication to establish a secure and authenticated connection.

Authenticator validation focuses on assessing the capability of O-RAN components, such as O-DUs, O-RUs, and cloud entities, to fulfill the authenticator role, managing access requests from supplicants. The testing process involves configuring a Remote Authentication Dial-In User Service (RADIUS) server with necessary certificates and employing a test tool to emulate supplicant requests at varying levels of correctness regarding identity and certificates. Scenarios range from correct certificates and identities to incorrect ones, including the use of non-EAP TLS authentication methods. Expected outcomes are straightforward: authentication success, when correct credentials are provided, and failure otherwise, highlighting the authenticator's ability to enforce secure access protocols effectively.

Supplicant validation addresses the ability of O-RAN components to act as supplicants, requesting network access. Similar to authenticator validation, this process involves setting up a RADIUS server and using a test tool as an emulated authenticator. The O-RAN component, configured for 802.1X on its fronthaul interface, attempts to authenticate itself with various credential configurations. This mirrors real-world scenarios where devices must prove their identity to gain network access, ensuring that only authorized entities can connect. The expected results mirror those of the authenticator validation, focusing on the supplicant's ability to present valid credentials for authentication successfully.

4.7.3.5 Validation of O-RAN-Component-Specific Security Protection Mechanisms

The security testing for the O-RU focuses primarily on validating the security mechanisms related to SSH and TLS protocols on the M-plane interface. Through the SSH on M-plane interface test, the O-RU is evaluated for its adherence to robust protocol implementations, specifically verifying that it supports only SSHv2 along with strong cryptographic algorithms for various security functions as dictated by O-RAN specifications. This setup involves configuring the O-RU as an SSH server and utilizing a client equipped with an SSH audit tool to rigorously test server-side implementations. The objective is to confirm the O-RU's capability to support secure SSH communications, adhering to specified cipher suites and security protocols.

Similarly, the TLS on M-plane interface test assesses the O-RU's implementation of TLS protocols, checking for compliance with the support of TLS v1.2 and/or v1.3 and their associated protocol profiles. This test extends to include mutual authentication using both valid and invalid client certificates, thereby ensuring the O-RU's effectiveness in establishing secure TLS-based connections. Alongside, the O-RU is also subjected to 802.1X Supplicant Validation tests, which are essential for confirming its role and reliability as a supplicant in the network access control mechanism. Together, these security tests are indispensable for guaranteeing that the O-RU can securely manage data transmission over the network, crucially supporting the O-RAN's security framework and protecting against potential cybersecurity threats.

Security testing for Near-RT RIC focuses on ensuring robust security mechanisms across various interfaces and API communications, primarily through the implementation of security protocols such as IPsec, TLS, mTLS, and OAuth 2.0. The IPsec on the E2 interface test is designed to verify that communications between the Near-RT RIC and E2 nodes are secured, providing confidentiality, integrity, replay protection, and data origin authentication as stipulated in O-RAN specifications. This entails a detailed test setup where the Near-RT RIC platform is configured with IPsec for the E2 interface, ensuring that data in transit is adequately protected against potential threats and vulnerabilities.

Furthermore, the security of transactional APIs, which are crucial for the interaction between the Near-RT RIC and its xApps, is assessed through TLS and mTLS tests. These tests confirm the support for TLS to safeguard message confidentiality and integrity, and mTLS for mutual authentication via X.509v3 certificates, thereby enhancing the security of API communications. Additionally, the OAuth 2.0 test for transactional APIs checks for the implementation of the OAuth 2.0 authorization framework within the Near-RT RIC, ensuring that access to APIs is securely managed and controlled. Through these comprehensive security tests, the Near-RT RIC's adherence to O-RAN security requirements is thoroughly validated, ensuring that it can perform its critical functions securely and efficiently while protecting against a broad spectrum of cyber threats.

The Non-RT RIC plays a crucial role within the O-RAN architecture. To secure its operation, particularly concerning the R1 and A1 interfaces, the implementation of OAuth 2.0 is mandated following zero-trust principles. This includes tests for OAuth 2.0 Resource Owner/Server capabilities, ensuring that the Non-RT RIC can authorize or deny service requests based on OAuth 2.0 protocols, alongside OAuth 2.0 Client tests, verifying the Non-RT RIC's ability to request and obtain permissions to access resources securely. Additionally, the Non-RT RIC framework's support for OAuth 2.0 reinforces its capability to securely manage service requests, emphasizing the importance of robust authorization mechanisms in safeguarding the Non-RT RIC's operations against potential threats.

On the R1 interface, which facilitates communication between the Non-RT RIC framework and rApps, security tests focus on TLS and mTLS to ensure confidentiality, integrity, and authentication of data in transit. Similarly, OAuth 2.0 tests validate the authorization controls for accessing services and resources. For the A1 interface, connecting the Non-RT RIC and Near-RT RIC, tests are conducted to confirm the implementation of TLS for data protection and mTLS for mutual authentication, alongside OAuth 2.0 for secure authorization. These comprehensive security tests across the Non-RT RIC, and its critical interfaces, aim to verify adherence to O-RAN security specifications, ensuring secure and reliable communication within the O-RAN network's components.

The SMO is responsible for overseeing and managing the RAN elements and functions. To ensure the security of these operations, particularly over the O1 interface, which facilitates the management of various O-RAN components, the implementation of the Network Configuration Access Control Model (NACM) plays a pivotal role. NACM enforces role-based access control, aligning with the zero-trust security principles by ensuring that only authenticated and authorized users can access specific network configurations and operations based on their roles. This model not only enhances the confidentiality, integrity, and authenticity of the communications over the O1 interface but also supports the least privilege access control, significantly mitigating the risk of unauthorized access and potential security threats.

The validation process for NACM on the O1 interface involves setting up a secure NETCONF session over TLS, ensuring that only clients with valid certificates can initiate connections. By meticulously verifying the session establishment and mapping to the correct NETCONF user, alongside checking the NACM enforcement control settings and rule sets for predefined groups, the test ensures comprehensive adherence to O-RAN's stringent security requirements. Such a thorough validation process aims to confirm that the O-RAN component supports the necessary security measures for the NETCONF over TLS sessions and enforces NACM rules accurately, contributing to the robust security of the O-RAN network managed by the SMO.

The O-Cloud security testing entails the implementation of comprehensive security tests across various interfaces and functionalities, such as the O2 interface, which connects the SMO framework to the cloud infrastructure. These tests are designed to validate the effectiveness of security controls in place, including mTLS for ensuring the authenticity of communications, TLS for confidentiality to prevent sensitive data exposure, and integrity checks to ensure that data is not altered during transmission. Additionally, antireplay measures are assessed to prevent the malicious reuse of intercepted data, while authorization tests ensure that only properly authenticated and authorized entities can access or perform actions on the system.

Furthermore, the resilience of the O2 interface against DoS attacks is tested to ensure that the O-Cloud can maintain operational integrity under adverse conditions. The virtualization layer's security is also scrutinized through tests for secure authentication and authorization, verifying that only legitimate users and applications can interact with the cloud APIs and that their actions are within their permission scope. Network policies are evaluated for their effectiveness in permitting or denying traffic between virtual machines or containers, ensuring that communication is restricted according to predefined security policies. Last, efforts to identify and remediate vulnerabilities and insecure configurations within the O-Cloud components are undertaken, alongside validating the system's logging and monitoring capabilities to detect and respond to security incidents, thus reinforcing the O-Cloud's defense mechanisms against potential security threats.

4.7.3.6 Security Testing of xApps and rApps

The security testing framework for xApps within the Near-RT RIC underscores a focused effort to ensure the integrity and authenticity of these applications before their deployment and operation. Central to this security validation are the steps involved in the signing and subsequent verification of xApps, a procedure designed to authenticate the source of the software and to verify its integrity prior to integration within the network environment. This process is pivotal for maintaining a secure network, as it prevents the introduction of tampered or malicious software that could compromise the O-RAN system's performance or security. By employing standardized test cases, such as "STC-9-9.5-001: Software Image Signing" and "STC-9-9.5-002: Software Signature Verification," the framework ensures a systematic approach to verifying the digital signatures of xApps, thus validating their legitimacy and safeguarding the network against potential vulnerabilities introduced through software deployment.

The security testing for rApps, integral components within the Non-RT RIC framework, encompasses rigorous processes to assure both the authenticity and the authorization of these applications within the O-RAN network. Key to this testing is the practice of signing and verifying rApps, as well as managing their authorization through OAuth 2.0 protocols. By enforcing strict controls on the signing and verification, the O-RAN ensures that only legitimate and verified rApps are integrated into the network, effectively mitigating the risk of malicious software infiltration. This process not only validates the origin and integrity of the rApps but also confirms their compliance with the specified security requirements, safeguarding the network against potential vulnerabilities.

4.7.3.7 Validation of Miscellaneous Components

The security tests for common application life cycle management focus on ensuring the authenticity and integrity of application packages within the O-RAN system. These tests are crucial for protecting the network against compromised software that could potentially harm the network's operation or security. The process involves careful validation of application packages, starting from their creation by the application provider to their deployment by the service provider.

Initially, the application provider signs the application package to guarantee its authenticity and integrity. This signature serves as a seal, attesting that the package has not been tampered with since its signing. Upon receiving the application package, the service provider is responsible for verifying this signature to confirm the package's legitimacy. This verification process is a critical step in the onboarding of applications onto the network, ensuring that only verified and trusted software is allowed entry.

Following the verification of the application provider's signature, the service provider adds its own signature to the application package. This additional layer of endorsement signifies the service provider's approval of the package for onboarding into its image repository and future instantiation. The final phase of the security test occurs during the instantiation of the application, where both the application provider's and service provider's signatures are validated once more. Successful validation at this stage means that the application package has maintained its integrity and authenticity throughout its journey from creation to deployment. Conversely, failure to validate the signatures halts the instantiation process, serving as a protective measure against the introduction of unauthorized or altered software components into the network. This comprehensive approach to application package validation underscores the importance of stringent security practices in the management of software life cycle within the O-RAN architecture.

4.8 Major Takeaways of the Chapter

Security of O-RAN is a critical topic. Compared to a traditional RAN, O-RAN has an increased threat surface due to more nodes and open interfaces. Here are the key takeaways of the chapter.

- Development and implementation of remediation strategies are required to address identified vulnerabilities.
- Specific security protocols and configurations are required to protect the O-RAN architecture and ensure consistent and effective security measures.
- Various security controls and mechanisms must be implemented across O-RAN components including encryption, authentication, and access control measures.
- Detailed security test cases are specified by O-RAN to validate the implementation and effectiveness of security controls within the O-RAN system.
- O-RAN focuses on the rigorous validation of security implementations against the defined security specifications, ensuring that its components meet stringent security standards.
- O-RAN components prioritize secure communication through protocols like OAuth 2.0, TLS, and mTLS, ensuring encrypted and authenticated interactions across different interfaces.
- The O-RAN system implements strong access control measures, including role-based access controls and authorization checks, adhering to zero-trust principles to safeguard against unauthorized access.
- O-RAN employs a detailed security testing framework that includes protocol validation, application package signing, and verification to ensure system integrity and security.

- The O-RAN security framework focuses on resilience against DoS attacks, highlighting the importance of maintaining service continuity.
- Across the O-RAN architecture, zero trust principles guide the implementation of security measures, including least privilege access and stringent authentication mechanisms.

References

3GPP (2017). *Security Architecture and Procedures for 5G System*. TS 33.501, March 2017.

3GPP (2018). *Security Assurance Methodology (SECAM) and Security Assurance Specification (SCAS) for 3GPP Virtualized Network Products*. TS 33.818, September 2018.

Abdalla, A.S. and Marojevic, V. (2024). End-to-end O-RAN security architecture, threat surface, coverage, and the case of the open fronthaul. *IEEE Communications Standards Magazine* 8 (1): 36–43.

Chaudhuri, A. (2015). *Best Practices for Mitigating Risks in Virtualized Environments*. Cloud Security Alliance, USA.

IEEE-std-1588-2019 (2023). IEEE standard for a precision clock synchronization protocol for networked measurement and control systems. In: *IEEE Std 1588-2019 (Revision of IEEE Std 1588-2008)*, 1–499. 16 June 2020. doi: https://doi.org/10.1109/IEEESTD.2020.9120376.

IETF (2005). *Security Architecture for the Internet Protocol*. RFC 4301, December 2005, https://datatracker.ietf.org/doc/html/rfc4301 (accessed 30 July 2024).

IETF (2006). *The Secure Shell (SSH) Protocol Architecture*. RFC 4251, January 2006, https://datatracker.ietf.org/doc/html/rfc4251 (accessed 30 July 2024).

IETF (2009). *Transport Layer Security (TLS) Transport Mapping for Syslog*. RFC 5425, March 2009, https://datatracker.ietf.org/doc/html/rfc5425 (accessed 30 July 2024).

IETF (2012). *The OAuth 2.0 Authorization Framework*. RFC 6749, October 2012, https://datatracker.ietf.org/doc/html/rfc6749 (accessed 30 July 2024).

IETF (2015a). *Using the NETCONF Protocol over Transport Layer Security (TLS) with Mutual X.509 Authentication*. RFC 7589, June 2015, https://datatracker.ietf.org/doc/html/rfc7589 (accessed 30 July 2024).

IETF (2015b). *Recommendations for Secure Use of Transport Layer Security (TLS) and Datagram Transport Layer Security (DTLS)*. RFC 7525, May 2015, https://datatracker.ietf.org/doc/rfc7525 (accessed 30 July 2024).

IETF 4210 (2005). *Internet X.509 Public Key Infrastructure Certificate Management Protocol (CMP)*. RFC 4210, IETF, September 2005, https://datatracker.ietf.org/doc/rfc4210/ (accessed 30 July 2024).

Liyanage, M., Braeken, A., Shahabuddin, S., and Ranaweera, P. (2023). Open RAN security: Challenges and opportunities. *Journal of Network and Computer Applications* 214: 103621.

NIST SP 800-190 (n.d.). *Application Container Security Guide*. https://nvlpubs.nist.gov/nistpubs/specialpublications/nist.sp.800-190.pdf (accessed 29 July 2024).

O-RAN Working Group 11 (2024a). *O-RAN Security Threat Modeling and Remediation Analysis*. O-RAN.WG11.Threat-Model.O-R003-v06.00. https://specifications.o-ran.org/specifications (accessed 28 July 2024).

O-RAN Working Group 11 (2024b). *O-RAN Security Requirements and Controls Specifications*. O-RAN.WG11.SecReqSpecs.0R003-v7.00. https://specifications.o-ran.org/specifications (accessed 28 July 2024).

O-RAN Working Group 11 (2024c). *O-RAN Security Protocols Specifications*. O-RAN.WG11.SecProtSpecs-R003-v07.00. https://specifications.o-ran.org/specifications (accessed 24 July 2024).

O-RAN Working Group 11 (2024d). *O-RAN Security Test Specifications.* O-RAN.WG11.SecTestSpecs-v05.00. https://specifications.o-ran.org/specifications (accessed 24 July 2024).

O-RAN Working Group 11 (2024e). *O-RAN Study on Security for Near Real Time RIC and xApps.* O-RAN.WG11.Security-Near-RT-RIC-xApps-TR.0-R003-v05.00 00. https://specifications.o-ran.org/specifications (accessed 29 July 2024).

O-RAN Working Group 11 (2024f). *O-RAN Study on Security for Non-RT RIC.* O-RAN.SFG.Non-RT-RIC-Security-TR-v01.00. https://specifications.o-ran.org/specifications (accessed 29 July 2024).

O-RAN Working Group 11 (2024g). *O-RAN Study on Security for AI/ML.* O-RAN.WG11.TR-AIML-Security-Analysis.O-R003-v01.00. https://specifications.o-ran.org/specifications (accessed 29 July 2024).

O-RAN Working Group 11 (2024h). *O-RAN Study on Security for O-Cloud.* O-RAN.WG11.O-CLOUD-Security-Analysis-TR.0-R003-v06.00. https://specifications.o-ran.org/specifications (accessed 29 July 2024).

O-RAN Working Group 11 (2024i). *O-RAN OAuth2.0 Security.* O-RAN.WG11.OAuth2.0-Security-TR.0-R003-v02.00. https://specifications.o-ran.org/specifications (accessed 29 July 2024).

O-RAN Working Group 11 (2024j). *O-RAN Study on Security for Fronthaul CUS-Plane.* O-RAN.WG11.Security_Fronthaul_CUSPlane-TR.0-R003-v01.00. https://specifications.o-ran.org/specifications (accessed 29 July 2024).

Polese, M., Bonati, L., D'oro, S. et al. (2023). Understanding O-RAN: Architecture, interfaces, algorithms, security, and research challenges. *IEEE Communications Surveys & Tutorials* 25 (2): 1376–1411.

5

O-RAN Ecosystem

Design, development, integration, testing, deployment, maintenance, optimization, and promotion of Open Radio Access Network (O-RAN) require a comprehensive ecosystem. This chapter discusses the overall ecosystem of O-RAN from two perspectives—an internal ecosystem and an external ecosystem. The internal ecosystem of O-RAN carries out work within the scope of the O-RAN ALLIANCE focus streams. The external ecosystem includes Standards Developing Organizations (SDOs), industry alliances, and government entities. Some constituents of the external ecosystem such as SDOs are of existential importance to O-RAN; without these components, O-RAN simply cannot exist. In contrast, some constituents of the external ecosystem such as government entities help make the overall ecosystem vibrant to promote the deployment of O-RAN, realizing the benefits of O-RAN for consumers, enterprises, and government entities. This chapter discusses roles of prominent constituents of the internal ecosystem as well as the external ecosystem of O-RAN.

5.1 Internal and External O-RAN Ecosystems

The overall O-RAN ecosystem can be viewed as an internal O-RAN ecosystem and an external O-RAN ecosystem as shown in Figure 5.1. Note that such ecosystem categorization is one possible approach. The internal ecosystem includes various groups working on different aspects of O-RAN, while an external ecosystem includes organizations that are not formally part of O-RAN but play an important role in the design, deployment, maintenance, optimization, use, and promotion of O-RAN.

The internal ecosystem of O-RAN consists of the entities that actively work on three focus streams of O-RAN. These focus streams include specification efforts, testing and integration efforts, and O-RAN Software Community (OSC). As mentioned in Chapter 1, the specification efforts involve extending RAN specifications such as the Third-Generation Partnership Project (3GPP) 5G specifications to achieve openness and intelligence. Work groups involved in the specification efforts address different aspects of O-RAN and define the overall O-RAN specifications. The testing and integration efforts involve entities that facilitate testing and integration of O-RAN implementations to ensure interoperability of products and compliance with O-RAN specifications. The OSC of the O-RAN ALLIANCE, in cooperation with the Linux Foundation, oversees the release and the development of open-source software for O-RAN. Section 5.2 discusses the internal O-RAN ecosystem.

Fundamentals of O-RAN, First Edition. Nishith D. Tripathi and Vijay K. Shah.

Companion website: www.wiley.com/go/tripathioran

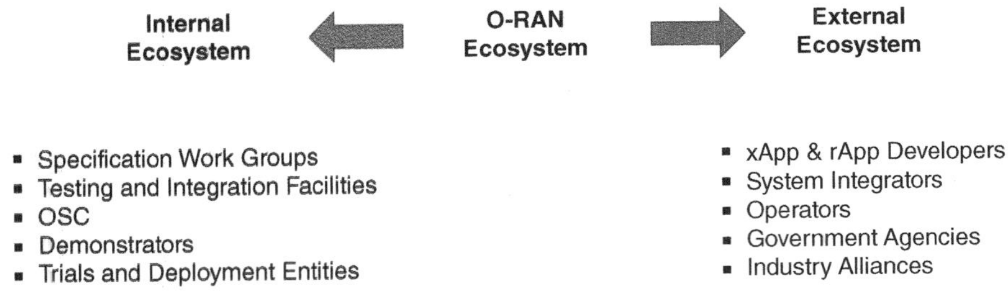

Figure 5.1 Overall O-RAN Ecosystem.

The external O-RAN ecosystem consists of diverse entities such as developers of extended Applications (xApps) and remote Apps (rApps), system integrators, operators, government agencies, and industry alliances/partners. One of the O-RAN goals is to promote intelligence in the RAN through non-real-time RIC and near-real-time RIC. These RICs provide the framework, where third-party developers can develop xApps and rApps. xApps and rApps significantly increase the value of O-RAN for operators and customers of operators. System integrators help create an integrated O-RAN-based system that integrates individual components of O-RAN possibly developed by different entities to construct a system that manages RAN operations effectively. System integrators rely on the testing and integration entities of the internal O-RAN ecosystem. Some operators may be actively contributing to the development of O-RAN specifications and hence would be considered to be part of the internal O-RAN ecosystem. However, some operators may not be involved in such specification efforts but deploy O-RAN. Such operators can be viewed as constituents of the external O-RAN ecosystem. Government agencies often promote the use of O-RAN. For example, the US Department of Defense (DoD) plans to use 5G O-RAN to modernize its communications systems while ensuring supply chain resilience for the RAN (Department of Defense 2022). Some industry organizations and academic institutes may not actively contribute to O-RAN specifications but help promote O-RAN by carrying out O-RAN research, prototyping O-RAN, and developing O-RAN savvy workforce.

Section 5.2 describes the internal O-RAN ecosystem, while Section 5.3 discusses the external O-RAN ecosystem. See Chapter 7 for another perspective about some elements of the O-RAN ecosystem.

5.2 Internal O-RAN Ecosystem

Section 5.2.1 discusses the O-RAN work groups that contribute to the development of O-RAN specifications. Focus and research groups of O-RAN are introduced in Section 5.2.2. Testing and integration entities are briefly described in Section 5.2.3. Demonstrators and field testing entities are discussed in Section 5.2.4. The role of the OSC is narrated in Section 5.2.5.

5.2.1 Work Groups of O-RAN Specification Efforts

As mentioned in Section 1.4.3 in Chapter 1, the technical O-RAN work is performed by O-RAN Work Groups (WGs), O-RAN Focus Groups (FGs), and Research Groups. As of early 2024, there are 11 WGs, 3 FGs, and 1 Research Group. Each WG focuses on a certain aspect of the O-RAN architecture. The Focus and Research Groups deal with the topics that are overarching the technical WGs or relevant for the entire O-RAN.

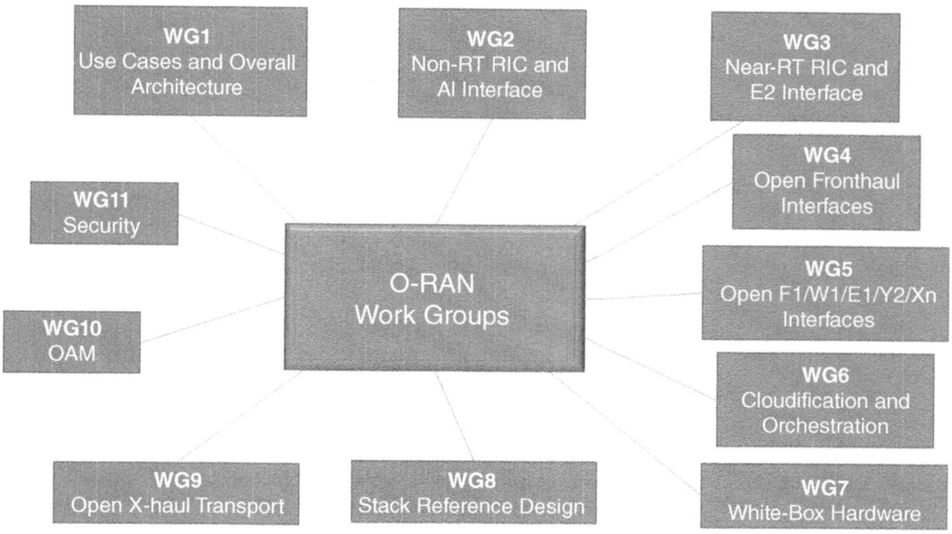

Figure 5.2 O-RAN Work Groups in a Nutshell.

Figure 5.2 mentions the scope of each of the 11 WGs. Work products created by these WGs provide a glimpse of the type of work performed by the O-RAN WGs.

WG1 is the Use Cases and Overall Architecture WG. WG1 defines the overall O-RAN architecture and use cases. WG2 is the non-real-time RAN Intelligent Controller and A1 interface WG and supports non-RT intelligent radio resource management, higher layer procedure optimization, policy optimization in RAN, and transfer of AI/ML models to near-RT RIC. WG3, the near-real-time RIC and E2 interface WG, defines the near-RT RIC architecture and data collection and actions over the E2 interface. WG4, the open fronthaul interfaces WG, defines open fronthaul interfaces to enable multi-vendor DU-RU interoperability. WG5 is the open F1/W1/E1/X2/Xn interface WG that defines 3GPP-compliant multi-vendor profile specifications for F1/W1/E1/X2/Xn interfaces and proposes any 3GPP specification enhancements. WG6, the Cloudification and Orchestration WG, enables decoupling of RAN software from the underlying hardware platforms to produce technology and reference designs that leverage commodity hardware platforms. WG7 is the White-box Hardware WG that specifies and releases a complete reference design to foster a decoupled software and hardware platform. WG8, the Stack Reference Design WG, develops the software architecture, design, and release plan for the O-CU and the O-DU based on O-RAN and 3GPP specifications for the NR protocol stack. WG9 is the open X-haul transport WG that specifies transport equipment, physical media, and control/management protocols associated with the transport network. WG10, Operation, Administration, and Maintenance (OAM) WG, specifies the OAM requirements, the OAM architecture, and the O1 interface. WG11, Security WG, addresses security aspects of O-RAN.

5.2.1.1 WG1 Work Products

Examples of the work products or documents created by WG1 are listed in self-explanatory Table 5.1 below; see O-RAN ALLIANCE (2024b) for the full list of the WG1 documents.

5.2.1.2 WG2 Work Products

Examples of the work products or documents created by WG2 are listed in self-explanatory Table 5.2 below; see O-RAN ALLIANCE (2024b) for the full list of the WG2 documents.

Table 5.1 Example WG1 Work Products.

Document Name	Document Scope
O-RAN Architecture Description	Specifies the overall architecture of O-RAN
O-RAN Slicing Architecture	Describes O-RAN slicing related use cases, requirements and architecture
O-RAN Use Cases Analysis Report	Describes potential O-RAN use cases at a very high level
O-RAN Use Cases Detailed Specification	Describes selected O-RAN use cases in further details to facilitate relevant O-RAN Work Groups to define requirements for associated O-RAN functions and interfaces
O-RAN Network Energy Saving Use Cases Technical Report	Defines overall use cases for network energy savings
O-RAN Massive MIMO Use Cases Technical Report	Captures the outcome of the WG1 UCTG Massive MIMO pre-normative phase work
O-RAN Operations and Maintenance Interface	Defines O-RAN OAM interface functions and protocols for the O-RAN O1 interface
O-RAN Operations and Maintenance Architecture	Identifies management services, managed elements, functions and applications supported in O-RAN, including the interworking between SMO and O-RAN O-Cloud management components
O-RAN Information Model and Data Models	Specifies the information model and the data models that are foundational for O-RAN's model-driven architecture and for the functions carried out over O-RAN interfaces
O-RAN Study on O-RAN Slicing	Provides common terminology for slicing, an overview of slicing activities in different SDOs (3GPP, ETSI, ONAP, GSMA, etc.), and a high level view of O-RAN slicing framework and architecture along with various deployment options

5.2.1.3 WG3 Work Products

Examples of the work products or documents created by WG3 are listed in self-explanatory Table 5.3 below; see O-RAN ALLIANCE (2024b) for the full list of the WG3 documents.

5.2.1.4 WG4 Work Products

Examples of the work products or documents created by WG4 are listed in self-explanatory Table 5.4 below; see O-RAN ALLIANCE (2024b) for the full list of the WG4 documents.

5.2.1.5 WG5 Work Products

Examples of the work products or documents created by WG5 are listed in self-explanatory Table 5.5 below; see O-RAN ALLIANCE (2024b) for the full list of the WG5 documents.

5.2.1.6 WG6 Work Products

Examples of the work products or documents created by WG6 are listed in self-explanatory Table 5.6 below; see O-RAN ALLIANCE (2024b) for the full list of the WG6 documents.

5.2.1.7 WG7 Work Products

Examples of the work products or documents created by WG7 are listed in self-explanatory Table 5.7 below; see O-RAN ALLIANCE (2024b) for the full list of the WG7 documents.

Table 5.2 Example WG2 Work Products.

Document Name	Document Scope
O-RAN A1 Interface: Transport Protocol	Describes the transport protocol of the O-RAN A1 interface
O-RAN A1 Interface: Type Definitions	Defines the data types for A1 policies and A1 enrichment information in a reusable and extensible way
O-RAN R1 Interface: General Aspects and Principles	Summarizes the R1 interface specification objectives and specifies the principles and procedures related to the O-RAN R1 interface
O-RAN Non-RT RIC Architecture	Provides the technical specification for the non-real-time ran intelligent controller (non-RT RIC) architecture
O-RAN R1 Interface: Use Cases and Requirements	Describes use cases and requirements for the O-RAN R1 interface
O-RAN Transport Protocols for R1 Services	Describes transport protocols for the O-RAN R1 interface services
O-RAN R1 Interface: Application Protocols for R1 Services	Contains a realization for the procedures identified in O-RAN R1 interface: general aspects and principles
O-RAN Non-RT RIC & A1/R1 Interface: Use Cases and Requirements	Describes O-RAN WG2 non-RT RIC use cases in detail including the required datasets, the flows, and the interactions within and outside of non-RT RIC with other O-RAN components
O-RAN A1 Interface: General Aspects and Principles	Specifies the general aspects and principles of the A1 interface
O-RAN A1 Interface: Use Cases and Requirements	Describes use cases for the O-RAN A1 interface
O-RAN A1 Interface: Application Protocol	Specifies the application protocol of the A1 interface
O-RAN A1 Interface: Test Specification	Specifies test cases for conformance testing and interoperability testing of the non-RT RIC and the near-RT RIC over the A1 interface
O-RAN AI/ML Workflow Description and Requirements	Provides details on AI/ML lifecycle management including model design, composition, and model runtime access to data and the model deployment solutions
O-RAN Non-RT RIC: Functional Architecture	Offers the rApp definition, the non-real time (RT) radio intelligent controller (RIC) framework functions and the R1 interface

5.2.1.8 WG8 Work Products

Examples of the work products or documents created by WG8 are listed in self-explanatory Table 5.8 below; see O-RAN ALLIANCE (2024b) for the full list of the WG8 documents.

5.2.1.9 WG9 Work Products

Examples of the work products or documents created by WG9 are listed in self-explanatory Table 5.9 below; see O-RAN ALLIANCE (2024b) for the full list of the WG9 documents.

5.2.1.10 WG10 Work Products

Examples of the work products or documents created by WG10 are listed in self-explanatory Table 5.10 below; see O-RAN ALLIANCE (2024b) for the full list of the WG10 documents.

Table 5.3 Example WG3 Work Products.

Document Name	Document Scope
O-RAN E2 General Aspects and Principles (E2GAP)	(Together with E2AP) describes the general architecture of Near-RT RIC and the main functions and procedures supported over the E2 interface, and the E2 application protocol of Near-RT RIC
O-RAN E2 Application Protocol (E2AP)	(Together with E2GAP) describes the general architecture of Near-RT RIC and the main functions and procedures supported over the E2 interface, and the E2 application protocol of Near-RT RIC
O-RAN E2 Service Model (E2SM)	Serves as the overall framework for the set of specialized E2SMs, each of which is dedicated to a specific RAN function
O-RAN E2 Service Model (E2SM) KPM	Defines the functions and protocols of the E2 interface, which connects the O-RAN Near-RT RIC to the underlying O-RAN nodes (i.e. E2 nodes)
O-RAN E2 Service Model (E2SM), RAN Control	Specifies the capabilities exposed over E2 interface to enable efficient control of RAN, including radio bear control, radio access control, connected mode mobility, etc.
O-RAN E2 Service Model (E2SM) Cell Configuration and Control	Aims at exposing configuration and control related processes on a cell-level basis
O-RAN Use Cases and Requirements	Details the functional and non-functional requirements on Near-RT RIC and E2 interface from the O-RAN use cases under study in O-RAN WG3
O-RAN Near-RT RIC Architecture	Specifies the Near-RT RIC internal architecture and functionalities, and the stage 2 definitions of Near-RT RIC APIs
O-RAN O1 Interface Specification for Near Real Time RAN Intelligent Controller	Defines O-RAN OAM interface functions and protocols for the O-RAN O1 interface for the Near RT RIC
O-RAN E2 Interface Test Specification	Introduces: additional conformance tests cases for RIC subscription delete request, miscellaneous subscription procedures for E2SM-KPM; IOT test methodology description and IOT cases for RIC global procedures and functional procedures of E2SM-KPM
O-RAN Near-RT RIC APIs Specification	Defines the signaling and data transport protocols for Near-RT RIC APIs
O-RAN Near-Real-time RAN Intelligent Controller E2 Service Model (E2SM), RAN Function Network Interface (NI)	Defines the functions and protocols of the E2 interface, which connects the O-RAN near-RT RIC to the underlying O-RAN nodes (i.e. E2 nodes)

5.2.1.11 WG11 Work Products

Examples of the work products or documents created by WG11 are listed in self-explanatory Table 5.11 below; see O-RAN ALLIANCE (2024b) for the full list of the WG11 documents.

5.2.2 Focus and Research Groups of O-RAN

The focus groups of O-RAN include Standard Development Focus Group (SDFG), open-source focus group (OSFG), Testing and Integration Focus Group (TIFG), Industry Engagement Focus

Table 5.4 Example WG4 Work Products.

Document Name	Document Scope
O-RAN Control, User and Synchronization Plane Specification	Specifies the control plane, user plane and synchronization plane protocols used over the fronthaul interface
O-RAN Management Plane Specification	Defines the management plane for the O-RAN open fronthaul based on the selected lower-layer split point as defined within the open fronthaul control plane, user plane and synchronization plane specification
O-RAN Management Plane Specification – YANG Models	Defines YANG models including common models, cross-WG common models, RU specific models, and RU specific models
O-RAN Conformance Test Specification	Defines tests that validate the conformance to the protocols defined for the lower layer split option 7.2x interface between the O-RU and the O-DU
O-RAN Fronthaul Interoperability Test Specification (IOT)	Specifies the interoperability testing (IOT) for O-DU and O-RU from different vendors connected using the O-RAN fronthaul (FH) interface
O-RAN Cooperative Transport Interface Transport Control Plane Specification	Describes the cooperative transport interface (CTI) and related message protocol when the fronthaul between the O-DU and O-RU is realized via a resource allocation-based transport network
O-RAN Cooperative Transport Interface Transport Management Procedures Specification	Describes the transport management procedures for the cooperative transport interface (CTI)
O-RAN Cooperative Transport Interface Transport Management Procedures Specification- YANG Models	Defines YANG models for the CTI

Group (IEFG), and Sustainability Focus Group (SuFG) (O-RAN ALLIANCE 2024j). The SDFG determines the standardization strategies and interfaces with other SDOs. The SDFG coordinates standardization-related activities and outputs across O-RAN WGs to ensure that the specifications developed by O-RAN WGs are timely and coherently submitted to relevant SDOs. The SDFG also tracks related activities of other SDOs/entities. The OSFG launched the OSC in 2019. Since the open-source software development activities are being carried out by the OSC, the OSFG is now a dormant group.

The TIFG defines the overall approach for testing and integration including coordination of test specifications across various WGs. The TIFG focuses on test and integration of specifications, creation of profiles to facilitate O-RAN commercialization, and specification of suitable processes for O-RAN integration and solution verification. Furthermore, the TIFG plans and coordinates the O-RAN PlugFests and defines the guidelines for third-party open testing and integration centres (OTICs). See Section 5.2.3 for a brief introduction to OTICs.

The IEFG aims to proactively engage with industry and the O-RAN ecosystem to promote and accelerate O-RAN-based technology adoption, proliferation, and continuous innovation. The SuFG focuses on optimizing energy consumption, reducing environmental impact, and creating more energy-efficient and environmentally friendly mobile networks.

Examples of the work products or documents created by TIFG are listed in self-explanatory Table 5.12 below; see O-RAN ALLIANCE (2024b) for the full list of the TIFG documents.

Table 5.5 Example WG5 Work Products.

Document Name	Document Scope
O-RAN NR C-Plane Profile	Further clarifies (i.e. profiles) for each RAN node the expected behavior (still within the scope of the 3GPP specifications)
O-RAN O1 Interface Specification for O-CU-UP and O-CU-CP- YANG Models	Defines YANG models for the O1 interface for the O-CU-UP and the O-CU-CP
O-RAN O1 Interface Specification for O-DU	Describes the supported management features including "start-up" installation, PNF software management, provisioning management, performance management, fault management and file management towards the O-DU via O1 interface
O-RAN O1 Interface Specification for O-DU- YANG Models	Defines YANG models for the O1 Interface specification for the O-DU
O-RAN O1 Interface Specification for O-CU-UP and O-CU-CP	Describes the supported management features including "start-up" installation, PNF software management, provisioning management, performance management, fault management and file management towards the O-CU-CP and O-CU-UP via O1 interface
O-RAN O1 Interface Specification for O-CU-UP and O-CU-CP- YANG Models	Defines YANG models for the O1 interface specification for the O-CU-UP and the O-CU-CP
O-RAN Interoperability Test Specification (IOT)	Specifies interoperability tests for RAN Node from different vendors implemented in accordance to the WG5 C-plane and U-plane profiles
O-RAN O1 Interface Specification for O-DU- Configuration Tables	Defines the O1 interface related O-DU configuration tables
O-RAN O1 Interface Specification for O-CU-UP and O-CU-CP- YANG Models	Defines YANG models for the O1 interface specification for the O-CU-UP and the O-CU-CP
O-RAN NR U-Plane Profile	Further clarifies (i.e. profiles) for each RAN node the expected behavior (still within the scope of the 3GPP specifications)
O-RAN Transport Specification	Clarifies the transport layer functions required/recommended for O-CU, O-DU, and eNB

At the time of this writing, there is a single research group, next-Generation Research Group (nGRG). The nGRG carries out research of open and intelligent RAN principles in 6G and future network standards. Chapter 8 provides an overview of selected 6G-specific topics addressed by the nGRG. Table 5.13 lists examples of research reports available on the O-RAN ALLIANCE website including 6G-related nGRG reports (O-RAN ALLIANCE 2024k).

Membership of the O-RAN ALLIANCE is not needed to access the O-RAN specifications. The O-RAN specifications can be obtained free of charge at O-RAN ALLIANCE (2024b). The O-RAN ALLIANCE has a tiered membership structure with the major membership categories being members, contributors (e.g., companies), and academic contributors (e.g., universities) (O-RAN ALLIANCE 2024f). All these membership categories can contribute to the development of O-RAN specifications. The membership is subject to an annual fee that depends on the size and annual revenue of the member entity. Operators are examples of members. Contributors are

Table 5.6 Example WG6 Work Products.

Document Name	Document Scope
O-RAN Acceleration Abstraction Layer General Aspects and Principles	Defines O-RAN O-cloud hardware accelerator interface functions and protocols for the O-RAN AAL interface
O-RAN Acceleration Abstraction Layer Common API	Focuses on the common API related to common aspects of accelerator abstraction from an administration and AAL Application perspective
O-RAN Cloud Architecture and Deployment Scenarios for O-RAN Virtualized RAN	Introduces and examines different scenarios and use cases for O-RAN deployments of network functionality into cloud platforms, O-RAN cloudified NFs and O-RAN physical NFs
O-RAN O-Cloud Interface Conformance Test Specification	Specifies a test specification for test and validation of the O-Cloud interfaces/APIs
O-RAN O2 Interface General Aspects and Principles	Defines O-RAN O-cloud functions and protocols for the O-RAN O2 interface
O-RAN Cloudification and Orchestration Use Cases and Requirements for O-RAN Virtualized RAN	Introduces different use cases for O-RAN orchestration of virtualized RAN and the interfaces used for management and orchestration
O-RAN O2dms Interface Specification: Kubernetes Native API Profile for Containerized NFs	Outlines the deployment management services of the O2dms interface profile based on Kubernetes native API for containerized NFs
O-RAN O2dms Interface Specification: Profile based on ETSI NFV Protocol and Data Models	Specifies the O-RAN O-cloud's DMS interface protocol and data model based on the ETSI NFV solutions
O-RAN Application Life Cycle Management (LCM) for Deployment Technical Recommendation	Contains deployment use cases and potential requirements
O-RAN O2ims Interface Specification	Defines O-RAN O-cloud IMS interface functions and protocols for the O-RAN O2 interface
O-RAN Acceleration Abstraction Layer High-PHY Profiles	Specifies how to accelerate the high-PHY functionality of an O-DU where the high-PHY acceleration function (AF) is hosted on an AAL-LPU that is assigned to the AAL application that is part of the O-DU
O-RAN O-Cloud Notification API Specification for Event Consumers	Describes a REST API that allows event consumers (EC) such as a vO-DU or CNF to subscribe to events/status from the O-cloud
O-RAN Acceleration Abstraction Layer FEC Profiles	Details the AAL specification consisting of the description of the interface, information models and requirements to implement an AALi, for the AAL_PUSCH_FEC and AAL_PDSCH_FEC profiles
O-RAN Acceleration Abstraction Layer AAL Profile – MUMIMO Precoder/BeamFormer Calculation	Defines an AAL acceleration profile to allow an AAL DU application to offload computation of precoding and beam-forming weights for MU-MIMO to an accelerator
O-RAN O2 General Aspects and Principles	Specifies the general aspects and principles associated with the O2 interface
O-RAN Cloud Platform Reference Designs	Provides an update of the O-cloud Reference Designs document to encompass both deployment scenario B and scenario C

Table 5.7 Example WG7 Work Products.

Document Name	Document Scope
O-RAN Outdoor Macrocell Hardware Architecture and Requirements (FR1) Specification	Specifies the hardware architecture and requirement as required by deployment scenario and base station classes specification
O-RAN Hardware Reference Design Specification for Outdoor Macrocell with Split Architecture Option 7.2	Describes the white box outdoor macrocell hardware reference designs of O-CU, O-DU, and O-RU based on deployment scenario and base station classes specification for O-RAN Split option 7-2x
O-RAN Network Energy Savings Procedures and Performance Metrics	Describes functional procedures for reducing O-RU or O-RU+O-DU energy consumption, and defining related power KPIs
O-RAN Hardware Reference Design Specification for Outdoor Micro Cell with Split Architecture Option 7.2	Specifies the hardware reference design as required by architecture and requirements for the outdoor microcell and deployment scenario and base station classes specifications
O-RAN Requirements and Reference Design Specification for Enterprise Microcell	Describes the hardware architecture, requirements and reference designs for the enterprise microcell deployment scenario based on deployment scenario and base station classes specification
O-RAN Deployment Scenarios and Base Station Classes	Specifies all deployment scenarios as identified by the operator group within O-RAN's Work Group 7
O-RAN Hardware Reference Design Specification for Indoor Picocell (FR1) with Split Architecture Option 8	Describes one white box hardware reference design example for indoor deployments (for indoor picocell FR1 with split architecture Option 8)
O-RAN Hardware Reference Design Specification for Indoor Picocell (FR1) with Split Architecture Option 7-2	Describes one white box hardware reference design example for indoor deployments (for indoor picocell FR1 with split architecture Option 7-2)
O-RAN Hardware Reference Design Specification for Fronthaul Gateway	Specifies a hardware reference design for the fronthaul gateway
O-RAN Hardware Reference Design Specification for Indoor Pico Cell with Fronthaul Split Option 6	Specifies a high-level architecture and hardware design for an indoor picocell using the split option 6 deployment scenario as specified in the deployment scenarios and base station classes document
O-RAN Outdoor Micro Cell Hardware Architecture and Requirements (FR1) Specification	Specifies system requirements and high-level architecture for the outdoor micro cell deployment scenario case as specified in the deployment scenarios and base station classes document
O-RAN Indoor Picocell Hardware Architecture and Requirement (FR1 Only) Specification	Describes the architecture and requirements for indoor picocell deployment scenarios as specified in O-RAN deployment scenarios and base station classes for white box hardware

Table 5.8 Example WG8 Work Products.

Document Name	Document Scope
O-RAN Stack Interoperability Test Specification	Defines interoperability test cases for O-CU and O-DU
O-RAN Base Station O-DU and O-CU Software Architecture and APIs	Defines the O-CU and O-DU software architecture, functional blocks, and API definitions

Table 5.9 Example WG9 Work Products.

Document Name	Document Scope
O-RAN Management Interfaces for Transport Network Elements	Describes best practices for O-RAN transport network element (TNE) management interfaces
O-RAN Xhaul Packet Switched Architectures and Solutions	Describes best practices for O-RAN transport based on end-to-end packet switching technology
O-RAN Xhaul Transport Testing	Defines the testing requirements for an Open Xhaul transport infrastructure
O-RAN Synchronization Architecture and Solution Specification	Describes best practices for O-RAN architecture and solution for Xhaul timing & synchronization
O-RAN WDM-Based Fronthaul Transport	Describes best practices for O-RAN fronthaul transport based on WDM technology
O-RAN Xhaul Transport Requirements	Defines the requirements for an Open Xhaul transport infrastructure

Table 5.10 Example WG10 Work Products.

Document Name	Document Scope
O-RAN O1 Interface Specification	Specifies management services for the O-RAN O1 interface and includes requirements, notifications and protocols
O-RAN Information Model and Data Models Specification	Specifies the information model and the data models that are foundational for O-RAN's model-driven architecture and for the functions carried out over O-RAN interfaces
O-RAN Information Model and Data Models Specification – Yang Models	Defines YANG models for the information model and the data models
O-RAN Operations and Maintenance Architecture	Identifies management services, functions and interfaces supported in O-RAN, including the interworking between SMO and O-RAN O-Cloud components
O-RAN Topology Exposure and Inventory Management Services Use Cases and Requirement Specification	Specifies the information model used to support TE&IV services within the SMO

the organizations that develop reference platforms and products to be tested and deployed by members. Contributors are companies with more than 500 employees or more than $50 Million annual revenue. Startup contributors have less than 50 employees. Small and medium enterprise contributors have less than 500 employees or less than $50 million annual revenue. Academic contributors are research or academic institutions.

5.2.3 Testing and Integration Entities

There are two aspects of testing and integration—development of test specifications and integration (and subsequent testing) of O-RAN components. Test specifications can be used to verify compliance of the O-RAN component to the O-RAN specifications. Interoperability test specifications are useful in validating interoperability of an O-RAN component with other O-RAN components that may be designed by the same entity or different entities.

Table 5.11 Example WG11 Work Products.

Document Name	Document Scope
O-RAN Security Requirements and Controls Specification	Specifies the security requirements and appropriate security controls per O-RAN interface and per O-RAN component
O-RAN Security Protocols Specifications	Provides security protocol specifications for an O-RAN compliant implementation
O-RAN Security Test Specifications	Validates the proper implementation of security requirements and protocols, emulates security attacks, and validates the effectiveness of security mitigation methods
O-RAN Security Threat Modeling and Risk Assessment	Identifies the various assets that are at risk, outlines the potential security threats that O-RAN system may face, along with security principles on how to mitigate these risks
O-RAN Study on Security for O-Cloud	(Technical report) analyzes the security of O-cloud components and interfaces
O-RAN Study on Security for Shared O-RU (Shared ORU)	(Technical report) provides the threat model and risk assessment for the shared O-RU
O-RAN Study on Security Log Management	(Technical report) considers the security log management as far as it applies to entities defined in the O-RAN architecture
O-RAN Study on Security for Near Real Time RIC and xApps	(Technical report) considers the security aspects of the near-RT RIC platform and xApps, as well as the associated network and management interfaces (E2, A1, O1) and APIs
O-RAN Security Threat Modeling and Remediation Analysis	(Technical report) identifies the various assets that are at risk, outlines the potential security threats that O-RAN system may face, along with security principles on how to mitigate these risks
O-RAN Study on Security for Application Lifecycle Management	(Technical report) considers security of O-RAN applications during their lifecycle phases of development, onboarding, and operation
O-RAN Study on Security for Service Management and Orchestration (SMO)	(Technical report) provides the threat model and risk assessment for the SMO
O-RAN Study on Security Log Management	(Technical report) considers the security log management as far as it applies to entities defined in the O-RAN architecture
O-RAN Study on Security for Non-RT-RIC	(Technical report) provides a security analysis for the non-RT RIC framework, rApps, R1 interface and A1 interface
O-RAN Study on Security for Fronthaul CUS-Plane	(Technical report) examines various aspects of securing CUS-plane on the open fronthaul interface between the O-DU and O-RU
O-RAN Study on Security for Artificial Intelligence and Machine Learning (AI/ML) in O-RAN	(Technical report) provides the threat model and risk assessment for AI/ML in O-RAN
O-RAN Study on Zero Trust Architecture for O-RAN	(Technical report) studies the applicability of a zero trust architecture (ZTA) to the O-RAN architecture

Table 5.12 Example TIFG Work Products.

Document Name	Document Scope
O-RAN Certification and Badging Processes and Procedures	Describes the certification and badging processes and associated detailed technical procedures for the open testing and integration centres (OTICs) and various other testing entities
O-RAN Certification and Badging Processes and Procedures – Summary report template	Defines a template for O-RAN certificate and badge summary report
O-RAN Certification and Badging Processes and Procedures – Test report template	Defines a template for TEST REPORT of O-RAN certificate, IOT badge and E2E badge
O-RAN Criteria and Guidelines of Open Testing and Integration Centre	Describes the essential criteria and guidelines (guiding principles) from process, organization, space and technical perspective on the qualified Open Testing and Integration Centre (OTIC)
O-RAN End-to-end Test Specification	Focuses on validating the end-to-end system functionality, performance, and key features of the O-RAN system as a black box
O-RAN End-to-End System Testing Framework Specification	Defines the end to end (E2E) testing framework for comprehensive E2E testing and validation of the open RAN (O-RAN) systems ensuring robust interoperability and operation with high performance as intended in real-world deployments and operation with high performance in real-world deployments

Table 5.13 Example nGRG Work Products.

Use Case Analysis on mmWave Antenna Distribution (mWAD)
Research Report on Cross-Domain AI
Architecture Principles for a Cloud-Friendly Future 6G RAN Architecture
O-RAN Native AI Architecture Description
Spectrum Sharing Based on Shared O-RUs
Research Report on Quantum Security
Research Report on Native and Cross-Domain AI: State of the Art and Future Outlook
O-RAN Towards 6G

Before the actual integration testing by an O-RAN certified entity is carried out, vendors of O-RAN components may carry out their own integration testing. For example, an O-DU vendor may carry out interoperability testing with an O-RU vendor. O-RAN PlugFests provide an opportunity to vendors to perform testing to ensure compliance with the O-RAN specifications and to verify interworking of their products with other products. O-RAN arranges Global PlugFests and third-party OTICs facilitate (i) compliance of vendor products to O-RAN specifications as well as (ii) interworking and integration of various O-RAN components.

The O-RAN ALLIANCE periodically organizes O-RAN Global PlugFests for testing and integration, where solution providers test, evaluate, and verify their products and solutions. Global

PlugFests are intended to facilitate progress of the overall O-RAN ecosystem (O-RAN ALLIANCE 2024c). O-RAN PlugFests help ensure the openness and interoperability of O-RAN solutions from different providers (O-RAN ALLIANCE 2024e). As an example, 60 companies and institutions participated in the O-RAN Global PlugFest Spring 2023 that was held from April 2023 to June 2023 in 10 labs across Asia, Europe, and North America. Information about Spring 2023 O-RAN Global PlugFest Spring 2023 can be found at O-RAN ALLIANCE (2024e), where four different PlugFest venues can be virtually visited. The tests carried out in PlugFests help validate O-RAN equipment conformance, performance, and interoperability. Issues can be identified and later resolved based on the learnings at PlugFests. PlugFests also provide vendors with opportunities to showcase new features, scenarios, and applications. For example, in Spring 2023 Global PlugFest, participants validated the security of end-to-end O-RAN architecture and demonstrated the ability of the RIC to support use cases such as QoS and energy efficiency (O-RAN ALLIANCE 2024e).

OTICs provide a collaborative, open, and impartial working environment for conformance, interoperability, and/or end-to-end testing of O-RAN products and solutions (O-RAN ALLIANCE 2024c, 2024g). OTICs provide reference setups and testbeds. OTICs can award O-RAN certificates and/or badges. OTICs can also host PlugFests. OTICs support demos, community events, and trials to support wide adoption of O-RAN specifications and promote the openness of the O-RAN ecosystem. Furthermore, OTICs also hold workshops or tutorials to increase O-RAN-related skills in the O-RAN ecosystem.

O-RAN certification and badging can be performed by only a qualified OTIC. The O-RAN certification and badging program is a comprehensive mechanism to instill confidence in O-RAN products and solutions. This program can minimize repetition of tests before commercial deployments and reduce test efforts for network operators. O-RAN certification also helps vendors promote their products and solutions. The program ensures certified interoperability among different vendors and acceptance from other vendors.

There are three distinct concepts in O-RAN certification and badging, an O-RAN certificate, an O-RAN Interoperability (IOT) Badge, and an O-RAN End-to-End (E2E) Badge. An O-RAN certificate indicates that a product is compliant to O-RAN certification. Such a certificate is awarded for one or more O-RAN interfaces based on O-RAN conformance tests. An O-RAN IOT Badge indicates interoperability of a pair of products connected via an O-RAN interface or O-RAN profiled 3GPP interface. This badge is awarded based on successful O-RAN interoperability tests. An O-RAN E2E Badge indicates that an end-to-end system or subsystem meets minimum requirements for functionality and security. Such a badge is issued upon successful completion of O-RAN end-to-end tests.

5.2.4 O-RAN Demonstrators and O-RAN Field Trials and Development Entities

In addition to test and integration demonstrations during PlugFests, there are other opportunities available to O-RAN demonstrators. Vendors can participate in a virtual exhibition, where they can demonstrate their O-RAN solutions (O-RAN ALLIANCE 2024c). For example, the O-RAN ALLIANCE has two main categories in the virtual exhibition hosted on its website (O-RAN ALLIANCE 2024d): (i) O-RAN demonstrations and (ii) Intelligent RAN Control Demonstrations. In the O-RAN demonstrations category, there are 45 virtual booths in open interface subcategory, 6 virtual booths in open-source subcategory, 6 virtual booths in open-source subcategory, 6 virtual booths in open-source subcategory, 6 virtual booths in open cloud subcategory, 17 virtual booths in open white-box subcategory, 16 virtual booths in open RAN stack subcategory, and 6 virtual booths in open xHaul subcategory. In the Intelligent RAN Control category, there are 28 virtual booths in

Intelligent Control subcategory and 22 virtual booths in Intelligent Management subcategory. As an example, a virtual booth may showcase a demonstration presented by a company at the Mobile World Congress in Barcelona in 2023.

Operators, in collaboration with their vendors, carry out field trials and commercial deployments. According to O-RAN ALLIANCE (2024h), there have been multiple commercial launches of O-RAN around the globe including the United States, Japan, Germany, Peru, and Italy.

5.2.5 O-RAN Software Community

The OSC is a collaboration between the O-RAN ALLIANCE and the Linux foundation (O-RAN ALLIANCE 2024i). The OSC's mission is to support the creation of open-source software for the RAN. The OSC aligns an open-source software reference implementation with the O-RAN ALLIANCE's open architecture to unify and accelerate the evolution and deployment of O-RAN. The OSC provides wireless technology support for essential patents.

The OSC comprises two project streams: (i) the OSC as a Linux Foundation project (supported and funded by O-RAN to lead the open-source implementation of the O-RAN specifications) and (ii) the O-RAN Specification Code Project as an O-RAN ALLIANCE project (to allow contributions that acknowledge essential patents). The Linux Foundation project leverages the Apache 2.0 License, while the Linux Foundation project leverages the O-RAN Software License. As specified by O-RAN ALLIANCE (2024i), Apache 2 is the standard license for open-source O-RAN software contributions, and the O-RAN License supports software that addresses RAN essential licensing and supports fair, reasonable, and nondiscriminatory terms. Both project streams are managed by the project's Technical Oversight Committee. The OSC is expected to enable the development of open-source software enabling modular, open, intelligent, efficient, and agile disaggregated radio access networks.

The OSC releases its software twice a year. The first OSC software release named Amber occurred in December 2019. Table 5.14 lists the OSC software releases with the release dates (OSC 2024a). The OSC software can be downloaded by visiting (OSC 2024a).

The OSC carries out its work through different projects as summarized in Figure 5.3 (OSC 2024b). The OSC codes are updated to reflect the features of a new release. The actual progress made by each project depends on the contributors to each project.

Table 5.14 A Snapshot of OSC Software Releases.

OSC Software Release Name	Release Date
Amber	November 2019
Bronze	June 2020
Cherry	December 2020
D Release	June 2021
E Release	December 2021
F Release	June 2022
G Release	December 2022
H Release	June 2023
I Release	December 2023

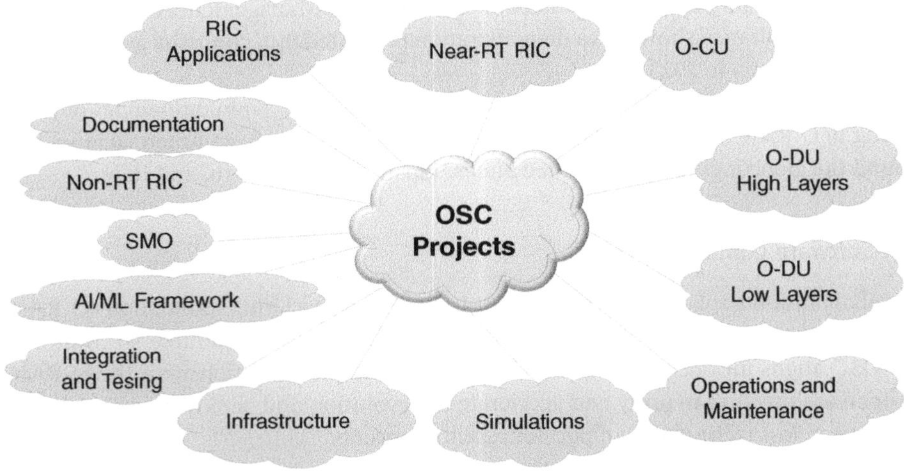

Figure 5.3 OSC Projects.

The RIC Applications Project includes open-source sample xApps and platform applications that can be used for integration, testing, and demonstrations.

The near-RT RIC Project develops software for the near-RT RIC platform to support xApps with limited support for O1, A1, and E2 interfaces.

The non-RT RIC Project develops software for the non-RIC platform to support rApps, the A1 interface, and the R1 interface. This project leverages and extends certain existing infrastructure from Open Network Automation Protocol (ONAP).

The goal of the O-CU Project is to support a basic E2 interface to enable integration testing between the near-RT RIC and the O-CU.

The O-DU High Project works on the RLC and the MAC functional blocks of the O-DU.

The O-DU Low Project works on the L1 functional block of the O-DU.

The SMO project aims to integrate different software artifacts of existing open-source projects creating a fully functional open-source SMO.

The Operations and Maintenance Project provides reference implementation according to the O-RAN WG1 OAM documents. This project also provides a common MnS-Consumer for development and module test purposes.

The Infrastructure Project defines the building blocks for the infrastructure to run O-RAN NF components.

The Simulations Project identifies simulators for testing O-RAN NF interfaces.

The Integration and Testing Project focuses on end-to-end testing and use case testing. This project also aims to establish a full end-to-end call flow using OSC components to carry out end-to-end integration tests and/or plug-and-play interoperability tests.

The Documentation Project prepares documentation on contributions and O-RAN SC contents.

5.3 External Ecosystem

As mentioned in Section 5.1, example constituents of the external O-RAN ecosystem are SDOs, developers of xApps and rApps, system integrators, operators, government agencies, and industry alliances/partners.

5.3.1 Standards Developing Organizations (SDOs)

As mentioned in Section 1.6 in Chapter 1, the specifications developed by the SDOs such as 3GPP, European Telecommunication Standards Institute (ETSI), Internet Engineering task Force (IETF), Cloud Native Computing Foundation (CNCF), OpenStack, and ONAP are important in designing and deploying O-RAN-based networks.

3GPP. Over the last few decades, the 3GPP has developed specifications for 3G, 4G, and 5G wireless technologies (3GPP 2024). The O-RAN ALLIANCE utilizes the 3GPP specifications of the 5G NG-RAN (and 4G LTE) and adds support for O-RAN-specific functions and interfaces. The 3GPP-defined gNB-CU-CP and gNB-CU-UP functions are implemented by the O-CU-CP and the O-CU-UP. The 3GPP-defined gNB-DU functions are implemented together by the O-DU and the O-RU. The gNB-CU-CP, the gNB-CU-UP, and the gNB-DU together constitute the gNB. Since a 5G User Equipment (UE) communicates with the gNB using the 5G NR air interface and since the O-CU-CP, the O-CU-UP, the O-DU, and the O-RU together represent the O-RAN-compliant gNB, O-RAN cannot exist without the 3GPP-defined specifications.

ETSI. ETSI develops global standards related to Information and Communications Technology-enabled systems, applications, and services deployed across all sectors of industry and society (ETSI 2024). Specific to 4G, 5G, and O-RAN implementations, ETSI standards related to virtualization are quite important. As mentioned in Chapter 1, O-Cloud can make use of ETSI-defined technologies such as Network functions Virtualization (NFV) in the O-Cloud to implement O-RAN nodes including the O-CU, the O-DU, the non-RT RIC, and the near-RT RIC. In particular, O-RAN ALLIANCE (2024m) lists the ETSI NFV standards useful for the O2 interface. These documents are related to Virtualized Network Function (VNF) packaging, Life Cycle Management (LCM) operations, and security for managing Virtual Machines and Containers. Examples of ETSI NFV standards relevant to the O2 interface implementation are (i) SOL001 standard to describe VNF, Physical Network Functions (PNF), and Network Service (NS) in Topology and Orchestration Specification for Cloud Applications (TOSCA), (ii) SOL002 standard for VNF/VNFC-level EM triggered scenarios such as LCM, Fault, Performance, and Configuration, (iii) SOL003 standard for VNF Package Management, LCM and Monitoring, (iv) SOL004 standard for VNF and PNF packages, (v) SOL005 standard for NS/PNF/VNF Package Management, LCM, and Monitoring, (vi) SOL006 standard to describe VNF, PNF, and NS in YANG, (vii) SOL007 standard is used for NS package, and (viii) SOL013 standard for interface security. In summary, ETSI plays an important role in the realization of O-RAN.

IETF. The IETF defines specifications in the form of Request for Comments (RFCs). Do not let the name RFC mislead you...although the name implies that the document is merely seeking comments, these are actually standards or specifications that are followed by compliant entities! Note that the 3GPP-defined 5G specifications and O-RAN specifications reuse relevant RFCs instead of reinventing the wheel. O-RAN nodes can interact with each other using IP. To ensure security of communications between the O-RAN nodes, IP-level security mechanisms can be applied. Examples of such mechanisms applicable to O-RAN include IPsec and Transport Layer Security (TLS) (O-RAN ALLIANCE 2024l). The 3GPP also supports the use of these mechanisms in a 5G network to provide security at the IP layer. The IETF is responsible for specifying these IP-based mechanisms. Examples of IETF specifications useful for O-RAN include Secure Shell (SSH), TLS, NETCONF, DTLS, IPsec, CMPv2, OAUTH 2.0, and cryptographic Ops.

CNCF. As mentioned in Chapter 1, Kubernetes can be used in the design and operation of an O-RAN platform such as the near-RT RIC platform (Open AI Cellular (OAIC) 2024). The near-RT RIC can be installed using two Kubernetes clusters. The first cluster called the RIC Cluster is used for deploying the near-RT RIC platform and applications, and the second cluster called the AUX cluster is used for deploying auxiliary functions. The RIC cluster consists of three major Kubernetes Systems of kube-system ns, ricinfra ns, and ricplt ns.

Kubernetes, abbreviated as K8s, is a portable, extensible, and open-source platform for managing containerized workloads and services (Kubernetes 2024). K8s is a container orchestration engine for automating deployment, scaling, and management of containerized applications. Google originally developed K8s and open-sourced it in 2014. The K8s project is hosted by the CNCF.

OpenStack. As mentioned in Chapter 1, since O-RAN is highly virtualized, OpenStack can be used to manage O-Cloud compute, storage, and networking resources through APIs or a dashboard. OpenStack is a widely deployed open-source cloud software. Example OpenStack functionalities include Infrastructure-as-a-Service, orchestration, fault management, and service management (OpenStack 2024).

ONAP. ONAP is an open-source platform that can be used to manage functions and/or components of a virtualized network. ONAP is a viable platform for automation and orchestration in an O-RAN implementation. ONAP can facilitate specific use cases such as SON and network slicing. As mentioned in Chapter 1, the sixth release of ONAP, called Frankfurt, focuses on support for 5G network slicing (Rethink 2024). ONAP supports the O-RAN O1 interface for fault, performance, and configuration management. New functions to support A1 Policies have been added to ONAP (Bonneau and Keeney 2023). Recall that A1 policies enable the non-RT RIC to provide policy guidance to the near-RT-RIC. Specifically, ONAP supports O-RAN's A1 interface, between (i) the orchestration layer implemented by ONAP or another Management and Orchestration (MANO) framework and (ii) the RIC.

5.3.2 Commercial Developers of xApps and rApps

xApps in the near-RT RIC and rApps in the non-RT RIC help O-RAN realize the goal of intelligence in the RAN. O-RAN provides the framework where the O-RAN functions such as the O-CU-CP, the O-CU-UP, the O-DU, and the O-RU can communicate with xApps and rApps. xApps and rApps enable service providers to manage and optimize various RAN operations depending on target use cases. Management of RAN operations has traditionally been proprietary. For example, the scheduling algorithm implemented in gNB-DU or equivalently the O-DU has been vendor-specific and hence not exposed to third parties for any significant modifications. However, in the O-RAN framework, it is possible for an xApp to influence the O-DU scheduler's operations through a suitable E2 Service Model (E2SM). In such case, an xApp in the near-RT RIC can modify the scheduler's behavior based on the measurements obtained from relevant E2 nodes.

The OSC has published several xApps as part of the xApp Store. Figure 5.4 mentions examples of such xApps developed by companies such as HCL, AT&T, CCMC, Samsung, and UTFPR.

Anomaly Detection. This HCL-developed xApp detects the presence of anomalous UEs in the network. This xApp works with the traffic steering xApp.

Bouncer. This is another HCL-developed xApp that is used for RIC benchmarking with the help of an E2 simulator communicating with the near-RT RIC. This xApp interacts with an E2 simulator using processes such as Sending Subscription Request, Receiving Subscription Response, Receiving RIC Indication, and Sending Control Message. The communication between the near-RT RIC and a suitable E2 node is stored in a file with relevant timestamps to determine throughput and control loop latency.

Load Prediction. This xApp has been developed by CCMC, which exploits ML to enhance the support for the traffic steering use case.

Figure 5.4 Examples of OSC-Published xApps.

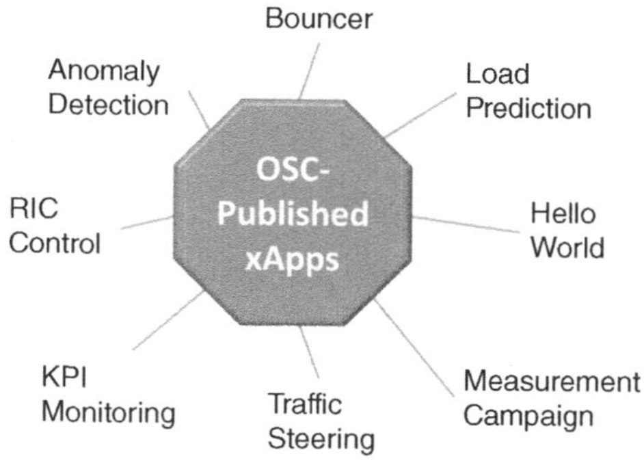

Hello World. This xApp is intended to help the xApp developers understand the APIs and libraries of the near-RT RIC to facilitate the development of xApps. The xApp demonstrates how an xApp uses the A1 and the E2 interfaces and performs read-write operations for a persistent database. This xApp has been developed in different languages such as C++ by AT&T, GO by Samsung, and Python by Samsung.

Measurement Campaign. This xApp calculates several metrics and KPIs based on X2 messages. For example, the xApp quantifies the number of EN-DC UEs, the duration of dual connections, and signal strength metrics (e.g., Reference Signal Received Power [RSRP], Reference Signal Received Quality [RSRQ], and Signal to Interference plus Noise Ratio [SINR]) of EN-DC UEs.

Traffic Steering Related xApps. A set of interacting xApps supports traffic steering. The traffic steering xApp (developed by AT&T and UTFPR) finds the UEs of interest based on the A1 policy received from the non-RT RIC, sends the prediction requests to the QoS Prediction (QP) xApp, receives the predictions from QP xApp, and steers the traffic by providing the control instructions to suitable E2 nodes. The QP Driver xApp, developed by AT&T and HCL, queries metrics from UE and Cell databases through the SDL layer for the UEs of interest and updates them to QP xApp. The QP xApp, developed by AT&T and HCL, predicts future throughput for UEs and makes the predictions available to the traffic steering xApp.

KPI Monitoring. The KPI Monitoring (KPIMON) xApp fills the cell metrics and the UE metrics into databases.

RIC Control xApp. This xApp acts as a Google Remote Procedure Call (GRPC) server that receives a GRPC request containing RIC control request parameters from a GRPC client. The xApp constructs a RIC CONTROL Request, encodes it in the ASN.1 format, and sends to a RAN/E2 node using RMR.

The xApps (and rApps) published by the O-RAN ALLIANCE can be viewed as a glimpse of O-RAN's potential in revolutionizing the management of the RAN. Third-party developers can create xApps and rApps that address the target use cases identified by the O-RAN ALLIANCE or even other use cases of interest.

5.3.3 System Integrators

While some RAN vendors may build all the O-RAN components, others may build only a subset of O-RAN components. Furthermore, O-RAN may be utilized to design a custom 5G network such as a private 5G enterprise network, an IIoT network, a utility network, or a military network. Custom networks often have service requirements that are quite distinct from the service requirements of nationwide commercial wireless networks. Hence, system integrators that integrate O-RAN and non-O-RAN components and meet target service requirements of custom networks are important for the growth of the O-RAN ecosystem.

5.3.4 Service Providers

As mentioned earlier in this section, some wireless service providers may not directly contribute to the development of O-RAN specifications but may deploy O-RAN. Such operators can be viewed as constituents of the external O-RAN ecosystem. These service providers realize the benefits of choosing O-RAN such as increased RAN supplier diversity, flexibility of mix-and-match of O-RAN components, supply chain resiliency, reduced costs, security, and efficient optimization of increasingly complex RAN. Acceptance and deployment of O-RAN by wireless service providers is critical to the survival and the growth of O-RAN. Hence, service providers are an essential component of the O-RAN ecosystem.

5.3.5 Government Entities

As mentioned in Section 5.1, government agencies often promote the use of O-RAN. For example, the US DoD plans to use 5G O-RAN in its communications systems (Dano 2003). Government agencies or departments that work across countries often promote O-RAN for enhanced security and supply chain resiliency. Examples of such international O-RAN collaboration include cooperation among the Quad countries of the United States, India, Australia, and Japan. Another example of intergovernmental collaboration includes the cooperation among the governments of the United States, the United Kingdom, and Germany.

Certain governments promote the deployment of O-RAN and offer incentives. For example, National Telecommunications and Information Administration (NTIA) in the United States launched a program to fund O-RAN security and testing research and development. Industry organizations such as Booz Allen Hamilton and universities such as Virginia Tech and Northeastern University have been granted awards by the NTIA (2024). Additional O-RAN funding opportunities are expected to become available in 2024 and beyond.

5.3.6 Industry Organizations

Some companies and organizations do not contribute to the development of O-RAN specifications but help promote O-RAN through conferences, seminars, and other events. For example, the IEEE has focused on O-RAN in some of its conferences such as IEEE SA ORAN SUMMIT (IEEE 2023).

5.3.7 Academic Institutions

As shown in Figure 5.5, universities can play a variety of roles as an external ecosystem constituent.[1] Examples of these roles include development of xApps and rApps, development of the O-RAN savvy workforce, and investigation and research of custom use cases using testbeds.

Universities can create xApps and rApps like commercial third-party developers. For example, Virginia Tech has developed an xApp that processes the positioning-related measurements such as

1 A university may also act as a constituent of the internal O-RAN ecosystem by contributing to the development of O-RAN specifications and serving as OTICs. Virginia Tech and Iowa State University serve as OTICs.

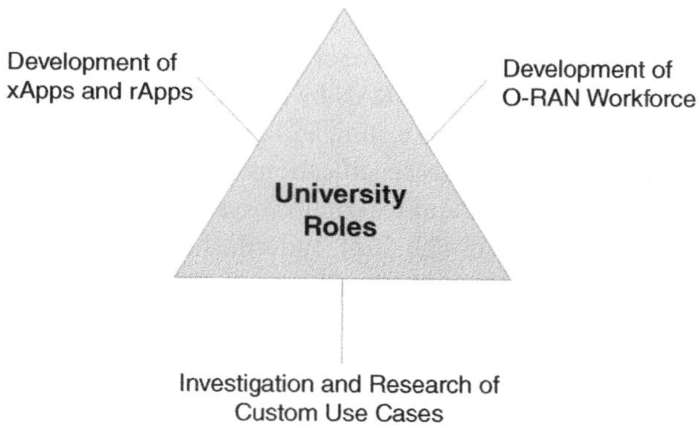

Figure 5.5 Roles of a University in an External O-RAN Ecosystem.

the propagation delays between the UE and the gNBs to determine the UE position (OAIC 2024). George Mason University and Mississippi State University have also developed xApps (e.g., interference detection) as part of the OAIC project (OAIC 2024).

Universities can contribute significantly to workforce development by creating O-RAN-focused educational materials. Students can be taught O-RAN fundamentals and get trained on how to develop xApps and rApps. For example, Virginia Tech has developed course materials on O-RAN to educate students of advanced wireless communications classes about O-RAN.

Research is an important aspect for a university. Students can get hands-on experience with O-RAN by working on O-RAN-related research projects. For example, Mississippi State University, Virginia Tech, and George Mason University have developed a platform called OAIC as part of an NSF-sponsored project to facilitate O-RAN research at universities and in the industry. The OAIC platform consists of two main components, OAIC-Control (OAIC-C) and OAIC-Testing (OAIC-T). OAIC-C is intended to facilitate development of xApps and rApps using suitable flavors of the platform. OAIC-T is intended to automate testing of AI-based applications to reduce manual operations and associated errors and accelerate the pace of research and innovations. The OAIC platform supports a variety of open-source software packages including software radio systems Radio Access Network (srsRAN) and Open Air Interface (OAI). Simulation-based environments are supported using ZeroMQ software and custom E2-like interface framework. Details of the OAIC platform can be found in OAIC (2024, Upadhyaya et al. 2023).

Examples of 5G testbeds suitable for 5G O-RAN research include Arena (sub-6 GHz indoors) (Northeastern University 2024), Colosseum (sub-6 GHz network emulator) (Coleman et al. 2024), AERPAW (Aerial outdoors) (AERPAW 2024), COSMOS (mmW outdoors and sub-6 GHz indoors) (COSMOS 2024), POWDER (sub-6 GHz outdoors & indoors) (POWDER 2024), and 5GENESIS (5G outdoors) (5GENESIS 2024).

5.4 O-RAN: A Deployment Perspective

As mentioned in Section 1.4 in Chapter 1, the O-RAN ALLIANCE is aiming for more intelligent, open, virtualized, and fully interoperable mobile networks (O-RAN ALLIANCE 2024a). The O-RAN ALLIANCE has developed a comprehensive set of specifications to enable a more competitive and diverse RAN supplier ecosystem and accelerate the pace of RAN innovations to increase efficiency, reduce costs, and enhance user experience. Recall from Section 1.4 in Chapter 1 that there are three focus streams of the O-RAN ALLIANCE: (i) the specification efforts carried out by several work groups to extend 3GPP 5G specifications to achieve openness and intelligence, (ii) the release and the development of open-source software for O-RAN by the OSC of the O-RAN ALLIANCE in cooperation with the Linux Foundation, and (iii) testing and integration efforts to ensure interoperability of products and compliance with O-RAN specifications.

In spite of the extensive work performed by the internal and external ecosystems of O-RAN, widespread deployments of O-RAN have not occurred as of Fall 2024. Indeed, there are significant challenges in deploying O-RAN. As mentioned in Section 8.1 in Chapter 8, there are a variety of challenges related to orchestration and OAM, performance, security, product development, Total Cost of Ownership, interoperability, RAN product performance, and multi-vendor operational complexity. These challenges need to be overcome to facilitate the growth of the ORAN ecosystem and make O-RAN successful. Chapter 8 explains potential solutions to the challenges faced by O-RAN.

Even though O-RAN specifications are quite comprehensive, interoperability of O-RAN nodes from different vendors is certainly challenging. Interoperability test specifications and PlugFests provide a good framework to address the issue of interoperability. The RAN is a very complex part of the wireless system with many RAN functions implemented in a proprietary manner. Hence, ensuring high performance requires smooth interworking among various O-RAN nodes. PlugFests and OTIC testing can be used to detect interoperability issues and enhance O-RAN specifications. However, interoperability testing needs to be comprehensive to cover all intended deployment scenarios and service use cases because of the enormous complexity of 3GPP 5G RAN specifications and O-RAN specifications. While tangible progress has been made on the open fronthaul interface to verify interoperability of the O-DU and the O-RU, widespread interoperability of O-RAN nodes from different vendors has not materialized as of Fall 2024. A significant amount of time and efforts would be needed before widespread O-RAN deployments using products of numerous vendors become a reality.

The O-RAN ALLIANCE emphasizes openness and intelligence for the next generation wireless networks and beyond (O-RAN ALLIANCE 2018). The initial focus in the industry has been on the openness aspect of O-RAN. The intelligence aspect of O-RAN would be even more challenging than the openness aspect of O-RAN. Indeed, there could be conflicts among different xApps/rApps and among xApps/rApps and proprietary RAN function implementations in the O-CU, the O-DU, and the O-RU.

Increased support for R&D and interoperability testing for the O-RAN system by the governments would be critical to ensure the long-term success of O-RAN. In the United States, the government agencies such as the DoD have been encouraging the R&D and deployment of O-RAN (Department of Defense 2022). The NTIA, which is part of the Department of Commerce, has funded numerous R&D initiatives and large centers and continued to support O-RAN activities to facilitate O-RAN deployments (Hazuda 2024) (NTIA 2024). The government and industry support for relevant O-RAN R&D and testing can help accelerate O-RAN deployments.

O-RAN success also depends on the acceptance of O-RAN by major service providers in the United States and around the globe. If major operators start using products from different O-RAN vendors and on a wide scale, it would be a huge O-RAN success story. Scattered O-RAN deployments or O-RAN deployments with only a handful of vendors may not be able to provide sustainable growth of the O-RAN ecosystem.

In summary, while the O-RAN ALLIANCE has built a strong foundation, and the governments have been providing support, investments of a significant amount of time and resources in R&D and testing and acceptance of O-RAN by major service providers are a key to O-RAN's success.

5.5 Major Takeaways of the Chapter

O-RAN has a comprehensive ecosystem with hundreds of organizations. Here are the major takeaways of this chapter.

- The overall O-RAN ecosystem can be viewed as an internal O-RAN ecosystem and an external O-RAN ecosystem.
- The internal ecosystem includes various groups working on various aspects of O-RAN, while an external ecosystem includes organizations that are not formally part of O-RAN but play a significant role in the design, deployment, maintenance, optimization, use, and promotion of O-RAN.
- The internal ecosystem of O-RAN consists of the entities that actively work on three focus streams of O-RAN. These focus streams include specification efforts, testing and integration efforts, and OSC.
- The external O-RAN ecosystem consists of diverse entities such as developers of xApps and rApps, system integrators, operators, government agencies, and industry alliances/partners.
- The technical O-RAN work is performed by O-RAN WGs, FGs, and Research Groups.
- There are 11 O-RAN WGs, WG1 to WG11. These WGs focus on various aspects of O-RAN such as the overall architecture, the non-RT RIC, the near-RT RIC, open fronthaul, 3GPP-compliant profiles, cloudification and orchestration, White-box hardware, stack reference designs, transport network, OAM, and security.
- The focus groups of O-RAN include SDFG, open-source focus group, TIFG, IEFG, and SuFG.
- The O-RAN ALLIANCE has a tiered membership structure with the major membership categories being members, contributors (e.g., companies), and academic contributors (e.g., universities).
- The O-RAN ALLIANCE periodically organizes O-RAN Global PlugFests for testing and integration, where vendors and providers test, evaluate, and verify their products and solutions.
- OTICs provide a collaborative, open, and impartial working environment for conformance, interoperability, and/or end-to-end testing of O-RAN products and solutions.
- The OSC is a collaboration between the O-RAN ALLIANCE and the Linux foundation. The OSC's mission is to support the creation of open-source software for the RAN. The OSC aligns an open-source software reference implementation with the O-RAN ALLIANCE's open architecture to unify and accelerate the evolution and deployment of O-RAN. The OSC typically releases its software twice a year. The OSC carries out its work through different projects.
- The specifications developed by the SDOs such as 3GPP, ETSI, IETF, CNCF, OpenStack, and ONAP are important in designing and deploying O-RAN-based networks.

References

3GPP (2024). *About 3GPP*. https://www.3gpp.org/about-us (accessed 18 July 2024).

5GENESIS (2024). *5GENESIS—5th Generation End-to-End Network. Experimentation, System Integration, and Showcasing*. https://5genesis.eu/ (accessed 18 July 2024).

AERPAW (2024). *Aerial Experimentation and Research Platform for Advanced Wireless*. https://aerpaw.org/ (accessed 18 July 2024).

Bonneau, M. and Keeney, J. (2023). *O-RAN A1 Policies in ONAP*. April 2023, https://wiki.onap.org/display/DW/O-RAN+A1+Policies+in+ONAP (accessed 18 July 2024).

Coleman, D. et al. (2024). *Overview of the Colosseum: The World's Largest Test Bed for Radio Experiments*. https://www.jhuapl.edu/content/techdigest/pdf/V35-N01/35-01-Coleman.pdf (accessed 18 July 2024).

COSMOS (2024). *COSMOS—Cloud Enhanced Open Software Defined Mobile Wireless Testbed for City-Scale Deployment*. https://www.cosmos-lab.org/ (accessed 18 July 2024).

Dano, M. (2003). *US Military Bases Could Get Private 5G, Delighting Dish*. 14 December, https://www.lightreading.com/open-ran/us-military-bases-could-get-private-5g-delighting-dish#close-modal (accessed 18 July 2024).

Department of Defense (2022). DoD and National Spectrum Consortium Team for open RAN acceleration. Press release (3 June 3). https://www.defense.gov/News/Releases/Release/Article/3052013/dod-and-national-spectrum-consortium-team-for-open-ran-acceleration/ (accessed 18 July 2024).

ETSI (2024). *About ETSI*. https://www.etsi.org/about (accessed 18 July 2024).

Hazuda, N. (2024). Grant projects to unlock potential, secure the future of open networks. Press release (22 March). https://news.vt.edu/articles/2024/03/eng-ece-lingjia-liu-nishith-tripathi-ORAN-Innovation-Fund.html (accessed 18 July 2024).

IEEE (2023). *IEEE SA O-RAN Summit*. https://engagestandards.ieee.org/ORan-Summit-2023.html (accessed 18 July 2024).

Kubernetes (2024). *Overview*. https://kubernetes.io/docs/concepts/overview/ (accessed 18 July 2024).

Northeastern University (2024). *Arena: A 64-antenna SDR-based Ceiling Grid Testing Platform for Sub-6 GHz 5G-and-Beyond Radio Spectrum Research*. https://ece.northeastern.edu/wineslab/arena.php (accessed 18 July 2024).

NTIA (2024). *Award Recipients*. https://www.ntia.gov/program/innovation-fund/grant-programs/round-1-2023/award-recipients (accessed 18 July 2024).

OAIC (2024). *Open AI Cellular*. https://github.com/openaicellular/oaic (accessed 18 July 2024).

OpenStack (2024). *Software*. https://www.openstack.org/software/ (accessed 18 July 2024).

O-RAN ALLIANCE (2018). O-RAN: Towards an Open and Smart RAN. White Paper, October 2018, https://mediastorage.o-ran.org/white-papers/O-RAN.White-Paper-2018-10.pdf (accessed 18 July 2024).

O-RAN ALLIANCE (2024a). *About Us*. https://www.o-ran.org/about (accessed 18 July 2024).

O-RAN ALLIANCE (2024b). *O-RAN Specifications*. https://specifications.o-ran.org (accessed 18 July 2024).

O-RAN ALLIANCE (2024c). *O-RAN Ecosystem*. https://www.o-ran.org/o-ran-ecosystem (accessed 18 July 2024).

O-RAN ALLIANCE (2024d). *Virtual Exhibition*. https://www.virtualexhibition.o-ran.org/classic/generation/2023 (accessed 18 July 2024).

O-RAN ALLIANCE (2024e). *O-RAN Global PlugFest 2023*. https://plugfestvirtualshowcase.o-ran.org/2023/SPRING (accessed 18 July 2024).

O-RAN ALLIANCE (2024f). *Membership Info*. https://www.o-ran.org/membership-info (accessed 18 July 2024).

O-RAN ALLIANCE (2024g). *Overview of Open Testing and Integration Centre (OTIC) and O-RAN Certification and Badging Program*. https://oran-master.webflow.io/testing-integration#learn-otic (accessed 18 July 2024).

O-RAN ALLIANCE (2024h). *O-RAN Map*. https://map.o-ran.org/ (accessed 18 July 2024).

O-RAN ALLIANCE (2024i). *Open Software for the RAN*. https://www.o-ran.org/software (accessed 18 July 2024).

O-RAN ALLIANCE (2024j). *O-RAN Technical Work Groups*. https://www.o-ran.org/about#technical-workgroup (accessed 18 July 2024).

O-RAN ALLIANCE (2024k). *Research Reports*. https://www.o-ran.org/research-reports (accessed 18 July 2024).

O-RAN ALLIANCE (2024l). *Security Protocols Specifications*. O-RAN.WG11.Security-Protocols-Specification.O-R003-v05.00. https://specifications.o-ran.org (accessed 18 July 2024).

O-RAN ALLIANCE (2024m). *O2 Interface General Aspects and Principles*. O-RAN.WG6.O2-GA&P-R003-v03.00. The O-RAN ALLIANCE. https://specifications.o-ran.org (accessed 18 July 2024).

OSC (2024a). *Releases*. https://wiki.o-ran-sc.org/display/REL/Releases (accessed 18 July 2024).

OSC (2024b). *Projects*. https://wiki.o-ran-sc.org/display/ORAN/Projects (accessed 18 July 2024).

POWDER (2024). *POWDER—Platform for Open Wireless Data-driven Experimental Research*. https://powderwireless.net/ (accessed 18 July 2024).

Rethink (2024). *ONAP's New Release Focuses on 5G Slicing and O-RAN Integration*. June 2020, https://rethinkresearch.biz/articles/onaps-new-release-focuses-on-5g-slicing-and-o-ran-integration/ (accessed 18 July 2024).

Upadhyaya, P. et al. (2023). *Open AI Cellular (OAIC): An Open Source 5G O-RAN Testbed for Design and Testing of AI-based RAN Management Algorithms*. IEEE Network. doi: https://doi.org/10.1109/MNET.2023.3320933.

6

O-RAN Testing and Integration

Open Radio Access Network (O-RAN) is characterized by openness and intelligence. Openness of O-RAN is evident by the support for multiple open interfaces. Open interfaces enable a service provider to make use of O-RAN component products (i.e., the O-RAN radio unit [O-RU], the O-RAN Distributed Unit [O-DU], the O-RAN Central Unit [O-CU], and the RAN Intelligent Controllers [RICs]) from different vendors. Openness provides significant flexibility to a service provider to select vendors to meet its business objectives. However, integration of different O-RAN components is a nontrivial matter. Testing of integrated products is especially important to ensure proper operations of O-RAN in commercial deployments. This chapter discusses various aspects of O-RAN testing and integration. A brief overview of Open Testing and Integration Centers (OTICs) is given. O-RAN certification and badging program is also described. Global PlugFests that bring companies and institutions from multiple continents together are discussed. An overview of testing-related O-RAN specifications is given. Finally, examples of best practices and recommendations for O-RAN testing and integration are given.

6.1 O-RAN Testing and Integration in a Nutshell

The O-RAN ALLIANCE supports testing and integration of O-RAN solutions through the mechanisms as shown in Figure 6.1 (O-RAN ALLIANCE 2024e).

Test Specifications. O-RAN has defined comprehensive test specifications to facilitate testing of O-RAN components.

OTICs. OTICs provide an open, collaborative, vendor independent, and impartial working environment to support the O-RAN ecosystem.

Global PlugFests. The O-RAN ALLIANCE holds global PlugFests periodically (O-RAN ALLIANCE 2024b). These events facilitate progress of the O-RAN ecosystem through organized testing and integration. Companies of the O-RAN ecosystem participate in PlugFests to test, evaluate, and verify their products and solutions.

Certification and Badging. O-RAN certification and badging can be performed by qualified OTICs only. Such certification and badging help O-RAN vendors promote their products and provide an opportunity for certified interoperability among different vendors and acceptance from other vendors.

Fundamentals of O-RAN, First Edition. Nishith D. Tripathi and Vijay K. Shah.

Companion website: www.wiley.com/go/tripathioran

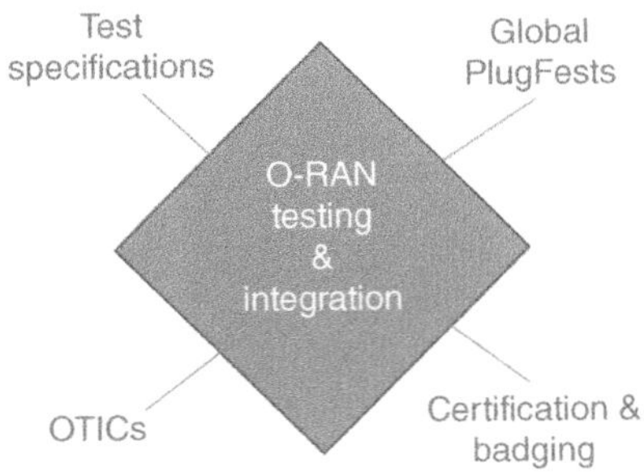

Figure 6.1 O-RAN Testing and Integration Mechanisms.

The rest of the chapter is organized as follows. Section 6.2 provides an overview of the test specifications developed by the O-RAN ALLIANCE. OTICs are discussed in Section 6.3. Section 6.4 describes global PlugFests. Section 6.5 summarizes certification and badging. Finally, key takeaways of the chapter are highlighted in Section 6.6.

6.2 Test Specifications

Test specifications facilitate several types or scopes of testing such as conformance certification, interoperability Test (IOT) badging, and end-to-end (E2E) badging (O-RAN ALLIANCE 2024g). For example, conformance certification involves compliance verification of a device under test (DUT) using an O-RAN interface or an implementation of O-RAN reference design based on conformance test specifications. IOT badging involves interoperability assessment of a DUT pair using O-RAN interfaces based on IOT test specifications. Interoperability of an O-RAN component with other O-RAN components that may be designed by the same vendor or different vendors can be verified by executing test procedures defined in interoperability test specifications. Finally, E2E badging is E2E system integration assessment of a DUT using O-RAN interfaces based on E2E test specifications.

Conformance testing and interoperability testing are within the scope of individual work groups (WGs), while E2E testing is within the scope of the Testing and Integration Focus Group (TIFG). The TIFG defines the process for E2E testing and integration including coordination of test specifications across various WGs. The TIFG develops test and integration specifications for E2E testing, creates profiles to facilitate O-RAN commercialization, and specifies processes for O-RAN integration and solution verification.

6.2.1 Conformance and Interoperability Testing

Table 6.1 lists test specifications for conformance and interoperability testing for O-RAN components and interfaces (O-RAN ALLIANCE 2024a). *A brief overview of the test specifications mentioned in Table 6.1 is given in Sections 6.2.1.1 to 6.2.1.8.*

6.2.1.1 O-RAN A1 Interface: Test Specification

The Work Group 2 (WG2) of the O-RAN ALLIANCE has specified test cases for conformance testing and interoperability testing of the non-real-time RIC (Non-RT RIC) and the near-real-time RIC (Near-RT RIC) over the A1 interface (O-RAN ALLIANCE 2024j). Test cases and methodology for A1 Policy (A1-P) consumer, A1-P producer, A1-enrichment information (A1-EI) consumer, and A1-EI producer are specified separately. Furthermore, A1-related test cases for interoperability

Table 6.1 Specifications for Conformance and Interoperability Testing.

Specification	Scope	Work Group
O-RAN A1 interface test specification	Conformance testing and interoperability testing of the Non-RT RIC and the Near-RT RIC over the A1 interface	WG2
O-RAN E2 interface test specification	Test configurations and test cases for E2 interface testing including conformance testing of the Near-RT RIC over the E2 interface, conformance testing of the E2 nodes over the E2 interface, and interoperability testing with Near-RT RIC and E2 nodes over the E2 interface	WG3
O-RAN (fronthaul) conformance test specification	Tests to validate the conformance to the protocols defined for the lower-layer split interface between the O-RU and the O-DU	WG4
O-RAN fronthaul interoperability test specification	Interoperability testing (IOT) for the O-DU and the O-RU from different vendors connected using the O-RAN fronthaul (FH) interface	WG4
O-RAN interoperability test specification	Interoperability tests for RAN nodes from different vendors implemented using the WG5 C-plane and U-plane profiles	WG5
O-RAN cloud conformance test specification	Tests for the O-Cloud conformance to WG6 specifications	WG6
O-RAN O-Cloud interoperability test (IOT) specification	O-Cloud IOT specification including definition of the interoperability testing methodology and a set of tests for the O-Cloud notification API	WG6
O-RAN stack interoperability test specification	Interoperability test cases for the O-CU and the O-DU	WG8

between the Non-RT RIC and the Near-RT RICs are specified. *Details of the entrance criteria, the procedure, and the expected result are described for each specified test case.* The procedures specified for the A1 application protocol (A1AP) are utilized to define the test cases to validate conformance and interoperability. The tester identifies relevant test cases based on an implementation statement related to a DUT. The policy types or enrichment information (EI) types to be used in a given test case are agreed between the testing entities as described in the test entrance criteria of the test case. For conformance testing, simulators are used for testing A1 procedures. These simulators are capable of generating Hypertext Transfer Protocol (HTTP) requests and responses for GET, PUT, POST, and DELETE operations. Flexibility in configuring URI, headers, and body for these HTTP requests and responses helps create multiple test cases.

Conformance Testing of the Non-RT RIC In support of conformance testing, the Non-RT RIC implements A1-P consumer and/or A1-EI producer functionality and relevant A1AP procedures. The Non-RT RIC also supports one agreed policy type and/or one agreed EI type. Similarly, the test simulator implements A1-P producer and A1-EI consumer that both have HTTP client and HTTP server capabilities and have flexibility to generate, receive, and validate HTTP messages for all the A1 procedures. Furthermore, the test simulator logs all the message content during the testing.

Multiple test cases are defined for the conformance testing of the Non-RT RIC (O-RAN ALLIANCE 2024j). For each of these test cases, the following details are specified: test description and applicability, test entrance criteria, and test methodology, including initial conditions, procedure, and expected result. The test cases are related to the following test case scenarios for A1-P Consumer: Query Policy Type test scenarios, Create Policy test scenarios, Query Policy test scenarios, Update Policy test scenarios, Delete Policy test scenarios, and Feedback Policy test scenarios. The test cases are related to the following test case scenarios for A1-EI Producer: Query EI types test scenarios, Create EI job test scenarios, Query EI jobs test scenarios, Update EI job test scenarios, Delete EI job test scenarios, Status of EI jobs test scenarios, and Deliver EI job result test scenarios.

Conformance Testing of the Near-RT RIC In support of conformance testing, the Near-RT RIC implements A1-P Producer and A1-EI Consumer functionality and the relevant A1AP procedures. The Near-RT RIC also supports one agreed policy type and/or one agreed EI type. On the other hand, the test simulator implements A1-P Consumer and A1-EI Producer that have HTTP client and HTTP server capabilities and can generate, receive, and validate HTTP messages for the A1 procedures. Furthermore, the test simulator logs the message content during the testing.

For the Near-RT RIC's A1-P Producer conformance test cases, the following test scenarios are considered: Query Policy Type test scenarios, Create Policy test scenarios, Query Policy test scenarios, Update Policy test scenarios, Delete Policy test scenarios, and Feedback Policy test scenarios. Furthermore, for the Near-RT RIC's A1-EI Consumer conformance test cases, the following test scenarios are considered: Query EI types test scenarios, Create EI job test scenarios, Query EI jobs test scenarios, Update EI job test scenarios, Delete EI job test scenarios, Status of EI jobs test scenarios, and Deliver EI job result test scenarios.

Interoperability Testing Between the Non-RT RIC and the Near-RT RIC The Non-RT RIC and the Near-RT RIC act as the DUT or the system under test (SUT). Note that the Non-RT RIC and the Near-RT RIC may be from different vendors or from the same vendor. The test methodology is compatible with production ready devices. The Non-RT RIC and the Near-RT RIC need to have a minimal set of matching functionalities to facilitate testing of A1-P and A1-EI procedures. For example, as an entry criterion for interoperability testing, the Non-RT RIC and the Near-RT RIC have at least one matching policy type and/or EI type. Examples of test tools and equipment needed for testing include real or simulated Service Management and Orchestration (SMO) O1 functionality, real or simulated user equipments (UEs) and Radio Access Network (RAN) (i.e., E2 nodes), real or simulated fourth-generation (4G)/fifth-generation (5G) core network, and a tap interface and protocol analyzer. A protocol analyzer connected to the A1 interface utilizes a tap to collect information and analyzes packets passively to validate the target test case.

For this interoperability testing, the SUT is a Non-RT RIC and a Near-RT RIC. The Non-RT RIC and the Near-RT RIC are also DUTs. The Non-RT RIC and the Near-RT RIC can be functions that are under development or finalized commercial products. The Non-RT RIC and the Near-RT RIC need to have passed the conformance test scenarios. To validate A1 procedure messages captured between the Non-RT RIC and the Near-RT RIC, an ability to disable encryption is needed. Furthermore, A1-P types and EI types used in test scenarios are agreed and configured in the DUTs. Note that specified A1 interoperability testing does not require usage of a specific policy type or EI type, because specific behavior is not validated; only the application protocol aspects and procedures are validated. Interoperability test cases for A1-P are defined for the following test scenarios: Query

Policy Type test scenarios, Create Policy Type test scenario, Query Policy test scenarios, Update Policy test scenarios, Delete Policy test scenarios, and Feedback Policy test scenarios. Additionally, interoperability test cases for A1-EI are defined for the following test scenarios: Query EI types test scenarios, Create EI job test scenarios, Query EI jobs test scenarios, Update EI job test scenarios, Delete EI job test scenarios, Status of EI jobs test scenarios, and Deliver EI job result test scenarios.

6.2.1.2 O-RAN E2 Interface Test Specification

The Work Group 3 (WG3) of the O-RAN ALLIANCE has specified test methodology and test cases for conformance testing and interoperability testing of the Near-RT RIC and the E2 nodes over the E2 interface (O-RAN ALLIANCE 2024k).

Conformance Testing for the Near-RT RIC The test configuration for E2 conformance testing of the Near-RT RIC can utilize an E2 node simulator, an O-RU simulator, and a UE simulator. A real O-RU with associated M-plane support and a real or test UE may also be used instead of their simulated forms. Configuration of the E2 nodes and the Near-RT RIC for the conformance testing is performed via the O1 interface.

In one example configuration, the test setup may use separate Near-RT RIC, O-CU control plane (O-CU-CP), O-CU user plane (O-CU-UP), and O-DU. In another example configuration, the E2 node simulator may support a range of different RAN configurations including (i) stand-alone (SA) O-CU-CP connected to one or more SA O-CU-UP and one or more SA O-DUs, (ii) combined O-CU-CP and O-CU-UP connected to one or more SA O-DUs, (iii) combined O-CU-CP/O-CU-UP may present either a common E2 interface or individual E2 interfaces corresponding to the individual O-RAN components, and (iv) combined O-CU-CP, O-CU-UP, and O-DU. A given O-RAN component may initiate an independent E2 connection or a shared E2 interface toward the Near-RT RIC.

In support of conformance testing, the Near-RT RIC implements the E2 interface and test case-related E2AP procedures. Note that the Near-RT RIC may support all or a subset of the E2 service models (i.e., E2SM-KPM, E2SM-NI, and E2SM-RC). The E2 node simulator simulates the E2 functionalities of the E2 node and has flexibility to generate, receive, and validate messages for all the E2 global procedures and certain or all of the E2 service models. The test simulator logs the message content during the testing. The test simulator can generate E2 node-initiated procedures, receive Near-RT RIC-initiated procedures, and generate responses for successful or unsuccessful operations. The E2 node simulator may also simulate suitable UE functions from the Near-RT RIC point of view.

Numerous test cases for conformance testing of the Near-RT RIC are defined. These test cases are related to E2AP Global Procedures, E2AP Functional Procedures, E2AP Functional Procedures for E2SM-KPM, and E2AP Functional Procedures for E2SM-NI. For each test case, details such as test purpose, test entrance criteria, and test methodology (including initial conditions, procedure, and expected results) are specified.

- Test Cases for E2AP Global Procedures: E2 Setup, RESET, Error Indication, RIC Service Update, E2 Node Configuration Update, E2 Connection Update, and E2 Removal
- Test Cases for E2AP Functional Procedures: RIC Subscription, RIC Subscription Delete, RIC Subscription Delete Required, RIC Indication, and RIC Control
- Test Cases for E2AP Functional Procedures for E2SM-KPM: RIC Subscription E2SM-KPM and RIC Indication E2SM-KPM
- Test Cases for E2AP Functional Procedures for E2SM-NI: RIC Subscription E2SM-NI

Conformance Testing for E2 Nodes The test setup for the conformance testing of E2 nodes may utilize a Near-RT RIC simulator, O-RU simulator, core simulator, and UE simulator. The test setup may use real equipment instead of simulators. In one example configuration, separate O-CU-CP, O-CU-UP, and O-DU may be used in the test setup.

In another example configuration, the E2 node may support a range of RAN configurations including (i) SA O-CU-CP connected to one or more SA O-CU-UPs and one or more SA O-DUs, (ii) combined O-CU-CP and O-CU-UP connected to one or more SA O-DUs, and (iii) combined O-CU-CP, O-CU-UP, and O-DU. Note that the combined node may present a common/shared E2 interface or individual E2 interfaces corresponding to the individual O-RAN components.

In the test setup, the DUT could be of any E2 node such as (i) the O-CU-CP, the O-CU-UP, or the O-DU; (ii) combined O-CU-CP, O-CU-UP, and O-DU; (iii) combined O-CU-CP and O-CU-UP; (iv) O-DU; and (v) O-RAN next-generation-evolved Node B (O-eNB).

In support of conformance testing, the E2 node implements E2AP procedures relevant to the target test case. The E2 node may support all or a subset of the E2 service models (i.e., E2SM-KPM, E2SM-NI, and E2SM-RC). The Near-RT RIC simulator simulates the E2 functionalities of the Near-RT RIC and can generate, receive, and validate messages for the E2AP procedures and one or more of the E2 service models. The test simulator also logs the message contents during the testing. The test simulator can generate the Near-RT RIC-initiated procedures, receive the E2 node-initiated procedures, and generate responses for successful or unsuccessful operations.

Numerous test cases for conformance testing of the E2 node are defined. These test cases are related to E2AP Global Procedures and E2AP Functional Procedures. For each test case, details such as test purpose, test entrance criteria, and test methodology (including initial conditions, procedure, and expected results) are specified.

- Test Cases for E2AP Global Procedures: E2 Setup, RESET, Error Indication, RIC Service Update, E2 Node Configuration Update, E2 Connection Update, and E2 Removal
- Test Cases for E2AP Functional Procedures: RIC Subscription, RIC Subscription Delete, RIC Subscription Delete Required, RIC Indication, and RIC Control

Interoperability Testing Between the Near-RT RIC and the E2 Node The test setup for interoperability testing between the Near-RT RIC and the E2 nodes can utilize a Non-RT RIC simulator, core simulator, O-RU simulator, and UE simulator. Real equipment can replace simulators in a different test setup. The Near-RT RIC and the E2 node can be configured via the O1 interface.

In one example configuration, the O-CU-CP, the O-CU-UP, and the O-DU may be separate entities. The E2 node may support RAN configurations including (i) SA O-CU-CP connected to one or more SA O-CU-UPs and one or more SA O-DUs, (ii) combined O-CU-CP and O-CU-UP connected to one or more SA O-DU, and (iii) combined O-CU-CP, O-CU-UP, and O-DU. The combined node may present a common/shared E2 interface or individual E2 interfaces.

The SUT for interoperability testing between the Near-RT RIC and the E2 node consists of the Near-RT RIC and the E2 node. The E2 node can be of any of the following type: (i) O-CU-CP, O-CU-UP, or O-DU; (ii) combined O-CU-CP, O-CU-UP, and O-DU; (iii) combined O-CU-CP and O-CU-UP; (iv) O-DU; and (v) O-eNB.

In support of interoperability testing, the Near-RT RIC and the E2 nodes implement the E2AP procedures related to target test cases. They may support one or more E2 service models such as E2SM-KPM, E2SM-NI, and E2SM-RC. In one approach, a TAP interface can be used on the E2 interface to passively capture and analyze packets to validate a given test case. In another approach, E2 logs from the SUT can be used for protocol analysis.

Several test cases for interoperability testing between the Near-RT RIC and the E2 nodes are defined. These test cases are related to E2AP Global Procedures and E2AP Functional Procedures. For each test case, details such as test purpose, test entrance criteria, and test methodology (including initial conditions, procedure, and expected results) are specified.

- Test Cases for E2AP Global Procedures: Near-RT RIC Support
- Test Cases for E2AP Functional Procedures for E2SM-KPM: RIC REPORT Service Styles and RIC Subscription Delete

6.2.1.3 O-RAN Fronthaul Conformance Test Specification

The Work Group 4 (WG4) of the O-RAN ALLIANCE has specified test methodology and test cases for conformance testing of the O-RU and the O-DU over the fronthaul (FH) for the C-plane and the U-plane (O-RAN ALLIANCE 2024l). Additionally, conformance testing for time synchronization of the O-RU and the O-DU for the S-plane includes both functional and performance tests. While functional tests validate the ability of the O-RU and the O-DU to achieve the proper synchronization management state, performance tests validate the conformance of the recovered clock for the O-DU or the O-RU.

Test configurations for the O-RU as the DUT, the O-DU as the DUT, and the shared cell are defined in O-RAN ALLIANCE (2024l).

O-RU Testing The O-RU is connected to the test equipment called test equipment O-RU (TER) during the O-RU conformance testing for the open FH. The TER connects to the FH interface as well as the radio interface. On the radio interface, the TER retrieves the downlink (DL) radio frequency (RF) signal from the DUT (i.e., the O-RU) transmitter and sends the uplink (UL) RF signal to the DUT receiver. The TER consists of (i) a CUSM-Plane Emulator (CUSM-E), (ii) a signal analyzer, and (iii) a signal generator. The RF test interface follows the Third Generation Partnership Project (3GPP) Long-Term Evolution (LTE) and New Radio (NR) specifications. The FH interface is based on the SA Ethernet.

O-DU Testing The O-DU is connected to the test equipment called test equipment O-DU (TED) during the O-DU conformance testing for the open FH. Two cases are supported for the O-DU testing. In the case of SA O-DU as the DUT, the TED wraps around the O-DU DUT via its F1 interface toward the combined Core + O-CU emulator, the front haul interface toward the O-RU + UE emulator, and the O1 or EMS interface toward the test configuration entity. In the case of combined O-CU + O-DU from the same vendor as the DUT, the TED wraps around the combined O-CU + O-DU DUT via the NG (SA) or S1 and X2 non-stand-alone (NSA) interfaces toward the core emulator, the front haul interface toward the O-RU + UE emulator, and the O1 or EMS interface toward the test configuration entity.

Shared Cell Testing Two shared cell configurations defined as shared cell modes are fronthaul multiplexer (FHM) mode and cascade mode (O-RAN ALLIANCE 2024l). The FHM mode makes use of a device called the fronthaul multiplexer (FHM), but the cascade mode does not utilize the FHM.

In support of the testing of the FHM, the FHM is connected to the test equipment called test equipment fronthaul multiplexer (TEM). The TEM surrounds the DUT (i.e., the FHM) and connects to the northbound and southbound ports of the FHM. At the northbound interface, the TEM acts as an O-DU-Emulator and connects to the DUT via the FH interface. Furthermore, at the southbound interface, the TEM acts as an O-RU + UE emulator. Multiple UEs can be attached to each of the emulated O-RUs in the TEM.

For conformance testing, test cases related to FH M-plane measurements, O-RU UC-plane measurements, S-plane conformance tests, O-DU UC-plane measurements, and shared-cell are described in detail in O-RAN ALLIANCE (2024l).

Fronthaul M-Plane Conformance Tests The O-RAN FH M-plane protocol conformance tests are described in detail in O-RAN ALLIANCE (2024l). As discussed in Chapter 2, the M-plane utilizes the NETCONF network element management protocol and YANG data modeling language. The NETCONF protocol makes use of a client and server model to exchange information and execute RPCs. An O-DU supports an NETCONF client, while an O-RU and an FHM support an NETCONF server. A given conformance test for the M-plane may be applicable to one or both of the O-RU and the O-DU.

Table 6.2 lists the test scenarios for the M-plane test cases. For each test case of a test scenario, details such as test description and applicability, test entrance criteria, and test methodology specifying initial conditions, procedure, and test requirement (expected result) are specified in O-RAN ALLIANCE (2024l). Furthermore, the applicability of the test case to the O-RU and the O-DU is also specified.

O-RU U-Plane and C-Plane Measurements This group of test scenarios include test scenarios for the O-RU as the DUT that implements the U-plane and C-plane O-RAN option 7.2x protocols. Conformance tests for four scenarios are defined: (i) conducted FDD tests for FR1 radios, (ii) nonconducted (i.e., over-the-air [OTA]) FDD tests for FR1 radios, (iii) conducted TDD tests for FR1 radios, and (iv) nonconducted (i.e., OTA) tests for TDD FR1 and FR2 radios. Both NR and LTE are included. Example tests include beamforming, compression of IQ data, and delay management. See O-RAN ALLIANCE (2024l) for details on the tests.

S-Plane Conformance Tests The O-RAN WG4 has described tests that validate the S-plane functional and performance conformance of the O-DU and the O-RU using FH interface and the M-plane

Table 6.2 Test Scenarios for the M-Plane Test Cases.

ID	Test Scenario Description	Test Numbers
1	Transport and Handshake	3.1.1.1 to 3.1.1.8
2	Manage Alarm Requests	3.1.2.1
3	M-Plane Connection Supervision	3.1.3.1 and 3.1.3.2
4	Retrieval of O-RU's information elements	3.1.4.1 and 3.1.4.2
5	Fault Management	3.1.5.1 and 3.1.5.2
6	O-RU Software Update	3.1.6.1 and 3.1.6.2
7	O-RU Software Activation	3.1.7.1 and 3.1.7.2
8	Access Control	3.1.8.1 to 3.1.8.10
9	External Input/Output Ports	3.1.9.1 and 3.1.9.2
10	O-RU Configurability	3.1.10.1 and 3.1.10.2
11	ALD Communications	3.1.11.1
12	Log Management	3.1.12.1 and 3.1.12.2
13	Connectivity Check	3.1.13.1 and 3.1.13.2
14	Energy Savings	3.1.14.1 to 3.1.14.3

interface (O-RAN ALLIANCE, 2024l). The S-plane functionality of the O-RU is determined based on status retrieved from the O-RU using the M-plane or SMO as appropriate. The S-plane performance is evaluated using OTA or conductive measurements of the O-RU synchronization signal. Similarly, the S-plane functionality of the O-DU is determined based on status retrieved from the O-DU using the SMO. The performance of the O-DU is evaluated using time/frequency error measurements on output signals such as 1PPS or FH SyncE + Precision Time Protocol (PTP) master ports in lower-layer split (LLS)-C1 and LLS-C2.

For the O-RU conformance testing, the stimulus is applied to the O-RU via an Ethernet interface by using equipment to generate an O-DU and/or fronthaul time reference. The FH interface carries the PTP and SyncE stimulus to test the O-RU. Furthermore, the FH carries the M-plane information to configure the DUT (i.e., the O-RU) and required C-plane and U-plane information to generate a test signal to measure time accuracy. Conductive or OTA performance measurements are made at the radio output. The setup for the conformance testing of the O-DU is similar to that for the O-RU.

A summary of the equipment useful for the S-plane testing is given below (O-RAN ALLIANCE 2024l).

CUSM-Plane Emulator. This emulator implements control, user, synchronization, and management planes to simulate a FH network connecting to the DUT such as the O-RU and the O-DU. It can test S-plane performance under stress with different noise profiles. The S-plane signals sent to the DUT and the measurement equipment are traceable to the reference signals received from the master time reference. The CUSM-plane emulator may consist of multiple components or may be a single component containing all the capabilities of the components. These components are time reference/S-plane, PRTC/T-GM, Global Navigation Satellite System (GNSS) emulator, PRTC emulator, SMO, M-plane, and UC-plane. The time reference/S-plane is the frequency and time reference and delivers frequency (e.g., SyncE, 10 MHz, and 1PPS), phase (e.g., 1PPS and a 10-ms frame boundary), and time (e.g., Time of Day [ToD] and PTP) signals to the Sync tester equipment and to various simulators/generators. The Primary Reference Time Clock (PRTC) /Telecom Grandmaster (T-GM) includes a PTP master clock and (potentially) a SyncE master function that propagates the time reference and allows addition of configured noise patterns. The GNSS emulator provides an RF signal that emulates a signal received by a GNSS antenna and also allows addition of configured noise patterns. The PRTC emulator provides, along with the PRTC status, the following signals to the O-DU: 1PPS per ITU G.703, 10 MHz, serial link carrying ToD (e.g., ITU-T G.8271 Annex A and NMEA). The SMO configures the CUSM-plane emulator to send configuration to the DUT relevant to the defined conformance tests. The M-plane translates the SMO configuration into the proper M-plane messages to correctly configure and report status from the DUT. The UC-plane implements the O-RAN UC-plane protocol and sends a stock data frame as UC-plane messages to the DUT.

Measurement Equipment. This equipment extracts the target synchronization information from the test signal transmitted by the DUT and measures this information to determine if the performance is within the limits. This equipment may be (i) a single piece of test equipment containing all of the required capabilities or (ii) created by connecting a signal analyzer and a performance measurement system.

For conformance testing of the O-RU, a signal extractor and a performance measurement equipment are needed. The signal extractor demodulates the time reference signal from the O-RU's DL radio interface and receives the RF signal from the DUT over the air or conductively. The performance measurement equipment performs error measurements of the timing and frequency output signals from the signal extractor.

For conformance testing of the O-DU, the PTP/SyncE/1PPS measurement equipment is used. This equipment extracts the target synchronization information from the Master PTP/SyncE Ethernet port or 1PPS output from the DUT and measures this information to determine if the performance of the DUT (i.e., the O-DU) is within the limits.

O-DU U-Plane and C-Plane Measurements These conformance test scenarios are defined for the O-DU that implements the U-plane and C-plane O-RAN option 7.2x protocols. These test scenarios utilize an SA O-DU or a combined O-CU + O-DU from a single vendor as the DUT. Main steps for the O-DU conformance testing for the FH include (i) configuration of the TED for test execution, (ii) configuration of the O-DU DUT via the O1 or the EMS interface, (iii) instructing of the UE emulator to initiate a connection toward (a) the O-DU DUT through the O-RU emulator via the open FH interface and (b) the rest of the emulated network via the F1 interface, (iv) verification of the successful connectivity of the UE emulator with the network and successful data transfer in the DL and the UL, and (v) observation and analysis of the FH data stream of the O-DU DUT. Examples of test scenarios include beamforming, compression of IQ data, and delay management.

Shared Cell Tests Conformance test scenarios for the devices supporting the FHM mode of shared cells are described in O-RAN ALLIANCE (2024l). The DUT in these tests is the FHM. The DUT is connected to TEM with O-RU capabilities. The FHM M-plane tests reuse many of the tests defined for the M-plane measurements. Example test scenarios for the FHM M-plane tests include transport and handshake in various IPv4 and IPv6 environments, performance management, software update, file management operations, and access control.

The FHM UC-plane tests are also defined for the DL and the UL. The FHM UC-plane DL test aims to ensure that the FHM can meet the requirement of the DL copy function to retrieve eCPRI messages from the north node as payload of Ethernet frames, copy them without any modifications as payload into Ethernet frames and send them toward configured south nodes. The purpose of the FHM UC-plane UL test is to ensure that the FHM can meet the requirement for the UL combine function to combine IQ data corresponding to the same radio resource element from each south node O-RU to eCPRI messages for the UL U-plane toward the north node O-DU.

6.2.1.4 O-RAN Fronthaul Interoperability Test Specification

The WG4 of the O-RAN ALLIANCE has specified test methodology and test cases for the interoperability testing (IOT) for the O-DU and the O-RU from different vendors or the same vendor, where the O-DU and the O-RU are connected using the O-RAN FH interface (O-RAN ALLIANCE 2024m). Various IOT profiles for DUTs are also defined in O-RAN ALLIANCE (2024m). An IOT profile is a high-level interoperability test configuration definition under which multiple detailed IOT profile test configurations may fit. The IoT profile allows groupings of test cases in broad categories.

The FH interoperability testing involves creation of a stimulus from the O-DU to the O-RU using an actual or emulated O-CU along with real or emulated core network, and measurement of the result at the output of the O-RU in the RF domain by an actual or emulated UE together with an RF signal/spectrum analyzer. Similarly, the FH interoperability testing in the reverse direction involves creation of a stimulus from the O-RU to the O-DU direction by an actual or emulated UE. In such case, the result is measured at the output of the O-DU by an actual or emulated O-CU with suitable real or emulated core network. Interoperability testing involves testing the FH interface in all the planes including the M-plane, S-plane, C-plane, and U-plane.

Key categories of interoperability test cases include (i) M-plane IOT tests, (ii) S-plane IOT test (both functional and performance tests), (iii) C/U-plane IOT tests, and (iv) C/U-plane delay management IOT tests. The M-plane FH tests focus on start-up (i.e., the O-RU start-up from the power-on of the O-RU to the availability of service). The S-plane FH tests focus on synchronization status detection and frequency and time error (performance). The C/U-plane FH tests focus on Radio Layer 3 C-plane establishment and initial radio U-plane data transfer, radio DL U-plane data transfer (DL throughput performance), and radio UL U-plane data transfer (UL throughput performance). The C/U-plane delay management FH tests aim to quantify minimum, intermediate, maximum, and excess latency.

For the IOT related to the FH M-plane, two architectural models are supported, a hierarchical model and a hybrid model. In the hierarchical model, the O-RU is entirely managed by a single O-DU using a NETCONF-based M-plane interface. In the hybrid model, one or multiple direct logical interfaces between the management systems and the O-RU are supported in addition to a logical interface between the O-DU and the O-RU. For the IOT involving FH synchronization, various synchronization options exist including LLS-C1, LLS-C2, LLS-C3, and LLS-C4. The system integrator identifies the test cases relevant to the specific deployment scenarios.

Interoperability tests are performed using testing tools that apply active stimulus and carry out passive monitoring and measurements of the DUTs.

Active stimulus testing tools include 5G NR O-CU or O-CU emulator, 4G LTE Master evolved Node B (MeNB) or MeNB emulator, and test UEs and/or UEs emulator and optional beamforming network.

The 5G NR O-CU or O-CU emulator may be a separate node or a combined node with O-DUs and provides Layer 2 and Layer 3 radio processing on the network side. The O-CU and O-CU emulator can be connected to a real or emulated core network. An external physical and/or logical connection between the O-CU (or the O-CU emulator) and the O-DU may be through a wireline or wireless medium.

The 4G LTE MeNB or MeNB emulator terminates the 3GPP EN-DC X2 interface with the 5G NR O-CU or the O-CU emulator. This emulator is essential when the DUTs such as the O-DU and the O-RU are configured to operate in the 5G NR NSA architecture option 3/3a/3x. This emulator may have a physical or logical connection with the test UE or UE emulator. The RF connection between the 4G LTE MeNB or MeNB emulator with the test UE or UE emulator may be conductive or OTA.

The test UEs and/or UEs emulator and optional beamforming network generate stateful device connections and traffic to validate the O-DU and the O-RU implementations of the O-RAN FH interface protocols. The test UEs require SIM cards that are pre-provisioned with subscriber profiles. The UE emulator requires a SIM profile configuration with the subscriber's profile. The UE emulator is useful in the test scenarios requiring sessions of multiple UEs, more flexibility, and configurability. The RF connection between the test UEs (or the UE emulator) and the O-RU is conductive or OTA.

6.2.1.5 O-RAN WG5 Interoperability Test Specification

The Work Group 5 (WG5) of the O-RAN ALLIANCE has specified test methodology and test cases for the interoperability testing (IOT) for the RAN nodes from different vendors connected via the X2, F1, and Xn interfaces (O-RAN ALLIANCE 2024n). The X2 interoperability testing focuses on the next-generation-evolved Node B (eNB) and EN-DC gNB (en-gNB) as the DUTs. EN-DC stands for E-UTRA- NR Dual Connectivity, where E-UTRA is Evolved Universal Terrestrial Radio Access and NR is New Radio. The F1 interoperability testing focuses on the disaggregated en-gNB and the next-generation Node B (gNB) (i.e., with the gNB-CU and gNB-DU) as the DUTs. The Xn interoperability testing focuses on the gNBs as the DUTs. The SUT for X2 interoperability testing consists of the eNB, the en-gNB, and the X2 interface. The SUT for F1 interoperability testing consists of the gNB-CU, the gNB-DU, and the F1 interface. The SUT for X2 interoperability testing consists of the gNB, the gNB, and the Xn interface.

A variety of testing tools are utilized for the interoperability testing of the X2, F1, and Xn interfaces (O-RAN ALLIANCE 2024n). The test UE and/or the UE emulator supports the test cases that require UE interactions. The core network or the core network emulator terminates the test UE and/or UE emulator's Non-Access Stratum (NAS) protocol and supports relevant core network procedures. The eNB or eNB emulator supports X2 and F1 interoperability testing. The Application Test Server supports user plane data transfer tests between the Application Test Server and Test UEs (and/or UE emulator). The protocol analyzer helps with verification of test results and troubleshooting. The Network Impairment Emulator is useful for the test cases requiring selective packet discards over the X2, F1, and Xn interface for X2, F1, and Xn interoperability testing, respectively.

The O-DU and O-RU emulator helps simplify connectivity to the en-gNB and the gNB with F1 split for X2 and Xn interoperability testing, respectively. Similarly, the O-RU emulator helps simplify connectivity to (i) the eNB and the en-gNB with open FH split for X2 interoperability testing and (ii) the gNB-DU and the gNB with open FH split for F1 and Xn interoperability testing, respectively. The test results and KPIs reporting tool and the test UE logging tool help simplify test results analysis.

The categories of X2 interoperability test cases are (i) X2 C-Plane IOT Test Cases (e.g., EN-DC X2 Setup, reset, Secondary Node Addition, and UE measurement transfer) and (ii) X2 U-Plane IOT Test Cases (e.g., Report Polling, Buffer discard indication, and Radio quality assistance information).

The categories of F1 interoperability test cases are (i) F1 C-Plane IOT Test Cases (e.g., F1 Setup, reset, Secondary Node Addition, paging, and UE context creation) and (ii) F1 U-Plane IOT Test Cases (e.g., Report Polling, Buffer discard indication, and Radio quality assistance information).

The categories of Xn interoperability test cases are (i) Xn C-Plane IOT Test Cases (e.g., Xn Setup, reset, Inter gNB Handover, and UE measurement transfer) and (ii) Xn U-Plane IOT Test Cases (e.g., Report Polling, Buffer discard indication, and Radio quality assistance information).

6.2.1.6 O-RAN Cloud Conformance Test Specification

The Work Group 6 (WG6) of the O-RAN ALLIANCE has specified test methodology and test cases for the O-Cloud conformance testing (O-RAN ALLIANCE 2024o). Since the cloud platform, the O-Cloud, may use different technologies such as virtual machines and containers, the O-Cloud needs to be tested based on the technology used. The SUT is the O-Cloud platform, which is the software part of the O-Cloud. The main architecture components of the O-Cloud platform software include synchronization, Acceleration Abstraction Layer (AAL) including AAL implementation and AAL Interface, and O2 interface support including Infrastructure Management Service (IMS) and Deployment Management Service (DMS).

The O-Cloud conformance testing includes O-Cloud Notification Application Programming Interface (API) test, O2 tests covering O2-IMS and O2-DMS, and synchronization.

O-Cloud Notification API Testing Recall that the O-Cloud provides a Notification Framework, where it acts as the event producer and exposes APIs to event consumers (e.g., applications such as the O-DU and the Cloud-native or Containerized Network Function [CNF]). The APIs enable the event consumers to subscribe to event notifications, delete subscriptions, get a list of subscriptions and detail of a subscription, and obtain notification by pull or push mode. Test cases for O-Cloud Notification APIs are described in O-RAN ALLIANCE (2024o). The SUT for the O-Cloud Notification API is the O-Cloud, which provides an O-Cloud Notification Event Producer. The test functions simulate the workloads (which are the applications running in the cloud) or other entities in the cloud that are interested in the events.

In support of the O-Cloud Notification API functional testing, several testing tools and simulators are recommended in O-RAN ALLIANCE (2024o). These include Postman, Curl, and HTTPie. Postman is an interface testing tool that is equivalent to a client. Postman can simulate various HTTP requests initiated by users, send the request data to the server, and obtain the corresponding response results to verify whether the result data in the response match the expected value. Postman acts as an event consumer and can also receive a notification from the event producer. Curl is a command-line tool for transferring data specified with URL syntax. Curl can be used as an event consumer. HTTPie is a command-line HTTP client with support for JSON, syntax highlighting, download functionality, and plugins. HTTPie utilizes a simple HTTP command and interacts with HTTP servers, RESTful APIs, and web services. HTTPie can be used as an event consumer.

Example test cases related to O-Cloud Notification API functional testing include (i) create a subscription resource, (ii) get a list of subscription resources, (iii) get detail of individual subscription resource, (iv) delete individual subscription resources, (v) event notification and notification sanity check, and (vi) event pull status notification.

O2 Tests (O2-IMS and O2-DMS) The O2 conformance testing verifies compliance of the O-Cloud to O-RAN O2 interface specifications. The SUT for O2 Test is an O-Cloud with O2 services to expose O-RAN O2 interfaces toward SMO. Two useful tools for O2 functional testing include the test tool and Robot. The test tool simulates SMO to issue O2 API request toward the O-Cloud. Robot is a generic test automation framework for acceptance testing and acceptance test-driven development. Robot is a keyword-driven testing framework and uses tabular test data syntax.

Example test cases for O2 IMS functional testing are related to subscriptions for inventory, and alarms and error messages such as security error and client error. The O2 DMS ETSI NFV Profile Test is an example test case for O2 DMS functional testing.

Synchronization Tests In O-RAN, time and frequency synchronization information can be distributed to the O-DU and the O-RU from the same source using various methods. Since the O-DU is a workload or an application of the O-Cloud, the O-Cloud needs to have the synchronization capability. The goal of the synchronization test is to validate that the O-Cloud is complaint to the O-RAN's open FH specifications. For O-Cloud synchronization testing, the O-Cloud node is the SUT, which hosts controller functions. Examples of two synchronization tests include Synchronization Test LLS-C1 and Synchronization Test LLS-C3.

6.2.1.7 O-RAN O-Cloud Interoperability Test (IOT) Specification

The WG6 of the O-RAN ALLIANCE has specified test methodology and test cases for the interoperability testing (IOT) of the O-Cloud regarding the O-Cloud Notification API, O2, and synchronization (O-RAN ALLIANCE 2024p).

Interoperability testing of the O-Cloud aims to verify the E2E functionality between the O-Cloud and peer devices connected by interfaces (e.g., O2 interface, Notification APIs, AAL interface, and synchronization interface). The SUT for O-Cloud interoperability testing consists of different components. The SUT for the O-Cloud IOT is the O-Cloud platform, the peer device, and the interfaces provided by the O-Cloud.

The categories of O-Cloud IOT test scenarios include O-Cloud Notification APIs IOT, AAL interface IOT, O2 interface IOT, and Synchronization IOT.

In the Notification APIs IOT test scenarios, one DUT is the O-Cloud, and another DUT is the O-CU/O-DU. In the AAL Interface IOT test scenarios, one DUT is AAL implementation (O-Cloud), and another DUT is the AAL Application (O-CU/O-DU). In the O2 Interface IOT test scenarios, one DUT is the O-Cloud, and another DUT is the SMO. In the synchronization IOT test scenarios, one DUT is the O-Cloud, and another DUT is the O-RU in case of the LLS-C1 architecture. Furthermore, in the synchronization IOT test scenarios, one DUT is the O-Cloud, and another DUT is the GM (which is a clock source) in case of the LLS-C3 architecture. In all these test scenarios, the SUT includes the DUTs and the interface between the DUTs.

As of late 2024, details of the O-Cloud Notification API IOT scenarios such as test purpose, test entrance criteria, initial conditions, and procedure are specified, but such test details for the AAL interface IOT scenarios, O2 interface IOT scenarios, and synchronization IOT scenarios are not specified (O-RAN ALLIANCE 2024p). See O-RAN ALLIANCE (2024a) for the latest WG6 specifications related to O-Cloud IOT scenarios.

6.2.1.8 O-RAN Stack Interoperability Test Specification

The Work Group 8 (WG8) of the O-RAN ALLIANCE has specified an extensive set[1] of (more than 75) interoperability (IOT) tests to verify stack interoperability among interfaces such as RLC-MAC, L1-L2, RRC-SDAP, SDAP-PDCP, RRC-PDCP, O1 interface, and E2 interface (O-RAN ALLIANCE 2024q).

Release 3 of the stack IOT specification focuses on functional behavior and limited performance of the O-CU and the O-DU in these scenarios: (i) Cell setup and activation, (ii) SystemInformation (MIB, SIB1, SIB2, and SIB3), (iii) RACH (with PRACH Format 0), (iv) Registration and De-Registration procedure, (v) Protocol Data Unit session establishment and modification, (vi) idle mode Paging, (vii) User Datagram Protocol (UDP) data transfer (DL/UL), (viii) O1 and E2 interface, and (ix) Beam Management. Furthermore, Release 3 of the stack IOT specification focuses on the following performance scenarios: (i) throughput under load condition in UL and DL and (ii) static configuration files are used to configure O-CU and O-DU.

In the stack IOT specification, the O-RU may be real or simulated. The O-DU or the O-CU may be simulated. The O-CU or the O-CU emulator may use a separate O-CU or a simulated O-CU based on the test setup. The O-DU or O-DU emulator is a separate O-DU or the simulated O-DU based on the test setup. The O-RU or O-RU emulator is a separate O-RU or simulated O-RU. The test UE or UE emulator can be a real UE or simulated one based on the test requirement. The Core Network or Core Network Emulator may use a real Core Network, or simulated core is used in the E2E test setup.

6.2.2 End-to-End Testing

As mentioned in Section 5.2.1 in Chapter 5, the TIFG has developed several specifications related to E2E testing, certification and badging, and OTICs (O-RAN ALLIANCE 2024a). To support testing, the TIFG has developed "O-RAN End-to-end Test Specification" that specifies how to validate the E2E system functionality, performance, and key features of the O-RAN system as a black box. The TIFG has also developed "O-RAN End-to-End System Testing Framework Specification" that describes the E2E testing framework for comprehensive E2E testing and validation of the O-RAN systems to ensure robust interoperability and operation with high performance in

1 More than 75 test cases have been identified. For each test case, details such as test purpose, initial conditions, test setup and configuration, and test procedure.

real-world deployments. The TIFG has also developed additional specifications that are related to certification and badging and OTICs.

6.2.2.1 O-RAN End-to-End Test Specification

The E2E test specification document developed by the TIFG of the O-RAN ALLIANCE specifies the methodology, procedures, configurations, and pass/fail criteria for the E2E testing of an O-RAN system (O-RAN ALLIANCE 2024i). That E2E test specification document treats the O-RAN system as a black box while enabling validation of the E2E system functionality, performance, and key features of O-RAN. The tests are suitable for both laboratory and field environments. Note that the O-RAN test specification specifies the test conditions, methodologies, and procedures but does not specify the test configuration (i.e., parameters) of the DUT/SUT. Such test configuration needs to be recorded as part of the test report to enable reproducing and/or benchmarking of the test if needed.

Figure 6.2 illustrates following network architectures supported for testing: (i) 4G LTE Option 1 with LTE-based radio network and Evolved Packet Core (EPC), (ii) 5G SA architecture Option 2 with NR-based radio network and the 5G Core (5GC), and (iii) 5G NSA architecture Option 3, Option 3a, and Option 3x. Note that these architectures are defined by the 3GPP and reused by O-RAN.

The Option 1 network architecture represents the baseline LTE-based network architecture, where the 4G LTE UE utilizes LTE-based radio interface to communicate with the LTE-based eNB. Furthermore, Option 1 architecture utilizes EPC as the core network. 4G NAS signaling is exchanged between the LTE UE and the EPC via the LTE eNB.

The Option 2 network architecture represents a 5G network architecture, where the 5G UE utilizes NR-based radio interface to communicate with the NR-based gNB. Furthermore, Option 2 architecture utilizes 5GC as the core network. 5G NAS signaling is exchanged between the 5G UE and the 5GC via the 5G NR-based gNB.

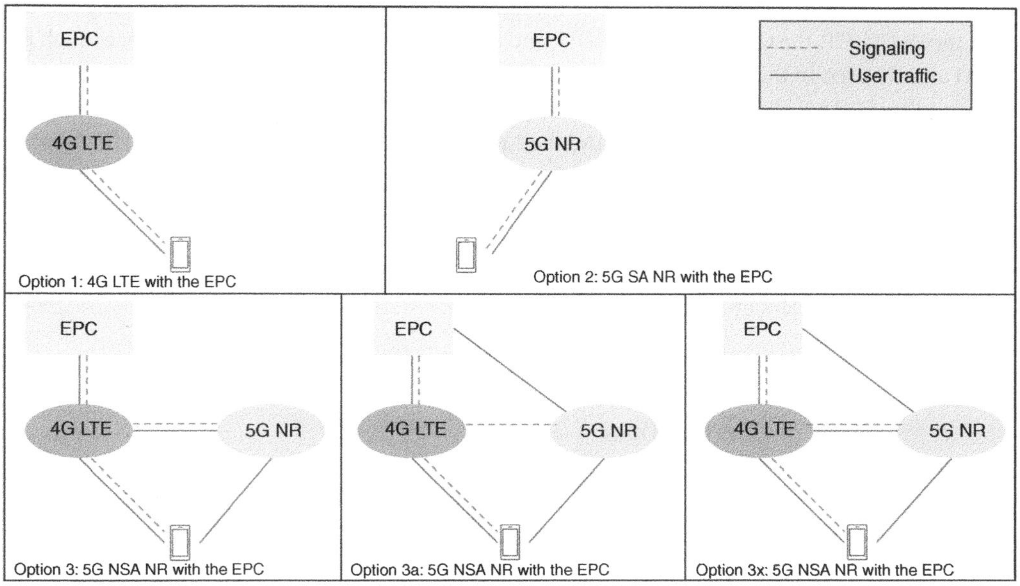

Figure 6.2 Network Architectures Supported for Testing.

The Option 3 network architecture represents a potential configuration for the 5G network architecture, where the 5G UE utilizes the LTE-based radio interface to communicate with the LTE-based eNB and NR-based radio interface to communicate with the gNB. Furthermore, Option 3 architecture utilizes EPC as the core network, and the EPC connects to the eNB for user traffic; there is no direct connectivity between the EPC and the gNB for user traffic. The user traffic between the core network and the gNB traverses the interface between the eNB and the gNB. 4G NAS signaling is exchanged between the 5G UE and the EPC via the eNB. The eNB and gNB exchange signaling messages.

The Option 3a network architecture represents another potential configuration for the 5G network architecture, where the 5G UE utilizes the LTE-based radio interface to communicate with the LTE-based eNB and NR-based radio interface to communicate with the gNB. Additionally, Option 3a architecture utilizes EPC as the core network, and the EPC connects to both the eNB and the gNB for user traffic. No user traffic regularly traverses between the eNB and the gNB. 4G NAS signaling is exchanged between the 5G UE and the EPC via the eNB. The eNB and gNB also exchange signaling messages.

The Option 3x network architecture represents yet another potential configuration for the 5G network architecture, where the 5G UE utilizes the LTE-based radio interface to communicate with the LTE-based eNB and NR-based radio interface to communicate with the gNB. Option 3x architecture also utilizes EPC as the core network, and the EPC connects to both the eNB and the gNB for user traffic. User traffic can regularly traverse between the eNB and the gNB. 4G NAS signaling is exchanged between the 5G UE and the EPC via the eNB. The eNB and gNB also exchange signaling messages. Among Option 3, Option 3a, and Option 3x, Option 3x is the most flexible network architecture option. Option 3x is widely used in commercial deployments.

Figure 6.3 shows the block diagram of the O-RAN SUT (O-RAN ALLIANCE 2024i).

Note that the O-RAN SUT can be viewed as an integrated black box from the perspective of the E2E testing. The E2E testing assumes that all involved O-RAN functions and interfaces are properly interoperating with one another. The testing of interoperability and conformance of the internal functions of the SUT is beyond the scope of the E2E testing. Hence, all the O-RAN components such as the O-CU-CP, the O-CU-UP, the O-DU, and the O-RU and all the O-RAN interfaces such as open FH and X2 used in the SUT should have been tested against their respective conformance and interoperability O-RAN specifications prior to the E2E testing. The E2E KPIs are defined across the E2E communication link between the UE and the Application/Traffic Server or another UE. As part of load and stress testing, the performance of the SUT near or exceeding full capacity is quantified. In support of such load and stress testing, additional traffic insertion points on the open FH or F1 interfaces may be utilized.

Table 6.3 summarizes test and measurement equipment and tools envisioned by the TIFG to be used in the tests (O-RAN ALLIANCE 2024i).

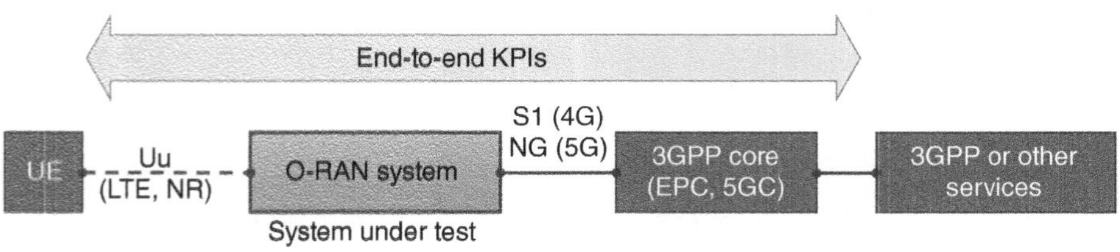

Figure 6.3 The O-RAN System as System Under Test.

Table 6.3 Test and Measurement Equipment and Tools for E2E Testing.

Equipment/Tool	Description
Real UE and/or UE emulator	A real UE or a UE emulator is used to establish stateful E2E connection and generate or receive data traffic. The UE is connected with the SUT via an RF cable or an over-the-air connection. A logging tool connected to the UE records measurements and KPI logs to facilitate test validation and reporting.
UE + O-RU emulator	This tool is used to establish stateful connections and generate or receive data traffic directly on the open fronthaul connection toward an O-DU within the O-RAN SUT. This emulator is used to create background traffic driven by multiple stateful UE protocol stacks to increase the load on the O-DU in addition to the E2E traffic applied on the Uu interface toward the SUT.
UE + O-DU emulator	This emulator is used to establish stateful connections with the O-CU and generate or receive data and signaling traffic directly on the F1 connection toward the O-CU within the O-RAN SUT. It is used to create background traffic driven by multiple UE connections to increase the load on the O-CU in addition to the E2E traffic applied at the Uu interface to the SUT.
4G/5G core or core emulator	This EPC/5GC or its emulator is used to terminate 4G/5G NAS sessions and support core network procedures required for SUT testing. It supports the E2E connection and data transfer between the real UE/UE emulator and the Application Server.
IMS core or IMS core emulator	The IMS core or its emulator supports VoLTE, ViLTE, VoNR, ViNR, and EPS Fallback. It interfaces with the EPC/5GC or a suitable emulator to setup dedicated bearers/QoS Flows to support these voice and video services.
Application/traffic server	This server is an endpoint for generation and/or termination of data traffic streams to/from real UEs/UE emulators. It generates data traffic for the services under test. It should be located as close as possible to the core/core emulator and connected to the core/core emulator via a high-capacity transport link.
Protocol analyzer	It is used for verification of test results and troubleshooting and root cause analysis of failed tests.
Network impairment emulator	This emulator is used for tests that require insertion of impairments like packet delay and/or jitter at the network interface such as open fronthaul.
RF attenuators and/or fading generator	RF attenuators are used to attenuate radio signals, while fading generators are used to simulate radio channel conditions such as urban, rural, and high-speed train.
RF shielded box/room	The RF shielded box or room is used for over-the-air (OTA) connectivity between the UE and the SUT in the lab environment.
Packet generation tool/DoS emulator	The combined packet generation tool and the denial-of-service (DoS) emulator is used for DoS traffic generation for security tests. The tool supports network traffic on network layers from 2 to 7 and conforms to network protocols including Ethernet, IP, UDP, TCP, PTP, eCPRI, TLS, and HTTP/HTTPS.
Packet capture tool	This tool captures samples of data traffic for validation, analysis, and troubleshooting. It can also be used to capture samples of legitimate traffic to create templates for fuzzing attacks. The tool captures network traffic on network layers from 2 to 7.
Network tap	This is a hardware or software device that provides access and visibility to the data flowing across a computer network.
Fuzzing tool	This tool is used for unexpected protocol input generation for security tests. The supports mutating and replaying of captured network traffic on network layers from 2 to 7.

(Continued)

Table 6.3 (Continued)

Equipment/Tool	Description
Vulnerability scanning tool	This tool is used for blind exploitation of well-known vulnerabilities during security tests. The tool utilizes cyclically updated database of known vulnerabilities based on common vulnerabilities and exposures (CVE) and supports scanning network services running on TCP/IP stack of protocols.
NFV benchmarking and resource exhaustion tool	The network function virtualization (NFV) tool is used for O-Cloud system performance measurement and resource exhaustion type of DoS attack generation. This tool may use public or private O-Cloud environment and supports testing of VNFs and CNFs.

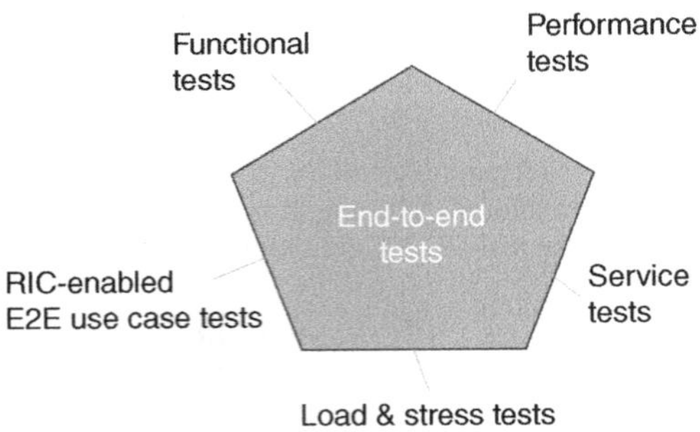

Figure 6.4 Types of End-to-End Tests.

Figure 6.4 mentions five types of E2E tests defined by the TIFG. Details of these tests can be found in O-RAN ALLIANCE (2024i).

6.2.2.2 Functional Tests

The functional tests evaluate and assess the functionality of the RAN from a network E2E perspective. This testing focuses on the end-user functionality based on 3GPP and O-RAN specifications. Pass–fail criteria, where applicable, are also defined for the functional tests. The functional tests include assessment of the following E2E functionality: (i) LTE/5G NSA Attach and Detach of a single UE, (ii) LTE/5G NSA Attach and Detach of multiple UEs, (iii) 5G SA Registration and Deregistration of a single UE, (iv) Intra-O-DU mobility, (v) Inter-O-DU mobility, (vi) Inter-O-CU mobility, (vii) 5G SA registration and deregistration for single network slices, (viii) 5G SA registration and deregistration for multiple network slices, (ix) idle mode intra-O-DU mobility, (x) idle mode inter-O-DU mobility, (xi) idle mode inter-O-CU mobility, (xii) 5G/4G inter-RAT mobility—5G to LTE handover, and (xiii) 5G/4G inter-RAT mobility—LTE to 5G handover. Details of functional tests such as test description, test setup and configuration, test procedure, and test requirements including expected results are given in O-RAN ALLIANCE (2024i) for each of the E2E functionalities mentioned here.

6.2.2.3 Performance Tests

These tests evaluate and assess the performance of the RAN from the network E2E perspective. The performance testing focuses on the end-user performance that is evaluated against the target and expected performance values. Furthermore, where feasible, the pass and fail thresholds for a given test are also defined. Examples of E2E performance test cases include (i) DL peak throughput, (ii) UL peak throughput, (iii) DL throughput in different radio conditions, (iv) UL throughput in different radio conditions, (v) bidirectional throughput in different radio conditions, (vi) DL coverage throughput (link budget), (vii) UL coverage throughput (link budget), (viii) DL aggregated cell throughput (cell capacity), (ix) UL aggregated cell throughput (cell capacity), (x) impact of FH latency on DL peak throughput, (xi) impact of FH latency on UL peak throughput, (xii) impact of midhaul latency on DL peak throughput, and (xiii) impact of midhaul latency on UL peak throughput. Calculations for the expected throughput and details of performance tests such as test description, test setup and configuration, test procedure, and test requirements including expected results are given in O-RAN ALLIANCE (2024i) for each of the E2E performance tests mentioned here.

6.2.2.4 Service Tests

These tests carry out testing of specific services. The services to be tested are classified as data services (e.g., web browsing and file upload/download), video streaming services, voice services, video calling services in the enhanced Mobile Broadband (eMBB) slice, and services supported using other slices such as ultra-reliable and low-latency communication (URLLC) and machine-type communication (mMTC). The impact of (i) scenarios such as handover and (ii) different radio conditions on the end user's service experience is also tested. Details of service tests such as test description, test setup, test methodology/procedure, and test expectation including expected results are given in O-RAN ALLIANCE (2024i). The test report includes information such as (i) throughput measured at different layers; (ii) measurements such as RSRP, SINR, and CQI; (iii) packet delay; and (iv) BLER.

6.2.2.5 Load and Stress Tests

These tests evaluate and assess the load and stress of the RAN from a network E2E perspective. These tests focus on the tolerability of the eNB/gNB under load based on 3GPP and O-RAN specifications. The eNB/gNB needs to handle various traffic patterns and loads that may occur in a commercial deployment. These tests may uncover problems that are otherwise difficult to observe and correct in the field. These tests also address special calling features such as emergency calls. Examples of load and stress tests include (O-RAN ALLIANCE 2024i) (i) simultaneous RRC_CONNECTED UEs, (ii) UE state transition rate testing, (iii) traffic load testing, (iv) traffic model testing, (v) long hours stability testing, (vi) multicell testing, (vii) emergency call, and (viii) Earthquake and Tsunami Warning System. Details of load and stress tests such as test description and applicability, test setup and configuration, test procedure, and test expectation including expected results are given in O-RAN ALLIANCE (2024i).

6.2.2.6 RIC-Enabled E2E Use Case Tests

These tests evaluate and assess the performance of a RAN with the Non-RT RIC and the Near-RT RIC optimizations from a network E2E perspective. These tests focus on (i) the end-user performance with user-specific optimizations by the Non-RT RIC and the Near-RT RIC or (ii) system-level performance with system-specific optimizations. Details of the test setup and test methodology for the RIC-enabled E2E use case tests are given in O-RAN ALLIANCE (2024i). For a given use case, system-level, cell-level, and UE-level test configurations and parameter settings and relevant KPIs to be measured are also specified. The initial RIC-enabled use case described by the TIFG in O-RAN ALLIANCE (2024i) is traffic steering with connected mode mobility control. The SA network architecture is assumed.

6.3 Open Testing and Integration Centers

OTICs provide a collaborative, open, and impartial working environment for conformance, interoperability, and/or E2E testing of O-RAN products and solutions (O-RAN ALLIANCE 2024e, 2024f). An organizer may submit an OTIC application any time. Details of the OTIC application process and qualifications are summarized in O-RAN ALLIANCE (2024f).

OTICs provide reference setups and testbeds. Furthermore, OTICs can award O-RAN certificates and/or badges. Additionally, OTICs can also host PlugFests that aim to accelerate the maturity of O-RAN products and solutions. OTICs support demos, community events such as speaker sessions, workshops, tutorials, and trials to support wide adoption of O-RAN specifications and promote the openness of the O-RAN ecosystem. Also, OTICs hold workshops or tutorials to increase O-RAN-related skills in the O-RAN ecosystem. Another important function of the OTIC is to provide feedback to the O-RAN community about the experiences with O-RAN specifications acquired during the testing to encourage and enable implementation-driven specifications.

Any O-RAN member or contributor that exhibits vendor neutrality can apply to host an OTIC. A given OTIC may be hosted by one or more hosts. All the activities of an OTIC are the full responsibility of the hosts(s). An OTIC provides selected services based on its capabilities and business objectives. As mentioned above, the O-RAN ALLIANCE has published specifications related to the criteria and guidelines for an OTIC (O-RAN ALLIANCE 2024h).

A given OTIC, as vendor-independent, open, and qualified physical space, may support one or more of the following (O-RAN ALLIANCE 2024h).

- Support adoption of O-RAN specifications and promote the openness of the O-RAN ecosystem through means such as demos, speaker sessions, workshops, tutorials, lab trials, and field trials.
- Demonstrate implementations and solutions based on O-RAN specifications via PlugFests and proof of concepts (PoCs).
- Test and verify the compliance of the RAN equipment to O-RAN interface specifications based on O-RAN conformance test specifications.
- Test and verify the interoperability of the RAN equipment from different vendors or the same vendor using O-RAN interface specifications based on O-RAN interoperability test specifications.
- Foster and develop the integrator's technical capabilities via workshops and tutorials.
- Conduct functional and performance (e.g., load and capacity) tests of both E2E systems and subsystems.
- Give the O-RAN ALLIANCE (e.g., O-RAN WGs and O-RAN contributors) feedback about the experiences of O-RAN specifications obtained during the testing activities.

The OTIC involves four distinct roles—the host, the partner, the client, and the observer (O-RAN ALLIANCE 2024h).

Host. The host is the founder and main sponsor of the OTIC. The host coordinates all the activities in OTIC. Host is an O-RAN Member or an O-RAN Contributor. A multi-host representative is designated as the main OTIC contact in case of a cohosted OTIC. The host provides (i) the space to conduct the tests and host the events, (ii) network connectivity, and (iii) test and measurement equipment, tools, and services. For example, space is needed for test rooms and server rooms. Examples of events include IOT and conformance testing, PlugFests, PoCs, and demos. The host allows the detailed results, blueprints, interface profiles, experiences, best practices, knowledge, lessons learned, and adopted processes to be shared per suitable agreements. The host regularly presents testing-related information at TIFG meetings.

Partner. This is an optional role; the host may perform all of the tasks of a partner. A partner is any company or organization with or without O-RAN ALLIANCE membership. A partner can provide, maintain, and/or operate test and measurement equipment and tools such as emulators, log tools, analyzers, and generators on behalf of the host on permanent or temporary basis per test. Furthermore, a partner can also provide and maintain space and services in the OTIC on behalf of the host. Additionally, a partner can design, set up, conduct, and evaluate the testing activities on behalf of the host.

Client. A client or customer is a company or organization with or without O-RAN ALLIANCE membership that produces O-RAN components and provides them to the OTIC for testing. The host or delegated partner of an OTIC signs a testing agreement with the client. The O-RAN ALLIANCE encourages clients to leave at least one representative product in the OTIC for a reasonable time period. The O-RAN ALLIANCE also encourages clients to leave more than one product in the OTIC and replace outdated products with newer products.

Observer. An observer or visitor is a company or organization with or without O-RAN ALLIANCE membership that is allowed to observe the testing activities in an OTIC for a limited period and/or a limited purpose. The role of an observer is expected to be more related to the O-RAN PlugFest.

According to the O-RAN ALLIANCE (O-RAN ALLIANCE 2024h), the OTIC physical lab needs to provide an open and collaborative environment while satisfying IPR protection concerns of OTIC participants. In an OTIC lab, participating companies collaborate onsite, and multiple testing activities can be hosted simultaneously. The space layout, architecture, and network setups in an OTIC lab need to be flexible and need to ensure restricted access to specific components by specific participants involved in specific testing activities. Furthermore, logical and physical separation between an OTIC lab and the rest of facilities of the host is needed to minimize any security risk.

Examples of OTICs include European OTIC in Berlin, European OTIC in Torino, European OTIC in Madrid, European OTIC in Paris, Auray OTIC and Security lab, Asia and Pacific OTIC in PRC, Kyrio O-RAN Test and Integration Lab, Japan OTIC, North American OTIC in NYC Metro Area/East (COSMOS), Asia and Pacific OTIC by ritt7layers, Asia and Pacific OTIC in Singapore, North American OTIC in the Boston Area (Northeastern University), North American OTIC in Washington, DC/Arlington, Virginia (Virginia Tech), North American OTIC in Central Iowa (ARA), North American OTIC in Salt Lake City (POWDER), North American OTIC at MITRE, Korea OTIC, North American OTIC in the Boston Area (UNH-IOL), and North American OTIC in the Raleigh-Durham Research Triangle Park Area (AERPAW).

6.4 Global PlugFests

O-RAN Global PlugFests are periodic events that are organized and cosponsored by the O-RAN ALLIANCE to enable efficient progress of the O-RAN ecosystem through well-organized testing and integration (O-RAN ALLIANCE 2024e). Vendors and providers participate in global PlugFests to test, evaluate, and verify their products and solutions.

In 2023, O-RAN PlugFests were held in Spring 2023 as well as Fall 2023. The O-RAN Global PlugFest Spring 2023 was held in 10 labs across the globe with participation from 60 companies or institutions (O-RAN ALLIANCE 2024c). *Information related to four PlugFest venues can be found at* O-RAN ALLIANCE (2024c): (i) O-RAN Global PlugFest hosted by Auray; (ii) O-RAN Global PlugFest hosted by Deutsche Telekom, EANTC, EURECOM, Orange, and Vodafone; (iii) O-RAN

Global PlugFest hosted by UNH-IOL, Kyrio, COSMOS Lab, and POWDER Lab; and (iv) O-RAN Global PlugFest hosted by Korea Telecom.

The O-RAN Global PlugFest Fall 2023 was held in 16 labs across the globe with participation from 52 companies or institutions (O-RAN ALLIANCE 2024d). *Information related to seven PlugFest venues can be found at* O-RAN ALLIANCE (2024d): (i) O-RAN Global PlugFest hosted by Japan OTIC; (ii) O-RAN Global PlugFest hosted by LG Uplus; (iii) O-RAN Global PlugFest hosted by SK Telecom; (iv) O-RAN Global PlugFest hosted by Digital Catapult; (v) O-RAN Global PlugFest hosted by EURECOM, i14y Lab supported by DT and EANTC, Orange, Telefonica, TIM, and Vodafone; (vi) O-RAN Global PlugFest hosted by CableLabs; and (vii) O-RAN Global PlugFest hosted by AT&T, DISH, COSMOS, NIST, POWDER, UNH-IOL, and Verizon and led by UNH-IOL.

A given PlugFest venue focuses on specific scope. In Fall 2023, the O-RAN PlugFest Fall 2023 hosted by Japan OTIC took place at Japan OTIC labs at Yokosuka Research Park (YRP) and labs at operator hosts (O-RAN ALLIANCE 2024d). A total of 33 companies and organizations from around the world participated in the Japan OTIC PlugFest. Examples of Japan OTIC achievements in Fall 2023 include (i) further expansion of open FH IOT, (ii) development of Non-RT RIC with multivendor rApps including energy savings, (iii) development of O-RU microsleep/O-DU scheduler optimization, (iv) confirmation of the validity of the intercell interference suppression, (v) Non-RT/Near-RT RIC A1/E2 Conformance Tests with IPv6 support, (vi) Near-RT RIC A1 Conformance Test and its automated execution, (vii) multivendor vRAN integration, (viii) robustness of open FH, and (ix) automated execution of test cases.

A total of 10 scenarios were tested by the Japan OTIC in the Fall 2023 PlugFest (O-RAN ALLIANCE 2024d). These scenarios included (i) multivendor vRAN expansion scenarios, (ii) O-RAN security, (iii) Non-RT RIC with multivendor rApps, (iv) vRAN integration including DMS deployment and vRAN Fault management, (v) verification of resource optimization for SLA assurance through RIC, (vi) Near-RT RIC conformance and test automation, (vii) RIC conformance, (viii) Non-RT RIC energy saving rApp-driven use cases, (ix) energy saving for O-RU, and (x) test automation. Note that these scenarios are reflected in the achievements mentioned above. See O-RAN ALLIANCE (2024d) for details of these scenarios.

6.5 Certification and Badging

The O-RAN certification and badging defines and unifies processes, procedures, templates, and data formats, which are provided to ensure sharing of the test results and repeatability of the executed tests (O-RAN ALLIANCE 2024f). The O-RAN certification and badging utilizes the tests defined in O-RAN test specifications produced by the TIFG. O-RAN certification and badging is performed by a qualified OTIC. The certification and badging program certifies interoperability among different vendors. O-RAN certification helps vendors promote their products and solutions.

As shown in Figure 6.5, there are three distinct concepts in O-RAN certification and badging (O-RAN ALLIANCE 2024f): O-RAN certification, O-RAN Interoperability (IOT) Badging, and O-RAN E2E Badging.

O-RAN certification aims to verify, using a set of O-RAN-defined conformance tests, that a vendor's product is compliant to O-RAN specifications. Upon successful completion of relevant tests, the O-RAN certificate is awarded to an O-RAN product for one or more O-RAN-specified or O-RAN-profiled interfaces such as O-RU open FH and O-DU O1. *O-RAN IOT badging* aims to prove interoperability of a pair of products connected via an O-RAN interface or O-RAN-profiled 3GPP interface using a set of O-RAN-defined interoperability tests. Upon successful completion of relevant tests, the O-RAN IOT badge is awarded to the pair of O-RAN products for an

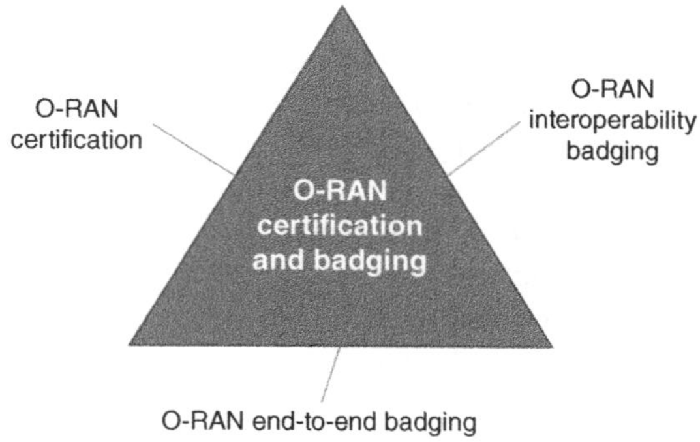

Figure 6.5 O-RAN Testing and Integration Mechanisms.

O-RAN-specified or O-RAN-profiled interface. For example, such badge is awarded to a specific vendor's O-RU and a specific vendor's O-DU for open FH. *O-RAN E2E badging* aims to demonstrate and validate, using O-RAN-defined E2E tests, that an E2E system or subsystem meets minimum requirements in terms of functionality and security and other mandatory E2E functionalities. Furthermore, O-RAN E2E badging provides an indication of E2E performance and optional E2E functionalities. The O-RAN E2E badge is awarded to all the products that constitute the E2E system or subsystem.

An O-RAN certificate or badge can be awarded to a product that represents (i) an O-RAN network function such as the Near-RT RIC, the O-CU-CP, the O-CU-UP, the O-DU, and the O-RU with O-RAN interfaces such as open FH, O1, and A1 or (ii) a 3GPP function such as the eNB, the gNB-CU, and the gNB-DU with O-RAN-profiled 3GPP interfaces such as Xn and F1.

The O-RAN ALLIANCE has defined the O-RAN certificates, IOT badges, and E2E badges for several products and interfaces (O-RAN ALLIANCE 2024f). O-RAN certificates may be obtained by a vendor for products such as (i) the O-RU with the open FH interface and (ii) the O-DU or the combined O-DU/O-CU with the open FH interface. O-RAN IOT badges may be obtained by a vendor for products such as (i) the O-RU and the O-DU (or the combined O-DU/O-CU) connected via the open FH interface, (ii) the eNB and the en-gNB connected via the X2 interface, (iii) the gNB-DU and the gNB-CU connected via the F1-C interface, and (iv) two gNBs connected via the Xn-C interface. O-RAN E2E badges may be obtained by a vendor for products such as the O-RU, the O-DU, and the O-CU or their combinations tested in the E2E system or subsystem.

O-RAN certificates or badges issued by an OTIC consist of the following (O-RAN ALLIANCE 2024f):

i) Logo of the O-RAN Certificate, the IOT Badge, or the E2E Badge
ii) Summary report containing a summary of O-RAN certification or badging testing
iii) Test report containing details of O-RAN certification or badging testing
iv) Detailed measurement results collected during certification or badging such as logs, screenshots, and packet captures

6.6 Major Takeaways of the Chapter

While openness in O-RAN provides significant flexibility to a service provider to select vendors to meet its business objectives, integration of different O-RAN components (especially from different vendors) is a nontrivial matter. Testing of integrated products is especially important to ensure proper operations of O-RAN in commercial deployments. Here are the major takeaways of this chapter.

- The O-RAN ALLIANCE supports testing and integration of O-RAN solutions through mechanisms such as test specifications, OTICs, global PlugFests, and certification and badging.
- O-RAN has defined comprehensive test specifications to facilitate conformance testing and interoperability testing of O-RAN components.
- OTICs provide an open, collaborative, vendor independent, and impartial working environment to support the O-RAN ecosystem.
- Global PlugFests are periodic events that facilitate progress of the O-RAN ecosystem through organized testing and integration, where companies participate to test, evaluate, and verify their products and solutions.
- Certification and badging involve O-RAN certification, O-RAN interoperability (IOT) badging, and O-RAN E2E badging.
- O-RAN certification aims to verify, using a set of O-RAN-defined conformance tests, that a vendor's product is compliant to O-RAN specifications.
- O-RAN IOT badging aims to prove interoperability of a pair of products connected via an O-RAN interface or O-RAN-profiled 3GPP interface using a set of O-RAN-defined interoperability tests.
- O-RAN E2E badging aims to demonstrate and validate, using O-RAN-defined E2E tests, that an E2E system or subsystem meets minimum requirements in terms of functionality and security and other mandatory E2E functionalities.

References

O-RAN ALLIANCE (2024a). *O-RAN Specifications*. https://specifications.o-ran.org (accessed 22 July 2024).

O-RAN ALLIANCE (2024b). *O-RAN Global PlugFest 2023*. https://plugfestvirtualshowcase.o-ran.org/2023/SPRING (accessed 22 July 2024).

O-RAN ALLIANCE (2024c). *O-RAN Global PlugFest 2023—Spring*. https://plugfestvirtualshowcase.o-ran.org/2023/SPRING (accessed 22 July 2024).

O-RAN ALLIANCE (2024d). *O-RAN Global PlugFest 2023—Fall*. https://plugfestvirtualshowcase.o-ran.org/2023/FALL (accessed 22 July 2024).

O-RAN ALLIANCE (2024e). *Testing & Integration*. https://www.o-ran.org/testing-integration (accessed 22 July 2024).

O-RAN ALLIANCE (2024f). Overview of Open Testing and Integration Centre (OTIC) and O-RAN Certification and Badging Program. White Paper, April 2023, https://assets-global.website-files.com/60b4ffd4ca081979751b5ed2/64561954d1802b0001099566_Overview%20of%20OTIC%20and%20O-RAN%20Certification%20and%20Badging%20Program_white%20paper_2023-04.pdf (accessed 22 July 2024).

O-RAN ALLIANCE (2024g). *Certification and Badging Processes and Procedures*. TIFG.Cert-Badge.0-R003-v08.00. https://specifications.o-ran.org (accessed 22 July 2024).

O-RAN ALLIANCE (2024h). *Criteria and Guidelines of Open Testing and Integration Centre*. TIFG.CGofOTIC.0-v05.00. https://specifications.o-ran.org (accessed 22 July 2024).

O-RAN ALLIANCE (2024i). *End-to-end Test Specification*. TIFG.E2E-Test.0-R003-v05.00. https://specifications.o-ran.org (accessed 22 July 2024).

O-RAN ALLIANCE (2024j). *A1 interface: Test Specification*. O-RAN.WG2.A1TS.R003-v04.00. https://specifications.o-ran.org (accessed 22 July 2024).

O-RAN ALLIANCE (2024k). *E2 Interface Test Specification*. O-RAN.WG3.E2TS-R003-v02.00. https://specifications.o-ran.org (accessed 22 July 2024).

O-RAN ALLIANCE (2024l). *Conformance Test Specification*. O-RAN.WG4.CONF.0-R003-v09.00. https://specifications.o-ran.org (accessed 22 July 2024).

O-RAN ALLIANCE (2024m). *Fronthaul Interoperability Test Specification (IOT)*. O-RAN.WG4.IOT.0-R003-v10.00. https://specifications.o-ran.org (accessed 22 July 2024).

O-RAN ALLIANCE (2024n). *Interoperability Test Specification (IOT)*. O-RAN.WG5.IOT.0-v09.00. https://specifications.o-ran.org (accessed 22 July 2024).

O-RAN ALLIANCE (2024o). *O-Cloud Conformance Test Specification*. O-RAN.WG6.O-CLOUD-CONF-Test-R003-v02.00. https://specifications.o-ran.org (accessed 22 July 2024).

O-RAN ALLIANCE (2024p). *O-Cloud Interoperability Test (IOT) Specification*. O-RAN.WG6.O-CLOUD-IOT-R003-v01.00. https://specifications.o-ran.org (accessed 22 July 2024).

O-RAN ALLIANCE (2024q). *Stack Interoperability Test Specification*. ORAN-WG8.IOT.0-R003-v09.00. https://specifications.o-ran.org (accessed 22 July 2024).

7

O-RAN Research and Development Initiatives

In the last few years, Open Radio Access Network (O-RAN) has received significant attention from government, industry, and academia. In this chapter, we outline a few key O-RAN Research and Development Initiatives led by industry, academia, and governments in the United States and across the world to propel the development and adoption of O-RAN for 5G and upcoming 6G networks.

7.1 Industry-Led Initiatives

There are several notable industry-led initiatives in O-RAN research and development. Figure 7.1 provides examples of such initiatives.

7.1.1 O-RAN ALLIANCE

As discussed in Chapter 1, the O-RAN ALLIANCE (O-RAN ALLIANCE 2024) is a worldwide community of mobile operators, vendors, and research and academic institutions with the mission to reshape radio access networks (RANs) to be more intelligent, open, virtualized, and fully interoperable. The O-RAN ALLIANCE was founded in 2018 by five mobile network operators (MNOs)—AT&T, China Mobile, Deutsche Telekom, NTT DOCOMO, and Orange. Specifically, the O-RAN ALLIANCE makes a number of efforts within the scope of O-RAN, such as O-RAN specifications, O-RAN software community (OSC) code, O-RAN demonstrations, and O-RAN testing and integration centers.

7.1.1.1 O-RAN Specifications

O-RAN ALLIANCE works on the standard specifications for individual parts of the O-RAN architecture, including O-RAN use cases, O-RAN non-real-time RAN intelligent controller (RIC) and A1 Interface, near-real-time RIC and E2 interface, open fronthaul interfaces, Open F1/W1/E1/X2/Xn interface, Cloudification and Orchestration, White-box Hardware, Stack Reference Design, Open X-haul Transport, OAM, O-RAN Test and Integration, and O-RAN Security.

7.1.1.2 O-RAN Software

OSC is a collaboration between the O-RAN ALLIANCE and Linux Foundation to support the creation of open software for O-RAN. It aims to accelerate the adoption of O-RAN technologies by providing reference software implementations, tools, and resources for developers. Specifically, the OSC engages in collaborative software development, addressing aspects such as open interfaces,

Fundamentals of O-RAN, First Edition. Nishith D. Tripathi and Vijay K. Shah.

Companion website: www.wiley.com/go/tripathioran

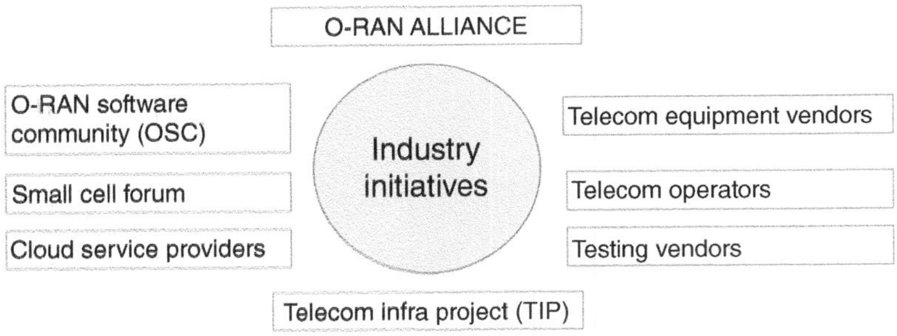

Figure 7.1 Prominent Industry-Led Initiatives.

orchestration, and management of O-RAN components. It encourages contributions from various organizations to build a robust open-source O-RAN software stack. See OSC Software (2024) to learn and contribute to the OSC software.

7.1.1.3 O-RAN Demonstrations

O-RAN ALLIANCE enables various O-RAN Virtual Exhibitions through events such as Mobile World Congress (MWC) that allow companies to demonstrate their O-RAN-based implementations. More information can be found in O-RAN Virtual Exhibition (2024).

7.1.1.4 Testing and Integration of O-RAN Implementations

The O-RAN ALLIANCE assists its members and contributors in testing and integrating their O-RAN implementations by providing a collaborative, open, and neutral testing environment via Open Testing and Integration Centers (OTICs). You can find the updated list OTIC centers throughout the world in O-RAN ALLIANCE OTIC (2024). See Chapter 6 for more details on OTICs and overall O-RAN testing and integration.

OTIC centers carry out a number of activities to advance O-RAN ecosystem, including the following:

- O-RAN Global PlugFests: These periodic events, organized and co-sponsored by the O-RAN Alliance, enable collaborative testing and integration, allowing vendors and providers to evaluate and validate their products and solutions effectively.
- O-RAN Certification and Badging: By awarding O-RAN certifications, these centers help vendors promote their O-RAN-based offerings, ensuring interoperability and fostering acceptance within the vendor community. For more information, refer to the O-RAN White Paper (2023).
- Comprehensive Testing: OTIC centers conduct conformance, interoperability, and end-to-end testing of O-RAN products and solutions to ensure they meet industry standards.
- Reference Setups and Testbeds: These facilities provide essential resources for developers to test and validate their implementations.
- Community Demos and Events: OTIC centers organize demonstrations and community trials to encourage widespread adoption of O-RAN specifications and highlight the ecosystem's openness.
- Workshops and Tutorials: These educational programs aim to enhance the technical skills of the O-RAN community, promoting growth and expertise.

7.1.2 Telecom Infra Project (TIP)

Founded in 2016, TIP is a global community of companies and organizations working on open and disaggregated solutions for telecom infrastructure. While not O-RAN-specific, TIP collaborates on projects related to O-RAN, aiming to accelerate innovation and address challenges in the telecom space. TIP hosts various project groups, including those focused on O-RAN solutions. These groups involve telecom operators, vendors, and other contributors working together to develop O-RAN and interoperable RAN technologies.

We give a concise overview of the TIP's OpenRAN™ **Project Group**, an initiative focused on creating vendor-neutral hardware, open interfaces, and software for RAN solutions spanning 2G, 3G, 4G, and 5G technologies. More details about TIP's OpenRAN™ project can be found in Telecom Infra Project (2024).

TIP OpenRAN™ aims to forge strong partnerships with organizations that share a vision for accelerating disaggregation and innovation in RAN technologies. The initiative seeks to unite the ecosystem for a comprehensive approach to developing next-generation RAN solutions. As illustrated in Figure 7.2. TIP OpenRAN™ will continue collaborating with various industry organizations, such as Global System for Mobile Communications Association (GSMA), Open Networking Foundation (ONF), and O-RAN ALLIANCE, through established liaison agreements. For instance, TIP and the O-RAN ALLIANCE have a Liaison Agreement that facilitates information sharing, referencing, and validation activities. According to this agreement, TIP will create specific subgroups dedicated to these activities. Participants in any TIP subgroup that discusses or references O-RAN Draft Specifications must also be members of O-RAN. Additionally, those involved in subgroups discussing O-RAN Final Specifications available on the O-RAN website must be O-RAN Adopters.

TIP's OpenRAN™ project has two major subgroups. See Figure 7.3.

1) **Component subgroups**
2) **Segment subgroups**

These subgroups are briefly discussed in Sections 7.1.2.1 and 7.1.2.2.

7.1.2.1 Component Subgroup

The component subgroup is divided into four groups—(i) radio unit (RU) subgroup, (ii) distributed unit (DU) and central unit (CU) subgroup, (iii) RAN Intelligence and Automation (RIA) subgroup, and (iv) RAN Orchestration and Management Automation (ROMA) subgroup.

- **RU Subgroup.** The RU subgroup aims to develop RU hardware based on an open and disaggregated architecture. This approach is expected to significantly reduce deployment and operational costs compared to traditional proprietary base stations in mobile networks. Collaborating with

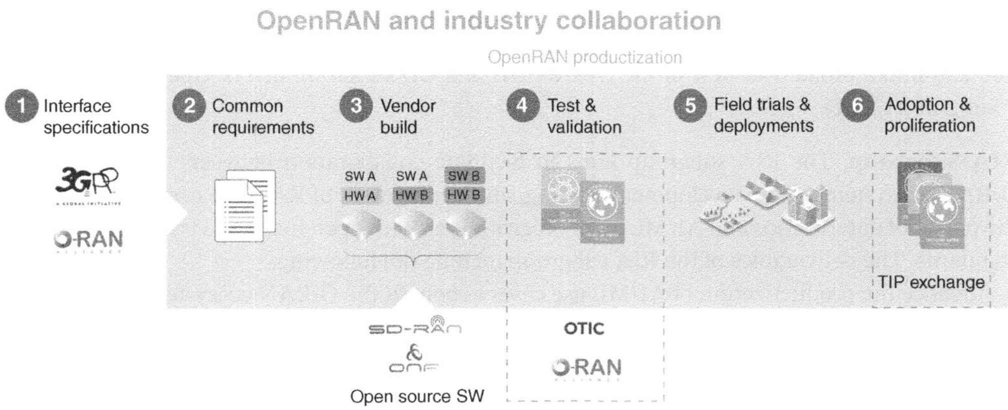

Figure 7.2 TIP O-RAN Liaisons (Telecom Infra Project 2024).

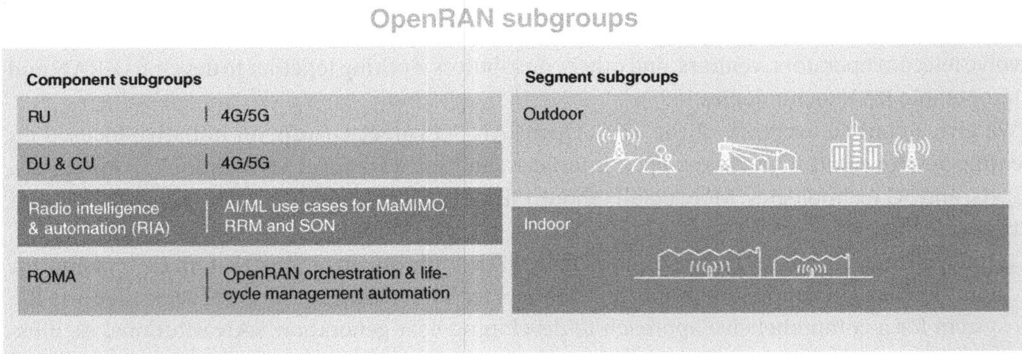

Figure 7.3 Open RAN Subgroups—(1) Component Subgroups and (2) Segment Subgroups (Telecom Infra Project 2024).

leading global operators and the O-RAN technology ecosystem, the group will align the requirements for radio units. The deliverables of this RU subgroup include:
- Various RU options suited for different deployment scenarios, such as single-band RU, multi-band RU, and massive Massive Input Massive Output (mMIMO) RU.
- Specifications for platform architecture and hardware/software requirements.
- Reference architecture and design specifications for the RU, focused on defined deployment use cases, including requirements and architectural design details.

You can find a broad overview of the O-RAN RU subgroup in TIP OpenRAN™ RU Subgroup (2021).

- **DU and CU Subgroup.** The DU and CU subgroup aims to design DU/CU hardware using an open and disaggregated architecture. This strategy seeks to substantially lower deployment and operational costs compared to traditional proprietary base stations in mobile networks. In collaboration with leading global operators such as T-Mobile USA, Vodafone, and Dish, the group will produce requirement documents for DU/CU. The deliverables will encompass:
  - Various DU/CU options tailored for different deployment scenarios, including split DU/CU and integrated DU/CU
  - Specifications for platform architecture and hardware/software requirements
  - Reference architecture and design criteria for both DU and CU

You can find a broad overview of the O-RAN DU and CU subgroup in TIP OpenRAN™ DU & CU Subgroup (2021).

- **RIA Subgroup.** The RIA subgroup aims to facilitate collaboration between MNOs and the O-RAN ecosystem in the development, testing, and deployment of RAN use cases that harness the power of data science and AI/ML technologies, along with open interfaces based on industry standards. The deliverables of the RIA subgroup include the following:
  - Advance the productization of AI/ML use cases to benefit the O-RAN ecosystem
  - Enable the O-RAN ecosystem to utilize AI/ML and data science technologies effectively
  - Launch testing and trials with MNOs and sharing the results with the O-RAN community

You can find a broad overview of O-RAN RIA subgroup in TIP OpenRAN™ RAN Intelligence and Automation (2021).

- **ROMA Subgroup.** ROMA is focused on gathering and standardizing MNO requirements for O-RAN orchestration and lifecycle management automation, while encouraging ecosystem partners to develop products and solutions that align with these requirements. The objectives of ROMA include:
  - Establish a unified set of use cases for O-RAN lifecycle management automation and orchestration that are agreed upon by multiple MNOs and members of the O-RAN ecosystem.
  - Create technical requirements for products and solutions that support the identified use cases, including necessary interfaces and data models.
  - Support product and solution development through lab testing, field trials, participation in TIP PlugFest, and certification on TIP exchange.
  - Facilitate large-scale O-RAN deployment with lifecycle management automation, including frameworks and tools for Continuous Integration (CI) and Continuous Deployment (CD).

7.1.2.2 Segment Subgroup

The segment subgroup includes (i) O-RAN Outdoor Macro subgroup and (ii) Indoor 5G NR Small Cell subgroup.

- O-**RAN Outdoor Macro Subgroup.** The Outdoor Macro subgroup aims to tackle the challenges associated with large-scale outdoor O-RAN deployments by establishing requirements, consolidating technology solutions, and creating playbooks tailored for macro-O-RAN configurations. The group will produce requirement documents for various outdoor macro segments, including rural, periurban (the transition zone between rural and urban areas), and urban settings. Key deliverables will include:
  - Reference architectures for RAN, encompassing components such as CU, DU, and RU.
  - Application Programming Interfaces (APIs) and interfaces designed for interoperability among multiple vendors, adhering to industry specifications.
  - Playbooks and test reports to guide implementation and validation.

You can find a broad overview of O-RAN Outdoor Macro subgroup in TIP OpenRAN™ Outdoor Macro (2021).

- **Indoor 5G NR Small Cell Subgroup.** The Indoor 5G NR Small Cell subgroup aims to tackle the challenges associated with large-scale indoor 5G NR small cell deployments by defining requirements for disaggregated indoor 5G NR small cells. The group will work alongside industry partners to expedite solution development and meet the requirements for 5G-scaled deployments. Key deliverables will include:
  - Reference architectures for 5G NR gNB, encompassing components such as CU, DU, and RU.
  - APIs and interfaces designed for interoperability among multiple vendors, in accordance with industry standards.

You can find a broad overview of the Indoor Small Cell subgroup here TIP OpenRAN™ Indoor Small Cell (2021).

7.1.3 Open Networking Foundation (ONF)

The Open Networking Foundation (ONF) is an operator-led consortium at the forefront of transformative network innovation. Established in 2011, ONF has become a recognized leader in open-source solutions for operators, initially focusing on Software-Defined Networking (SDN). With the backing of major partners like AT&T, China Unicom, Comcast, Deutsche Telekom,

Google, NTT Group, and Turk Telekom, the ONF is spearheading substantial advancements within the operator sector. For additional information, do check out ONF (2024).

Software-defined Radio Access Network (SD-RAN) is ONF's new exemplar platform for Third-Generation Partnership Project (3GPP) compliant SD-RAN, aligned with the O-RAN architecture. This cloud-native platform is built upon ONF's established, operator-approved systems, such as the Open Networking Operating System (ONOS) and Aether. Beginning with an ONOS-based near-real-time RAN intelligent controller (near-RT RIC), the platform aims to create open-source components for the control and user planes of the CU (specifically, the O-RAN central unit-control plane [O-CU-CP] and O-RAN central unit-user plane [O-CU-UP]) and the O-RAN distributed unit (O-DU) within the disaggregated RAN, in close collaboration with the O-RAN ALLIANCE and the O-RAN Software Community.

As illustrated in Figure 7.4, ONF has begun the developing an Exemplar Platform aligned with the O-RAN architecture, incorporating specific implementation choices ONF (2024).

- The solution will feature open-source implementations of the O-DU, O-CU-UP, and O-CU-CP.
- The O-CU-UP will be developed using P4.
- An open-source near-RT RIC implementation will be based on ONF's ONOS.
- The platform is expected to enhance the E2 interface to support scheduler control and network slicing, contributing these advancements back to O-RAN for integration into the specifications.
- It will be designed to interoperate with third-party RUs.
- The solution will utilize commercial off-the-shelf (COTS) and white-box P4-programmable switches.
- Aether5 will be utilized as the virtualization layer, Virtual Infrastructure Manager (VIM), and infrastructure management framework.

The architecture of the ONF SD-RAN Exemplar platform is depicted in Figure 7.5.

ONF is developing individual components of the exemplar platform in close coordination with the O-RAN Software Community. The different components are as follows:

- **Aether.** Aether is recognized as the leading open-source private 5G platform. It offers a cloud-optimized solution that runs on Kubernetes as native microservices, allowing for scalable deployment. Designed specifically for the demands of enterprise Industry 4.0 applications, Aether provides a comprehensive end-to-end solution that can be deployed with various RAN options, including O-RAN. Its high scalability allows users to begin with small deployments and expand into large distributed private networks that support both 4G and 5G connectivity. A pilot Aether network is operational at several sites today. More information on Aether can be found in ONF Aether (2024).

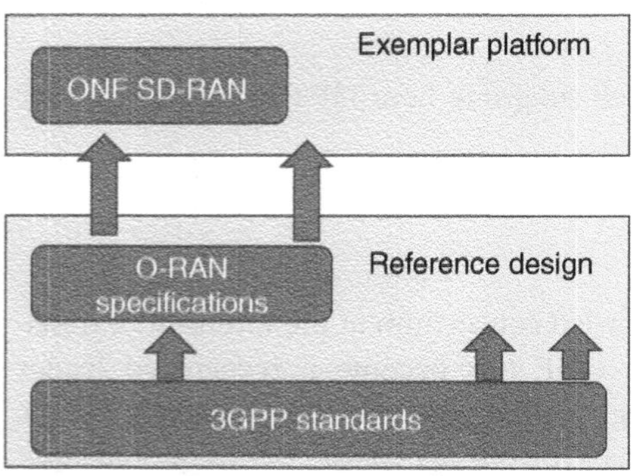

Figure 7.4 ONF's SD-RAN Platform—An Exemplar Platform for O-RAN Based on Specific Design Choices.

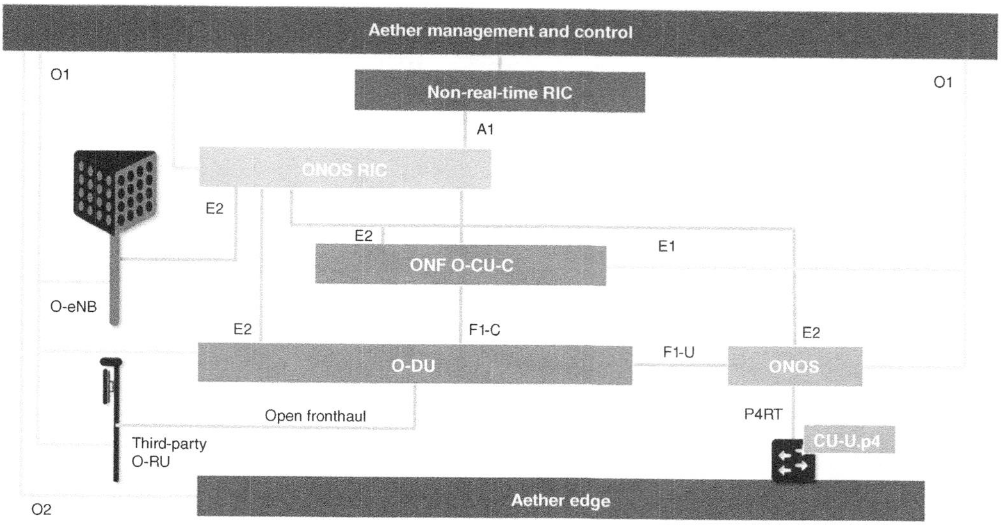

Figure 7.5 ONF's SD-RAN Exemplar Platform (ONF Aether 2024).

- **ONOS RIC.** The next-generation µONOS RIC will function as a logically centralized SD-RAN controller, utilizing a microservices architecture that comprises the following components:
  - *Certificate Manager*: Handles certificate authority (CA) functions and manages certificates to ensure secure interactions within the network environment.
  - *Topology Manager*: Keeps track of the inventory of network infrastructure devices and their connections, providing a comprehensive view of the network for both the platform and its applications.
  - *Configuration Manager*: Manages the issuance, tracking, rollback, and validation of atomic configuration operations using the gRPC Network Management Interface (gNMI) and gRPC Network Operations Interface (gNOI) across multiple network infrastructure devices to ensure consistent network performance.
  - *RAN Control Manager*: Configures mobile network infrastructure devices (including O-RU, O-DU, O-CU, and O-RAN-eNB) and manages radio resource management (RRM) and network optimization using O-RAN interfaces, all within specified latency constraints.
  - *Distributed Store*: Offers services to core systems and applications for data storage and access using various strategies, including Raft consensus and eventual consistency. The system will also provide client-side libraries for cache management and efficient data structures to enhance usability.

The high-level µONOS RIC architecture is illustrated in Figure 7.6 (ONF SD-RAN 2020).

The µONOS RIC supports multicluster operations, ensuring high availability and scalability. Its southbound interface is based on the O-RAN-specified E2 interface. ONOS features a distributed data store that maintains the Radio Network Information Base (R-NIB) and includes a messaging infrastructure, topology management, as well as control and configuration managers within a microservices framework. Furthermore, it is designed to facilitate conflict resolution to manage issues stemming from multiple RAN applications.

The µONOS RIC features open APIs to support third-party RAN applications, known as xApps, within the O-RAN architecture. These APIs allow xApps to connect with E2 Nodes (including O-DU, O-CU-C, O-CU-U, and all-in-one gNB) to the south, as well as with operators' OSS/BSS

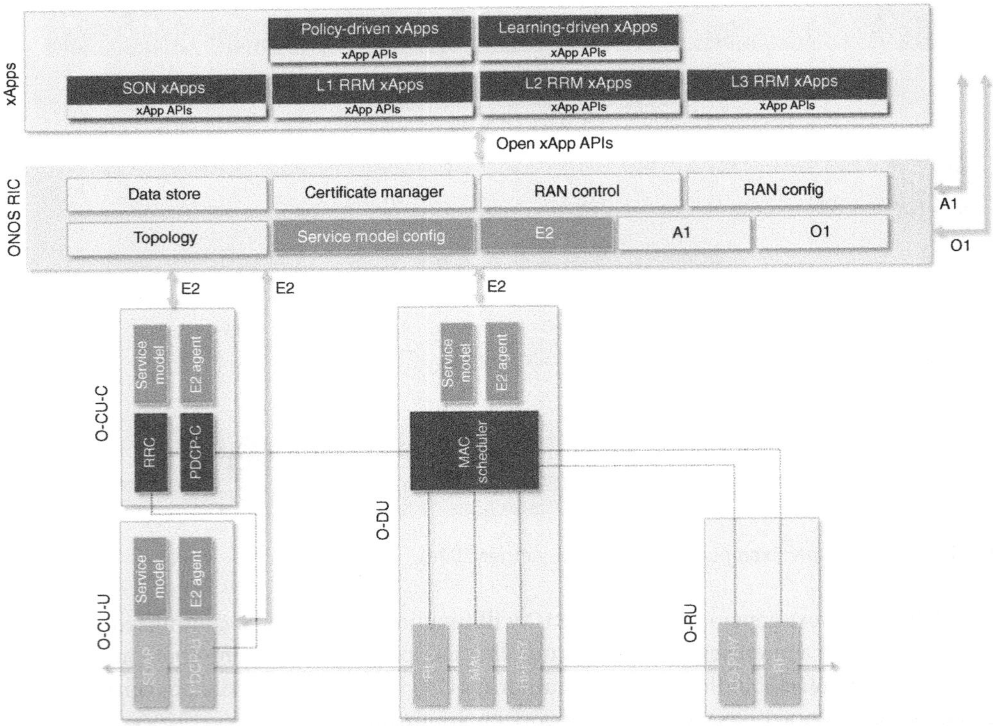

Figure 7.6 High-level µONOS RIC Architecture (ONF SD-RAN 2020).

platforms and the non-RT RIC to the north, in addition to interfacing with the distributed data store within the µONOS RIC. The range of applications will encompass basic RRM functions, Self-Organizing Network (SON) capabilities, machine learning-driven network optimization, and policy-based control and configuration, as illustrated in Figure 7.10.

Here are the key services of the µONOS RIC platform along with a summary of their functions.

- **E2 Termination**—Manages external ASN.1/SCTP connections to E2 nodes and acts as a relay, enabling applications to send and receive E2SM messages.
- **A1 Termination**—Processes external JSON/HTTP REST API requests from northbound orchestration systems and non-real-time applications
- **O1 Termination**—Handles external XML (YANG)/Network Configuration Protocol (NETCONF) requests, along with various operations-related tasks.
- **E2 Subscription Management**—Manages subscription requests for E2SM messages and assists in routing these messages between E2 termination nodes and applications.
- **R-NIB Service**—Monitors and disseminates information about E2 nodes and their interrelationships within the RAN environment.
- User Equipment-Network Information Base **(UE-NIB) Service**—Tracks and disseminates information regarding end nodes (UEs).
- **Storage Facilities**—Offers a distributed, scalable, and highly available system for applications to maintain their operational state.

Each of the services outlined above is (or will be) developed as an individual microservice, allowing for the deployment of multiple instances to achieve scalability and high availability.

The implementation of all services will be in Golang to align with other μONOS services and to take advantage of various μONOS patterns and libraries. A small exception might apply to certain elements of the E2 termination, which may incorporate specific C libraries for transcoding ASN.1 and protobuf messages.

Each of the "termination" services offers an external interface for interacting with the outside environment, utilizing the protocols and encodings defined by O-RAN standards. Additionally, they will feature an internal gRPC interface for use by RIC applications.

As of this writing, SD-RAN solution is focused on several key areas, as outlined in the following (SD-RAN Documentation 2024):

- Developing O-RAN compliant interfaces such as E2, O1, and A1, along with protocols including ASN.1, Stream Control Transmission Protocol (SCTP), and NETCONF.
- Implementing a clustered micro-ONOS architecture designed for high availability and performance.
- Integrating third-party vendor xApps.
- Creating an SDK that enables the portability of xApps across different RIC implementations.
- Facilitating integration with external CU and DU vendors.
- Developing ONF's own O-RAN compliant RU, CU, and DU to serve as a reference model
- Developing ONF's own O-RAN compliant RAN simulator to support scalable testing.
- Creating a complete end-to-end sdRan-in-a-Box (RiaB) solution for development, testing, and reference applications.
- Sharing insights and learnings related to new Service Models (SMs) and the app-sdk with O-RAN and the O-RAN Software Community (OSC).
- Committing to regular quarterly releases, complete with regression test suites on physical test pods.
- Collaborating with Deutsche Telekom at the O-RAN Open Testing and Integration Center (OTIC) in Berlin to ensure quality assurance and interoperability.
- Preparing for lab and field trials in partnership with operator collaborators.

For detailed information, refer to ONF SD-RAN (2020) and SD-RAN Documentation (2024).

7.1.4 Cloud Service Providers

Companies like Amazon Web Services (AWS), Microsoft Azure, and Google Cloud have shown interest in O-RAN by providing cloud-based solutions that complement O-RAN architectures. These providers offer cloud infrastructure and services that can support O-RAN deployments. By leveraging cloud resources, telecom operators can enhance the scalability, flexibility, and efficiency of their O-RAN implementations.

For instance, an AWS whitepaper describes how the entire O-RAN architecture, including, SMO, non-RT RIC, near-RT RIC, CU, and DU can be supported by various AWS services (AWS O-RAN evolution 2024). Figure 7.7 demonstrates how an O-RAN reference architecture can be hosted on AWS.

7.1.5 Major Telecom Vendors

Major telecom vendors such as Samsung, Nokia, and Ericsson have been actively involved in O-RAN initiatives, contributing to the development of open interfaces and solutions that adhere to O-RAN principles. These vendors actively participate in standardization efforts, contribute

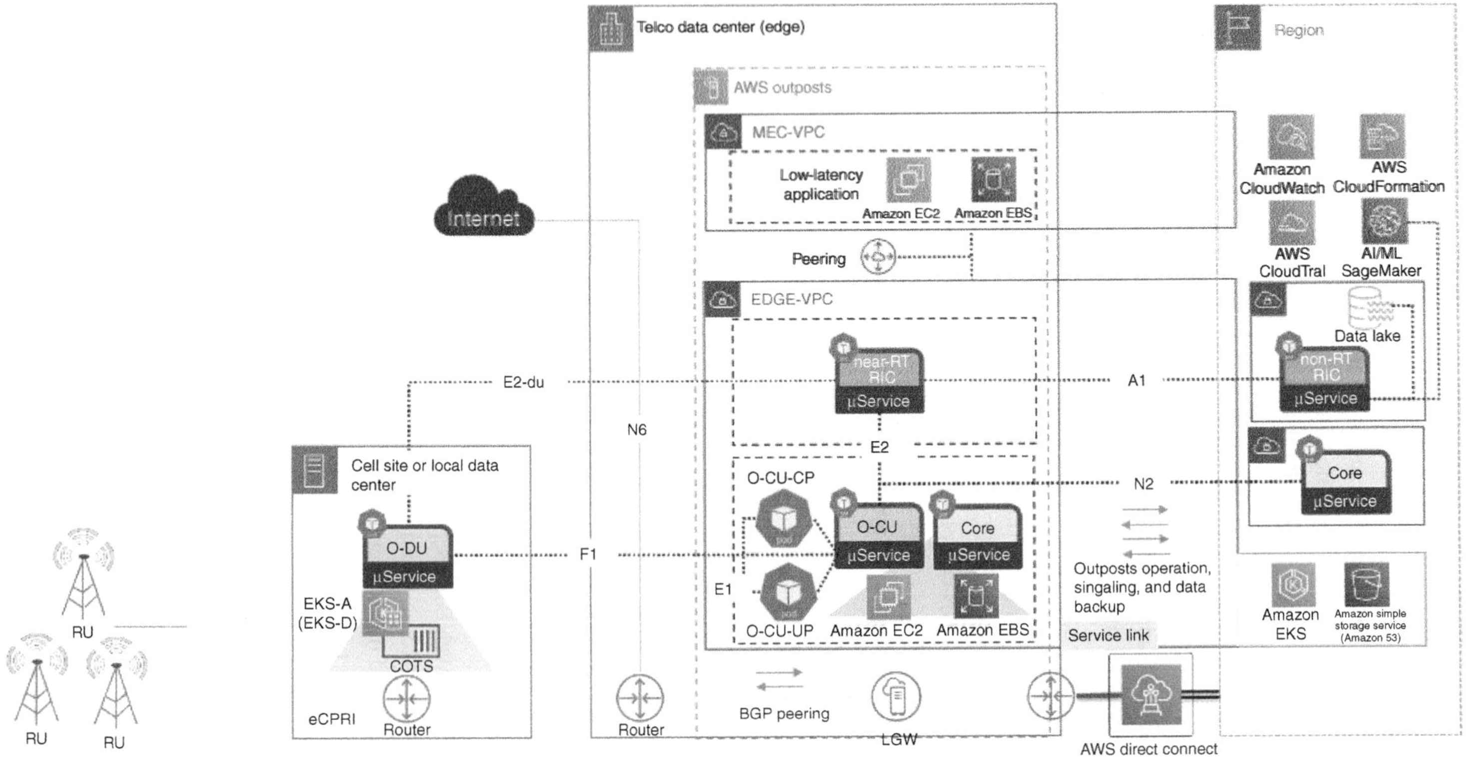

Figure 7.7 O-RAN Reference Architecture on AWS (AWS O-RAN evolution 2024).

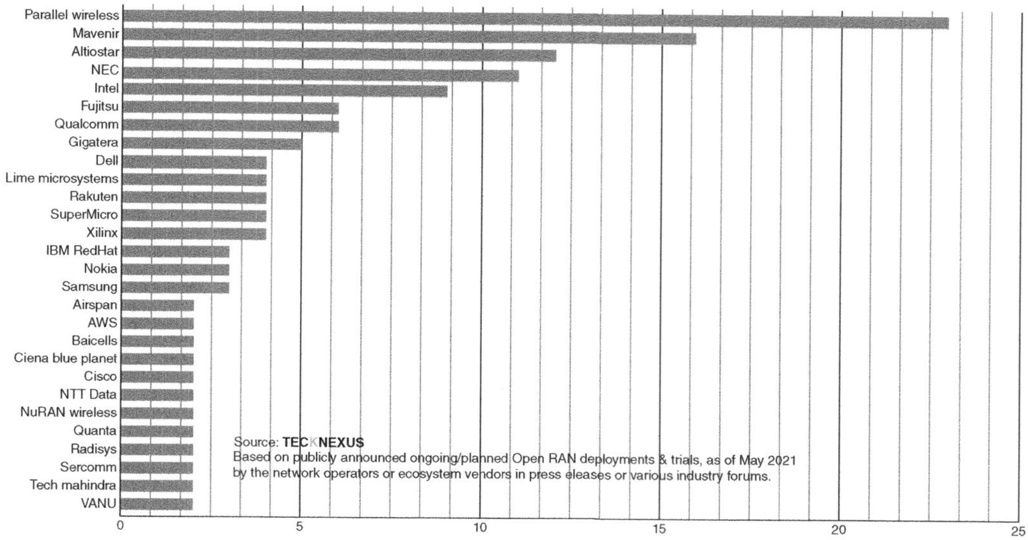

Figure 7.8 Top 28 O-RAN Telecom Vendors Across Categories (TeckNexus 2021).

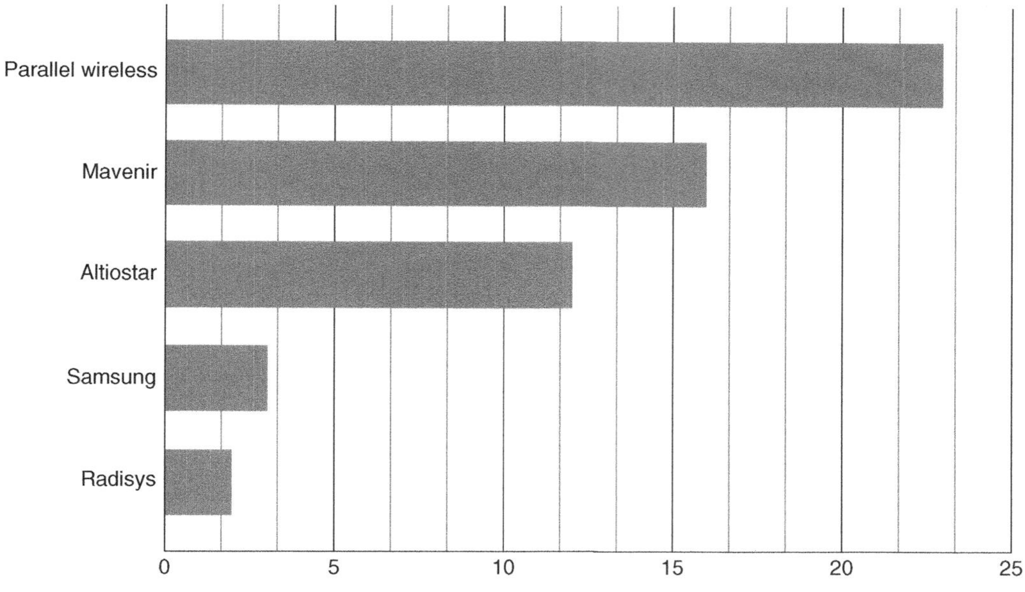

Figure 7.9 Top 5 Virtual RAN Vendors for Open RAN (TeckNexus 2021).

to O-RAN specifications, and provide hardware and software components that adhere to O-RAN architectures. Their involvement helps drive the adoption of O-RAN in the industry. In Figures 7.8–7.12, we present top 28 O-RAN vendors across categories, followed by top 5 Virtual RAN vendors, top 18 radio unit vendors, top 6 RAN Chip vendors, and top 7 system integrators for O-RAN, based on publicly announced ongoing/planned O-RAN deployments and trials as of May

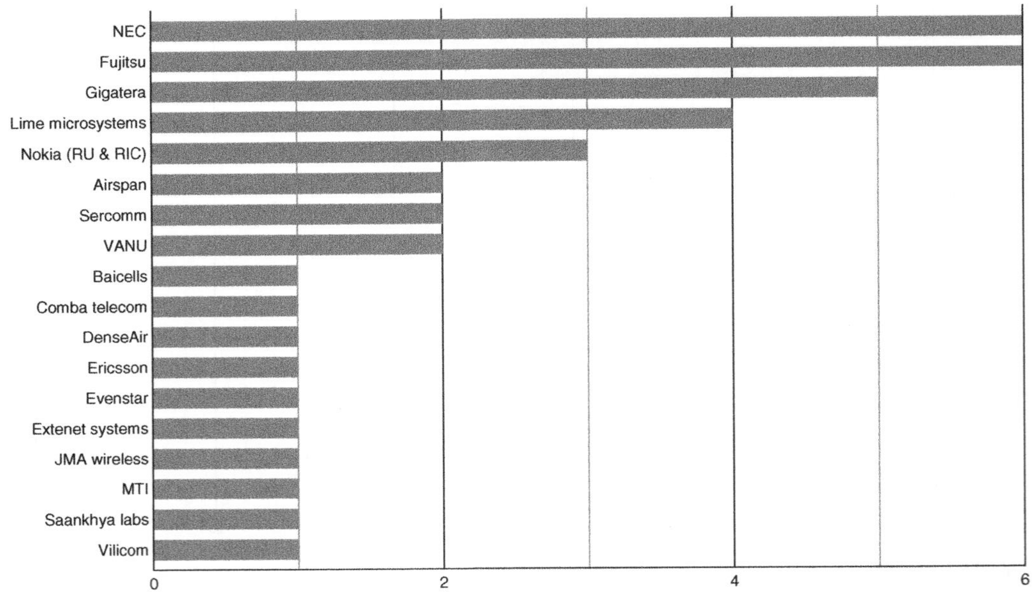

Figure 7.10 Top 18 Radio Unit Vendors for Open RAN (TeckNexus 2021).

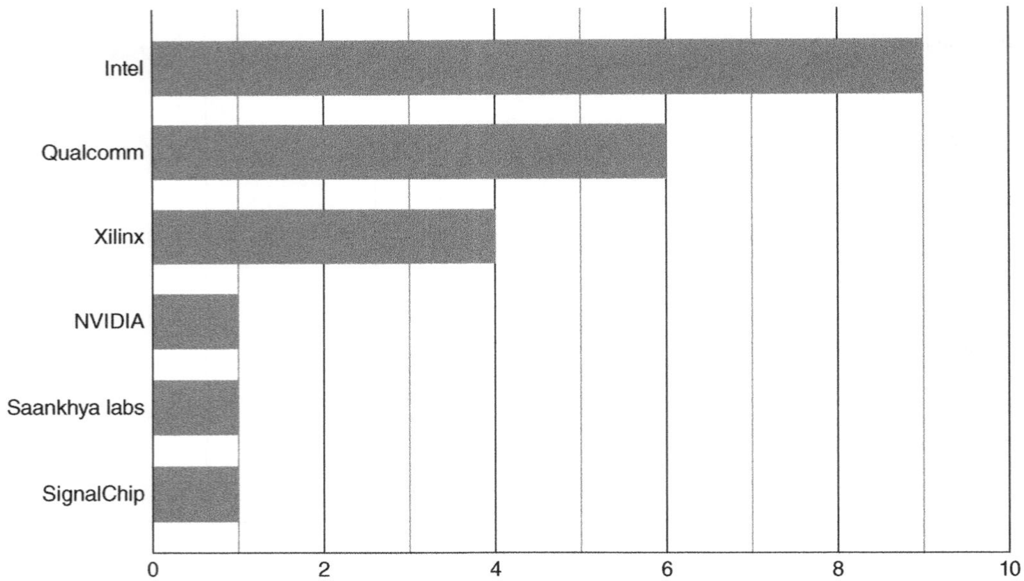

Figure 7.11 Top 6 RAN Chip Vendors for Open RAN (TeckNexus 2021).

2021 by the network operators or ecosystem vendors in press releases or various industry forums (TeckNexus 2021).

7.1.6 Major Telecom Operators

Some of the major O-RAN deployments happening globally, as per TeckNexus, STL Partners, and Voice&Data (TeckNexus 2021) (Economic Times 2023), are outlined below (see Figure 7.13).

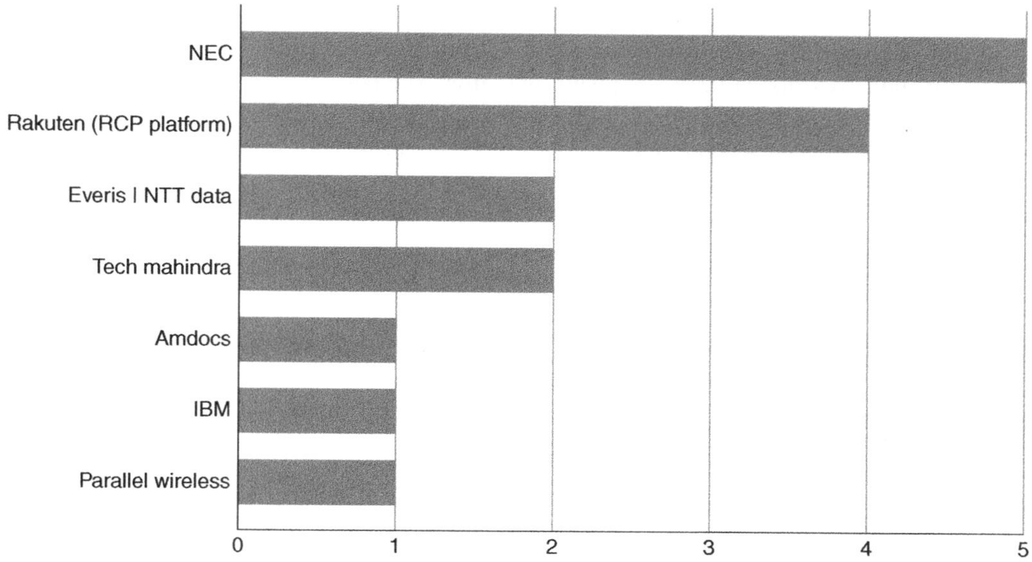

Figure 7.12 Top 7 System Integrators for Open RAN (TeckNexus 2021).

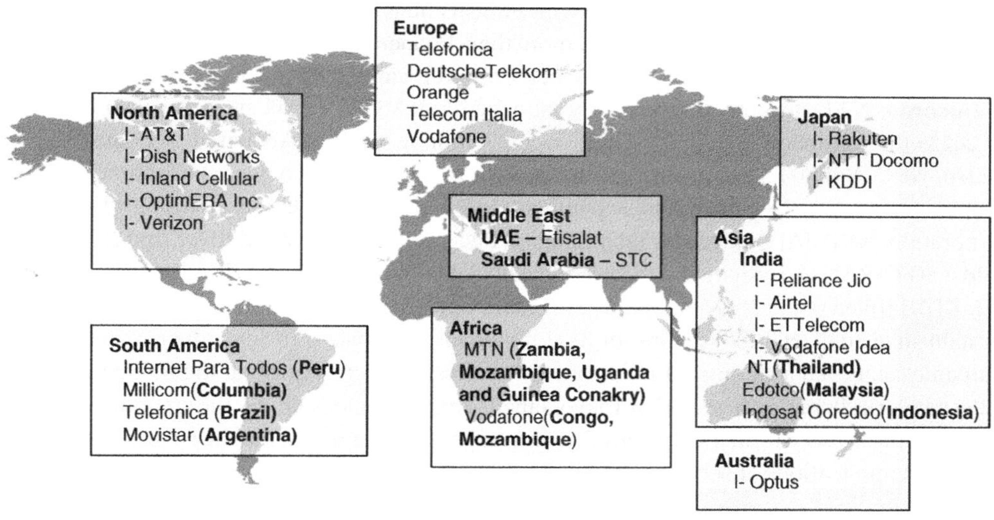

Figure 7.13 Open RAN Deployments and Trials Across the World (TeckNexus 2021).

7.1.6.1 North America

AT&T. AT&T has rolled out an O-RAN-based 5G network at various sites in Dallas, collaborating with Ericsson and Samsung. They have also performed several trials of O-RAN technology, demonstrating a millimeter wave 5G gNB and open fronthaul with support from CommScope and Intel. Additionally, AT&T recently entered into a $14 billion partnership with Ericsson to implement O-RAN in the United States. The aim is to have O-RAN sites operational by 2024 and to expand with multiple suppliers in 2025. By late 2026, AT&T projects that 70% of its wireless traffic will be handled through O-RAN systems (Nasdaq 2023).

Dish Network. Dish Network launched a cloud-native 5G broadband network in 2022 using O-RAN technology. They successfully tested a fully virtualized 5G Core Stand Alone (SA) system using the industry's first O-RAN-compliant radio from Microelectronics Technology (MTI), in collaboration with Nokia and Mavenir. Key vendors in their 5G ecosystem include Altiostar, Ciena Blue Planet, DigitalRoute, Fujitsu, Hansen Technologies, Intel, Matrixx, Mavenir, MTI, Nokia, Qualcomm, and VMware.

Inland Cellular. Inland Cellular has adopted a TIP O-RAN-based solution to successfully launch its 4G services and set the stage for future O-RAN-enabled 5G capabilities. The main vendors contributing to this 4G O-RAN deployment include Dell EMC, ExteNet Systems, Intel, and Parallel Wireless.

OptimERA Cellular. OptimERA focuses on O-RAN-based Long-Term Evolution (LTE) services, collaborating with parallel wireless to deliver broadband services.

Verizon. Key vendors include Ericsson, Nokia, and Samsung (Economic Times 2023).

T-Mobile. T-Mobile has not yet entered the O-RAN space (Nasdaq 2023).

7.1.6.2 Japan

Rakuten. Rakuten Mobile has launched commercial 4G and 5G services in Japan, utilizing O-RAN architecture. Its vendor ecosystem includes Altiostar, Cisco, Nokia, Intel, IBM Red Hat, OKI, Fujitsu, Ciena, NEC/Netcracker, Qualcomm, Mavenir, Qanta Cloud Technology, Sercomm, Tech Mahindra, Allot, Innoeye, VIAVI, Robin.io, Radcom, and Airspan. Rakuten has signed individual Memoranda of Understanding (MoU) with Fujitsu and NEC to collaboratively create O-RAN solutions. Rakuten is also discussing with more than 15 operators for the creation of a blueprint for O-RAN deployments based on the Rakuten Communications Platform (RCP).

NTT Docomo. NTT Docomo commercially launched O-RAN-based 5G sites in Tokyo in 2020. By 2023, the company stated they had over 27 million Open RAN-based 5G subscribers. Fujitsu, NEC, Nokia, and Samsung are involved in this initiative. They are collaborating with companies—such as Dell Technologies, Fujitsu, Intel, Mavenir, NEC Corporation, NTT DATA Corporation, NVIDIA, Qualcomm Technologies, Red Hat, VMware, Wind River, and Xilinx—to build a 5G O-RAN ecosystem and to expedite global deployment of O-RAN technology.

KDDI. KDDI rolled out O-RAN-compatible 5G radio units from Fujitsu and Samsung Electronics to establish virtualized base stations for its commercial 5G services. In September 2020, the two companies achieved a significant milestone by successfully demonstrating the first end-to-end (E2E) network slicing with an RIC in Tokyo, Japan. This technology allows the creation of multiple virtual networks within a single physical infrastructure, enabling each slice to be customized for distinct applications or services, thereby meeting a variety of operational needs.

7.1.6.3 India

Reliance Jio. Reliance Jio is constructing a 5G network in India utilizing O-RAN technology. The company has developed and tested its proprietary 5G radios, collaborating with manufacturers based in India. Key contributors to this ecosystem include Radisys (acquired by Jio) and Qualcomm, alongside other notable companies such as Saankhya Labs, Sterlite Technologies, Tejas Networks, and VVDN Technologies.

Airtel. Airtel has deployed O-RAN technologies from Altiostar in various cities to enhance its 4G services, utilizing a 5G-ready infrastructure that allows for a seamless upgrade to 5G.

ETTelecom. Airtel signed a deal with Mavenir to launch 5G O-RAN deployments in India, beginning with 2,300 sites and aiming to scale up to 10,000 sites. This move comes after extensive trials by India's second-largest telecom operator Mavenir, testing 4G and 5G O-RAN solutions.

Vodafone Idea. In India, Vodafone Idea has partnered with Mavenir to deploy O-RAN technology across various sites to support its 4G services.

7.1.6.4 Thailand

National Telecom Public Company Limited (NT). NT collaborated with Cisco, Mavenir, and 5GCT to launch an O-RAN-based private 5G network in Ban Chang in 2021. Mavenir supplied the cloud-native, open architecture private network solution, which includes 5G O-RAN, millimeter wave (mmWave) radios, and a 5G Core with Open API technology standards. This setup enables the state-owned telco TOT to launch its 5G services, while Cisco provides the necessary switching hardware and application services to support the "smart" features of the 5G solution.

7.1.6.5 Indonesia

Indosat Ooredoo. In December 2024, Nokia and Indonesia's leading mobile operator, Indosat Ooredoo Hutchison (IOH), announced a two-year extension to their partnership to expand IOH's 4G and 5G network across Indonesia. They aim to significantly improve operational performance as well as help advance Indonesia's rapidly evolving telecom sector. Indonesia has conducted field trials for O-RAN technology aimed at enhancing and expanding their video-grade 4G network to provide improved internet services. The trials received support from the TIP OpenRAN™ Project Group and Parallel Wireless.

7.1.6.6 Malaysia

Edotco. Together with TIP, Edotco is testing and implementing O-RAN 4G sites in some high-traffic areas across Malaysia. The company validated TIP OpenRAN™ in a Network as a service setting, offering wholesale services to mobile providers, beginning with Celcom Axiata and expanding over time. In Malaysia, the government is building and managing a national 5G wholesale network for use by existing 4G operators, including CelcomDigi in which Axiata has a 33% stake.

7.1.6.7 Australia

Optus. Australia's second-largest mobile operator has implemented O-RAN solutions from Parallel Wireless to deliver 4G mobile connectivity to customers in remote areas. The company has also launched 5G services in Australia, utilizing network equipment from both Nokia and Ericsson.

7.1.6.8 Europe

Deutsche Telekom, Orange, Telecom Italia, Telefonica, and Vodafone have entered into an MoU to collaborate with the O-RAN ALLIANCE, TIP, and policymakers. Their goal is to facilitate the deployment of O-RAN technologies across Europe, ensuring these solutions can compete effectively with traditional technologies.

Telefonica. Telefonica conducted numerous field trials, leading to initial rollouts in phase 1 across Spain, Brazil, Germany, and the United Kingdom. Telefonica aims to implement O-RAN in 50% of its network by 2025. Telefonica O2 has also launched various O-RAN projects to enhance network services in both rural areas and densely populated urban centers.

Deutsche Telekom. Deutsche Telekom has O-RAN-based 4G and 5G technology at multiple sites across Germany. Key vendors include Mavenir, Fujitsu, NEC, Nokia, Dell, and others.

In November 2024, Nokia signed a deal with Deutsche Telekom to roll out large-scale commercial O-RAN networks in Germany at more than 3,000 sites, supporting DT's strategy of increasing choice in supplier base and focusing on high-power, large-scale, multi-vendor network rollout.

Orange. In 2025, Orange plans to launch an exclusively O-RAN compliant network, which will undergo virtual testing. They believe that O-RAN will create opportunities for vendors while maintaining support for Nokia and Ericsson. Furthermore, in February 2023, Orange and Vodafone agreed to collaborate on building O-RAN networks in European rural areas where they both have mobile infrastructure, making this the first instance of shared O-RAN networks between the two firms in Europe.

Telecom Italia. Telecom Italia has rolled out 4G services in Italy and is actively participating with other companies in signing MoUs to advance the deployment of O-RAN technologies across Europe. Key vendors include JMA Wireless and MTI.

Vodafone. Vodafone introduced 4G services in Ireland through a partnership with Parallel Wireless leveraging system integrators to enhance its developing ecosystem with RAN automation vendors. The company has deployed O-RAN 4G sites in the United Kingdom in collaboration with Dell, Kontron, Lime Microsystems, and Mavenir. Additionally, Vodafone has conducted field trials of O-RAN technology for 4G services with Parallel Wireless and Lime Microsystems across Italy, Romania, and Spain. In the Netherlands, Vodafone Ziggo has tested O-RAN technology in partnership with Altiostar and NEC Europe. Furthermore, Vodafone has established O-RAN Research and Development labs throughout Europe. They launched their first O-RAN 5G site in Bath, UK, which marks the beginning of a planned rollout of 2,500 sites and signifies the start of the first large-scale O-RAN project in Europe (Economic Times 2023).

7.1.6.9 South America

Internet Para Todos (IPT). IPT is a wholesale operator in Peru, jointly owned by Telefonica, Facebook, and Latin American banks IDB Invest and CAF Bank. They have a commercial 4G mobile broadband service based on TIP O-RAN architecture. IPT provides connectivity and services to rural areas and communities through a revenue-sharing model. Key vendors involved in this deployment include Altiostar and Parallel Wireless.

Millicom Tigo. Millicom has rolled out 5G services in rural areas of Colombia using O-RAN technology through its partnership with Parallel Wireless. Parallel Wireless is responsible for uniting various partners for hardware, software, RIC, orchestration, and applications.

Telefonica Vivo. Telefonica Vivo has carried out trials of O-RAN technology for both 4G and 5G in Brazil, collaborating with Altiostar, GigaTera Communications, Intel, Mavenir, Parallel Wireless, Supermicro, and Xilinx. They are leveraging AWS Outposts as part of their infrastructure solutions.

Movistar. A brand of Telefonica, Movistar, has carried out an O-RAN proof-of-concept in Puerto Madryn, Argentina. Key vendors involved include Altiostar, GigaTera, Kontron, Quanta, and Red Hat. IBM serves as the system integrator overseeing the end-to-end deployment.

7.1.6.10 Africa

MTN (Zambia, Mozambique, Uganda, and Guinea Conakry, South Africa). MTN has launched more than 200 rural commercial sites powered by O-RAN technology. In Mozambique, the company performed O-RAN-based trials for 4G services, with key partners such as CrowdCell, Lime Microsystems, Parallel Wireless, and Mavenir. In Uganda and Guinea, MTN provides commercial 4G services using O-RAN technology, collaborating with Parallel Wireless, NuRAN Wireless, and VANU. They have executed O-RAN trials in South Africa in partnership with Parallel Wireless.

Vodafone (Democratic Republic of Congo, Mozambique). Vodacom has carried out trials using TIP OpenRAN™ technology in the Democratic Republic of Congo in collaboration with Parallel Wireless, and in Mozambique, they have tested O-RAN-based 4G services. Key vendors include CrowdCell, Lime Microsystems, Mavenir, and Parallel Wireless.

7.1.6.11 Middle-East

Etisalat. Etisalat, the UAE-based telecommunications operator, was the first to launch an Open Virtual Radio Access Network (Open vRAN) in January 2020. The company has plans to gradually expand the rollout of Open vRAN throughout the UAE. Key vendors involved in this deployment include Parallel Wireless, Altiostar, Cisco, and NEC. Etisalat is utilizing COTS hardware as a cost-effective alternative to traditional proprietary network equipment.

Etisalat operates in 16 markets throughout the Middle East, Asia, and Africa. Recently, they entered into an agreement with Parallel Wireless to implement an O-RAN compliant network in Afghanistan, collaborating with Intel and Supermicro to replace existing legacy applications. Intel will supply a third-generation Xeon processor to be used on Supermicro servers. Etisalat has been investigating the RCP for some time and has signed an MoU with Rakuten Mobile to create a reference blueprint for mobile broadband networks and O-RAN technology based on the RCP (RakutenMobile 2021).

STC (Saudi Arabia). STC tested O-RAN sessions in Saudi Arabia in December 2020, partnering with STC Solutions and Altiostar. This testing is in line with STC's DARE strategy for a promising digital future. Furthermore, STC entered into an MoU with Rakuten Mobile to explore various technologies, including O-RAN deployment options for greenfield and brownfield applications. The goal is to enhance the speed of mobile network service delivery using O-RAN technology.

7.1.7 Major Testing Vendors

Companies such as VIAVI Solutions, Spirent, and National Instruments (NI) have been actively involved in O-RAN initiatives, contributing to the development of testing equipment for performance, conformance, security, and interoperability aspects of O-RAN architecture.

VIAVI Solutions. VIAVI offers testing capabilities for individual network elements, including O-RU, O-DU, O-CU, and near-RT RIC, as well as for combined elements and full end-to-end open network validation. By adopting the fronthaul specifications from the O-RAN ALLIANCE and developing O-RAN compliant RAN and fronthaul testing practices, VIAVI has committed to advancing O-RAN development and deployment through various industry partnerships. Promoting interoperability testing methodologies that reduce reliance on built-in tests is crucial for managing costs associated with deployed network infrastructure. This also encompasses creating methods to assess O-RAN field performance. Figure 7.14 illustrates VIAVI's capabilities for end-to-end O-RAN testing.

Spirent. Spirent's O-RAN test solutions (Spirent 2024) offer the performance and flexibility necessary to support the evolving landscape of O-RAN technologies. These solutions assist global operators and equipment manufacturers in ensuring the highest quality user experience, providing confidence that their networks and devices are prepared for commercial launch with optimal performance to enhance business results. Key testing offerings include O-RAN End-to-End Testing, O-RU Testing, O-DU Testing, O-CU Testing, RIC Testing, and O-RAN Fronthaul Testing.

Keysight. Keysight's end-to-end O-RAN Architect solutions facilitate conformance, interoperability, performance, and security testing throughout the entire lifecycle. These suites are specifically

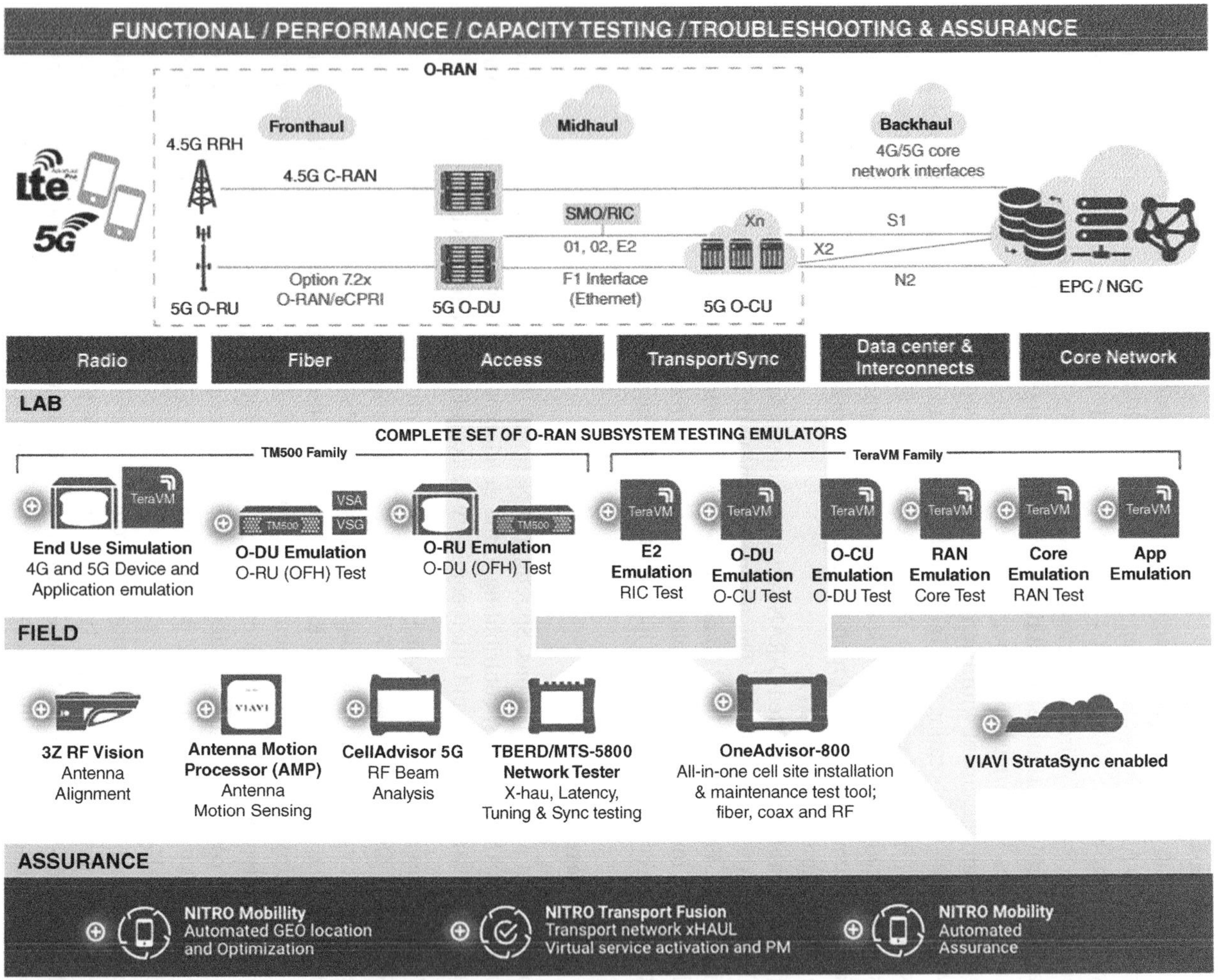

Figure 7.14 Overview of VIAVI's O-RAN End-to-End Testing Capability (VIAVI 2024).

designed for chipset manufacturers, software stack developers, network equipment providers, mobile operators, and Open Testing and Integration Centers (OTICs). For more details, visit (Keysight 2024).

National Instruments (NI). NI's O-RAN RU production test system (National Instruments 2024) helps tackle the complex challenges of developing and testing RUs in a standardized and disaggregated RAN architecture.

Rohde & Schwarz (RS). Rohde & Schwarz, in partnership with VIAVI Solutions, provides a comprehensive suite of O-RAN development and conformance testing solutions designed specifically for radio manufacturers (Rohde & Schwarz 2024). The company asserts that its solutions cover the entire product lifecycle and comply with the specifications established by TIP, 3GPP, O-RAN ALLIANCE Working Groups, and TIFG.

7.2 Government-Led Initiatives

Key US Government-led initiatives include the NTIA Wireless Innovation Fund, The RIC Forum, Department of Defense (DoD) Innovate Beyond 5G Program, DoD and the National Spectrum Consortium Effort on O-RAN Acceleration, DoD and Office in the Under Secretary of Defense (OUSD) 5G Challenge, and National Defense Authorization Act (NDAA). These initiatives are briefly described in Sections 7.2.1–7.2.5.

7.2.1 NTIA Wireless Innovation Fund

The Biden-Harris Administration seeks to enhance O-RAN through the $1.5 billion Public Wireless Supply Chain Innovation Fund (NTIA Innovation Fund 2024). Authorized under the FY 2021 NDAA, the fund received funding from the CHIPS and Science Act of 2022. Over the next decade, the Innovation Fund aims to foster wireless innovation, increase competition, and strengthen supply chain resilience. Its investments will create new opportunities for companies in the United States and partner nations.

The innovation fund will

- promote secure, open, and virtualized 5G+ technologies.
- facilitate the deployment of open and interoperable equipment.
- enable network integration using equipment from various manufacturers.
- develop criteria to determine compliance with open standards.

7.2.2 The RIC Forum

NTIA/ITS, in collaboration with the US Department of Defense FutureG Office, hosts a Forum where RAN RIC Application (App) vendors and universities can demonstrate the current state of the O-RAN RIC App market (RIC Forum 2024). This event serves as a platform to demonstrate practical RIC applications and their use cases, aimed at sharing technical knowledge. Presenters will showcase live demonstrations of RIC apps (rather than simulations) that measure and illustrate the impact of their xApps and rApps on a 5G O-RAN network.

7.2.3 DoD Innovate Beyond 5G Program

In early 2022, DoD's Innovate Beyond 5G (IB5G) program funded three new projects to advance 5G and O-RAN technology, out of which Open6G is related to O-RAN R&D efforts.

Open6G Project: Open6G is a collaborative initiative between industry and academia aimed at advancing 6G systems research focused on O-RANs (DoD Release 2022a). The initiative will concentrate on O-RAN research and the open-source implementation of 5G protocol stack features to facilitate the development of emerging applications beyond or enhancing 5G. Open6G will act as the DoD's central hub for the development, testing, and integration of trusted enhancements, fostering a NextG ecosystem for both industry and federal government efforts toward 6G technology objectives. Funded by a $1.77 million anchor award from IB5G, the project is managed by Northeastern University's Kostas Research Institute in collaboration with the Army Research Laboratory.

7.2.4 DoD and National Spectrum Consortium Effort on O-RAN Acceleration

In July 2022, the US Department of Defense, in partnership with the National Spectrum Consortium (NSC), issued a call for white papers to solicit feedback and ideas from industry and academia aimed at speeding up the development, manufacturing, and testing of 5G O-RAN technologies in the United States. The primary objective of this initiative is to identify challenges faced by both emerging and established companies in advancing the commercialization of O-RAN technologies in the United States. Further details can be found at DoD Release (2022b).

7.2.5 DoD and OUSD 5G Challenge

FutureG and the 5G Office of the Under Secretary of Defense for Research and Engineering (R&E) recently concluded the 2023 5G Challenge. Sponsored by FutureG and the 5G Office and organized by the NTIA within the Department of Commerce, the 5G Challenge aimed to accelerate the adoption of open interfaces, interoperable components, and multi-vendor solutions to foster the development of an open 5G ecosystem.

As per DoD Release (2023), O-RAN is a priority for the DoD because of its potential to support the Department's Joint All-Domain Command and Control goals for future military operations. Whereas a traditional RAN requires unified hardware and software, O-RAN gives combatant commanders more options to acquire or replace software and hardware used in military equipment and ensures more robust integration of systems and equipment on the battlefield. Principal Director for FutureG & 5G, Dr. Tom Rondeau, said "For the nation's defense and security, fully actualized O-RAN will allow the DOD the ability to illuminate and monitor networks to create zero-trust systems and provide real-time, relevant information to our commanders behind the lines and out in the field."

More information on the 5G Challenge can be found at 5G Challenge (2023).

7.2.6 National Defense Authorization Act (NDAA)

NDAA for fiscal year 2024 includes provisions for the deployment of private networks based on O-RAN at military installations (Fierce Networks 2023). The new NDAA mandates that the Department of Defense develop a strategy for implementing private wireless networks utilizing O-RAN architecture at these sites. Additionally, it allocates nearly $180 million for research, development, testing, and evaluation of "Next Generation Information Communications Technology (5G)."

7.3 Academia-Led Initiatives

Academia has been actively involved in O-RAN research and development, contributing to the advancement of O-RAN and intelligent RAN architectures. Academia-led initiatives often involve collaborative research projects, publications, and contributions to standards bodies. Some notable academia-led O-RAN initiatives are Open AI Cellular (OAIC) Platform, Colosseum, OpenRAN Gym, ColO-RAN, POWDER, Commonwealth Cyber Initiative (CCI) xG Testbed, AERPAW, Agriculture and Rural Communities (ARA) Testbed, 5GENESIS Consortium, and 5G-EmPOWER. These initiatives are briefly discussed below.

7.3.1 Open AI Cellular (OAIC) Platform

As a part of NSF-funded research project "*Open AI Cellular (OAIC): Prototyping Artificial Intelligence-Enabled Control and Testing Systems for Cellular Communications Research,*" Mississippi State University, Virginia Tech, North Carolina State University, and Worcester Polytechnic Institute together have developed a fully open-source software platform called OAIC that provides a 5G O-RAN testbed platform for building and testing AI-driven RAN controllers enabling cellular RAN research. More information about OAIC can be found on OAIC (2024). The OAIC framework consists of two primary components: OAIC-C (OAIC-Controllers), which facilitates the development of AI-controlled RAN functionalities, and OAIC-T (OAIC-Testing), designed for highly automated testing of AI-enhanced O-RAN-based networks. Refer to Upadhyaya et al. (2023) for a concise introduction to the OAIC platform.

The OAIC-C platform is built on an O-RAN-enabled 5G infrastructure, as illustrated in Figure 7.15. It leverages open-source 5G NR (New Radio) software implementations for User Equipment (UE), RAN, and Core Network, which are compatible with COTS Software-Defined Radio (SDR) testbeds and the OSC software library Release E. This allows for the implementation of Near-RT and non-RT RICs, as well as O-RAN interfaces such as E2, A1, and O1.

The deployment process consists of three steps: setting up the Near-RT RIC, configuring a 5G network using srsRAN with an E2 agent, and deploying an xApp. The Near-RT RIC is distributed across two Kubernetes clusters that orchestrate Docker containers for system deployment. Docker is used to run and manage applications in isolated environments (containers), simplifying dependency management and enhancing security through isolation. The first cluster focuses on deploying the Near-RT RIC platform and applications (RIC cluster), while the second cluster manages auxiliary functions (AUX cluster).

Testing of the UE and O-RAN connection can be conducted using standard network traffic and tools such as ping and iperf, provided they are connected to the correct IP endpoints. Additionally, OAIC-C includes prepackaged xApps, such as the KPI Monitoring (KPIMON) xApp, which is based on OSC's version, to showcase the use of containerized software in controlling the gNodeB (gNB). Moreover, the platform features innovative AI-controlled xApps, including a CNN-based interference classifier xApp and an ML-based radar detection xApp, as discussed in Stephenson et al. (2023) and Reus-Muns et al. (2023), respectively.

The OAIC-C project's source code (OAIC-GitHub 2024) and documentation (OAIC-Documentation 2024) are publicly accessible on GitHub and utilize Git submodules to effectively manage versions across multiple repositories. The repository features automatically updated documentation, along with installation and testing instructions, as well as GitHub Actions for

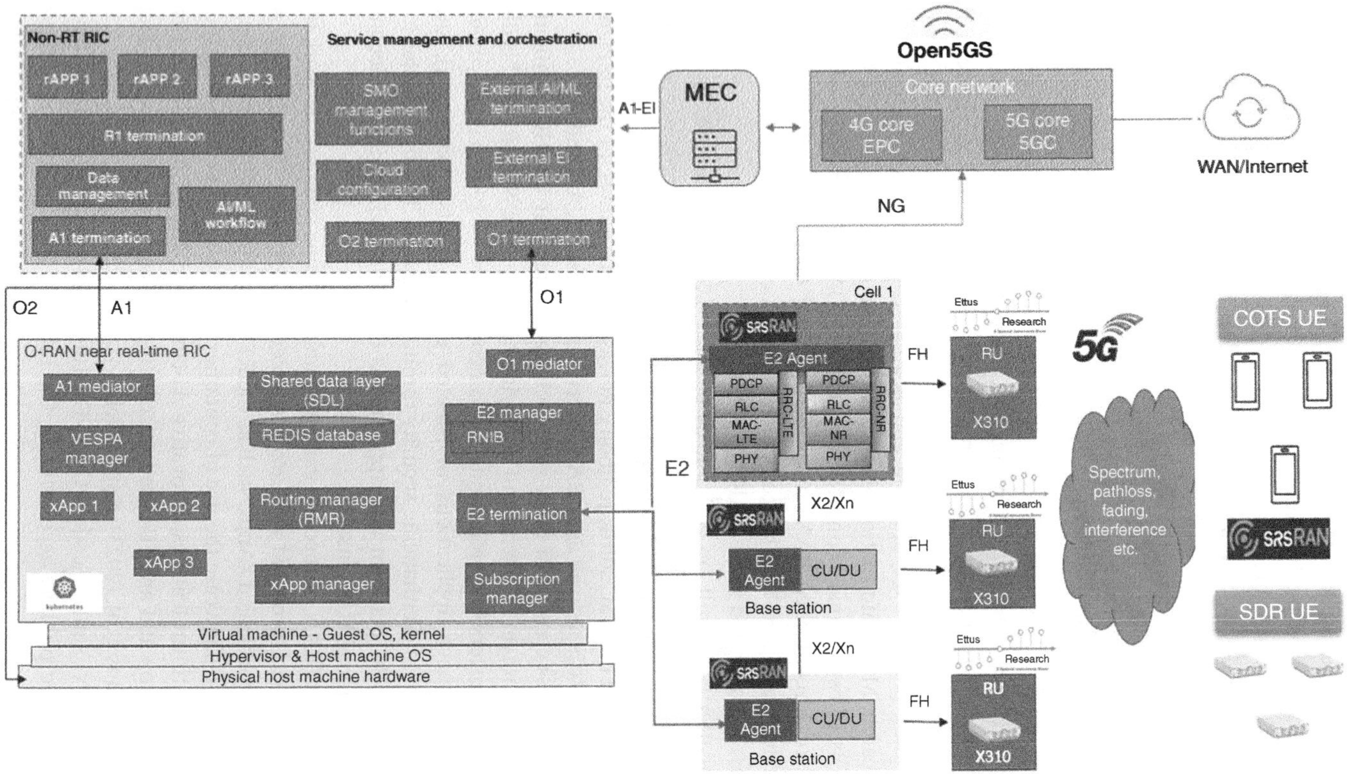

Figure 7.15 Overview of OAIC Controllers (OAIC-C) Platform.

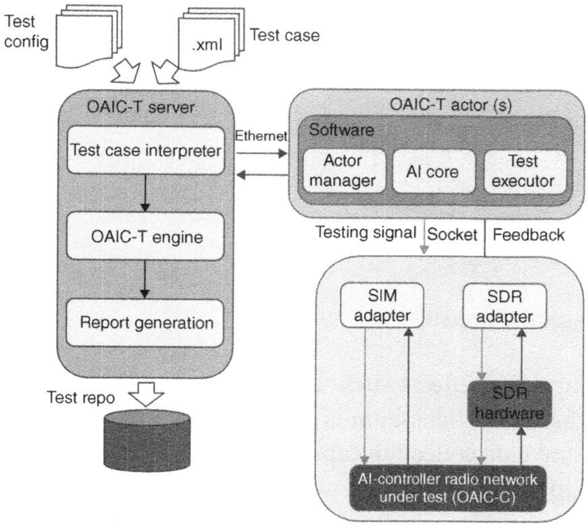

Server

Test script interpreter	Read test scripts in .xml, json, etc.
Server engine	Manage all actors, set up testing environment, and monitor test status.
Report generation	Generate testing reports and access test repository.

Actor(s)

Actor manager	Manage actor resources and running of test scripts.
AI Core	Provide AI capability, i.e., AI learner instances.
Test executor	Execute individual test actions defined in test script
SIM/SDR Adapter	Provide interfaces to the unit under test.

Figure 7.16 Overview of OAIC-T Platform.

continuous integration (CI) testing and continuous deployment (CD) of software packages. The OAIC-C GitHub organization welcomes new members, enabling them to stay updated on the latest pre-released software versions.

The OAIC-T framework facilitates automated, distributed, and AI-enhanced testing of xApps in O-RAN environments. It comprises three main components: (i) Test Input, which includes test configuration and script files; (ii) Server, which sets up the testing environment based on the configuration files and orchestrates test execution as defined in the scripts; and (iii) Actor(s), which carry out test actions as directed by the server, as illustrated in Figure 7.16.

The server oversees multiple remote testing actors, maintaining socket connections with each and monitoring their resource usage (such as CPU, memory, disk, and SDR hardware) and test status. Each actor consists of an actor manager, an AI core, and a test executor, along with two adapters for interfacing with the unit under test (i.e., a cellular radio network controller).

Actors can interact with the xApps/rApps under test using either the SIM Adapter or the SDR Adapter. The SIM Adapter functions as a testing xApp/rApp, which can be deployed in a non-RT, Near-RT, or RT RIC to send test data and receive responses from the RAN through socket communications. The SDR Adapter, on the other hand, operates as a User Equipment (UE), utilizing an SDR-based software suite (e.g., srsRAN) to transmit radio testing signals to the RAN via antennas.

The source code and documentation for OAIC-T 1.0 can be found on OAIC-Testing-GitHub (2024) and OAIC-Testing-Documentation (2024), respectively.

7.3.2 Colosseum

Currently located at Colosseum is the world's largest wireless network emulator, specifically designed to facilitate research and development of large-scale, next-generation radio network technologies within configurable and repeatable radio frequency (RF) and traffic environments. Figure 7.17 illustrates the Colosseum testbed infrastructure (Colosseum 2024).

It integrates 128 Standard Radio Nodes (SRNs), each outfitted with NI USRP X310 Software-Defined Radios (SDRs), along with a Massive Digital Channel Emulator (MCHEM) supported by a robust FPGA routing fabric. The MCHEM emulates real-world wireless RF channels

Figure 7.17 Colosseum (Colosseum 2024)/Northeastern University.

between the SRNs using FPGA technology, capturing effects such as fading and multipath for up to 256 × 256 independently customizable channels. Colosseum is accessible as a cloud-based platform, offering distinctive experimentation and data collection capabilities.

As an RF testbed, Colosseum can be utilized to

- Emulate various operational environments in real-time, incorporating high-fidelity ray-tracking to accurately simulate effects such as multipath and fading.
- Instantiate softwarized protocol stacks on a general-purpose infrastructure
- Carry out large-scale testing and data collection.
- Develop machine learning algorithms across various wireless communication techniques, including spectrum sharing, dynamic spectrum access, signal intelligence extraction, and optimized routing, by leveraging integrated high-performance computational resources.

7.3.3 OpenRAN Gym

OpenRAN Gym is an open-source toolkit designed for data collection and experimentation with AI in O-RAN environments. It utilizes various experimental platforms, including Colosseum (OpenRANGym 2024). The toolkit includes a softwarized RAN and a near-RT RIC, which can be deployed as containers within Colosseum to automate the data collection and testing of AI solutions for O-RAN. Figure 7.18 depicts a high-level overview of OpenRANGym architecture (OpenRANGym 2024).

7.3.4 ColO-RAN Platform

The ColO-RAN platform is a publicly accessible O-RAN testing framework that incorporates SDRs-in-the-loop (ColORAN 2024; Polese et al. 2022). Figure 7.19 depicts the overall architecture of the ColO-RAN Platform (ColORAN 2024). Leveraging the scale and computational power of

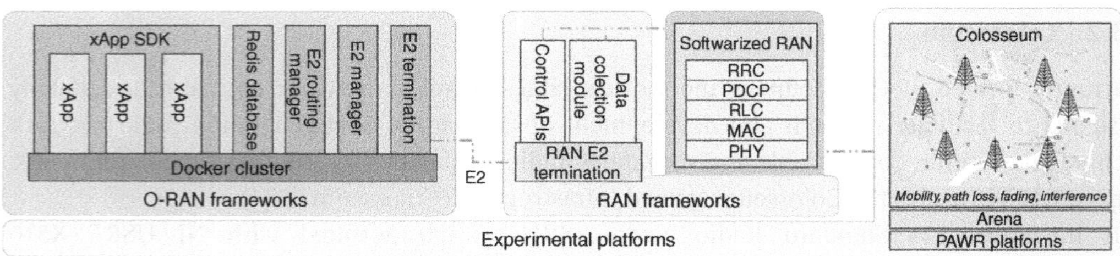

Figure 7.18 OpenRANGym Architecture (OpenRANGym 2024).

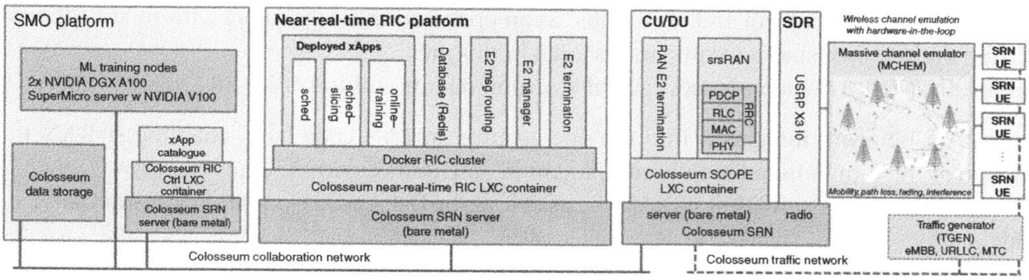

Figure 7.19 ColORAN Platform (ColORAN 2024).

the Colosseum wireless network emulator, ColO-RAN facilitates large-scale machine learning research using O-RAN components, programmable base stations, and a "wireless data factory." It features a near-real-time RIC, based on the O-RAN Software Community RIC, deployed on Colosseum, which manages a software-defined RAN through our SCOPE framework with a standard-compliant E2 termination (see Figure 7.19).

It also includes an SDK for the rapid integration and testing of AI/ML-based xApps for real-time RAN inference and control. At a high level, ColO-RAN xApps consist of two main components: (i) the SM connector, which manages communications to and from the near-RT RIC (e.g., facilitating interactions with base stations), handles ASN.1 message encoding/decoding, and executes queries to the RIC's Redis database; and (ii) the data-driven logic unit that performs tasks based on key performance metrics (KPMs) received from the RAN in real time, such as traffic prediction and base station control.

7.3.5 POWDER

The POWDER (Platform for Open Wireless Data-driven Experimental Research) is a large-scale wireless research testbed located at the University of Utah (POWDER 2024). It provides a unique platform for researchers to conduct experiments in wireless networking and communication. The testbed supports a wide range of experiments, including those related to 5G and beyond.

POWDER's O-RAN capabilities are designed to be flexible and open, enabling researchers to experiment with various network architectures, protocols, and technologies. Specifically, POWDER supports near-RT RIC, xApps, the O-CU subsystem, and open SMO software. It has integrated RAN slicing functionality into the srsRAN software, providing access to the RIC through a custom E2 RAN slicing service model. Additionally, POWDER has implemented the O-RAN KPM E2 service model in the srsRAN code, allowing xApps to retrieve RAN-related statistics. This implementation includes the NexRAN xApp, which facilitates policy-driven RAN slicing using these two service models.

NexRAN enables closed-loop control of RAN slicing within an O-RAN ecosystem. RAN slicing is integrated into the srsRAN open-source mobility stack and is accessible through a custom service model to the NexRAN xApp, which operates on the near-RT RIC from the O-RAN Software Community. The POWDER implementation of RAN slicing allows different slices to share the same frequency band, explicitly associates UEs with specific slices, and employs a slice-aware scheduler in the base station to manage RAN resources for each slice. The NexRAN xApp facilitates policy-driven closed-loop control of RAN slices by monitoring the current state of RAN elements using the standard O-RAN KPM service model with slice extensions and adjusting slice behavior through the custom slicing service model. NexRAN is designed for easy integration within the POWDER platform and is available as an open-source xApp, along with modifications to existing open-source stacks to support research in this area.

Specifically, NexRAN encompasses several key components within POWDER (2024):

- An O-RAN E2 agent integrated into the open-source srsRAN code base, enabling srsRAN to function as the foundational mobile component in an ORAN environment.
- A RAN slicing scheduler within srsRAN, accessible to the O-RAN ecosystem via a custom service model.
- A RAN slicing xApp that facilitates closed-loop RAN slicing control, serving as a top-to-bottom example of managing RAN functionality in O-RAN setting.
- Open-source software and a POWDER profile that promote easy repeatability, testing, and extension, providing essential building blocks for ORAN research and experimentation.

More information on NexRAN xApp can be found in Johnson et al. (2022). Within POWDER, researchers can explore and test O-RAN concepts, such as virtualized baseband units, intelligent algorithms for resource management, and open interfaces between network elements. This capability allows for experimentation with innovative approaches to RANs, fostering the development of more efficient and flexible wireless communication systems.

7.3.6 Commonwealth Cyber Initiative (CCI) xG Testbed

The *CCI xG Testbed* at Virginia Tech in Arlington, VA, contains a full suite of modern communication laboratory equipment for 5G/NextG wireless technology innovation, hardware prototyping, and SDR implementation (CCI-xG-Testbed 2024). Figure 7.20 depicts the xG testbed infrastructure. With 72 nodes, 4 multi-access edge computing (MEC), and O-RAN capability, this is one of the largest university-based SDR testbeds in the United States. These capabilities are supported by a staff of nine research engineers with hardware and software development and rapid prototyping expertise. The radio network is designed to be O-RAN and AI-powered compliant for at-scale and reproducible experimentation. The radio network offers users a choice of COTS radios (802.11x, LTE, 5G nonstand alone, and SA) and fully programmable SDRs (USRPs X310, X410, N310, B210, and B205-mini). The CCI xG Testbed provides end-to-end cellular stack programmability from a service management orchestrator, the core network, and the RAN, allowing users to create real-world wireless scenarios in a controlled, reproducible experimental setting. The CCI xG Testbed software allows experimenters to manipulate wireless interfaces

Figure 7.20 CCI xG Testbed (CCI-xG-Testbed 2024)/Commonwealth Cyber Initiative.

(such as the GnuRadio/USRP) to set up experiments and collect substantial performance data. There is also an SDN-based Ethernet and optical switching array with core and access switches with 180 ports to switch traffic from 72 radio node interfaces and the servers. The radio network backend system includes an OpenStack cluster for cloud computing experiments (enabling virtual machine and containers provisioning) and light-OpenStack version for edge computing based on the latest generation of Intel NUCs. In addition to the xG Testbed, Virginia Tech is currently building the PORTS5+ Testbed, a portable testbed that enables testing of mobility aspects of the 5G O-RAN.

7.3.7 AERPAW Testbed

Based at NC State, AERPAW—Aerial Experimentation and Research Platform for Advanced Wireless—is the first wireless research platform to study the convergence of 5G technology and autonomous drones. AERPAW is a five-year NSF PAWR project that started 1 September 2019. AERPAW supports experimentation with SDRs and unmanned aerial systems. A deployment map for AERPAW nodes can be seen in Figure 7.21.

AERPAW serves as an O-RAN Testing and Integration Center (OTIC) lab and possesses the capability to run O-RAN conformance tests and issue certificates and badges. The lab includes several test equipment such as Keysight S5040A O-RAN studio appliance, M9451A 8 GHz VXT transceiver and together with software and test automation packages for complete O-RU conformance test setup, Keysight ethernet Line Server Unit, and SDRv.4 with Air mosaic Software emulates up to 100 5G NR UE's and helps build E2E and IoT gNB tests. There are two IXIA X100-5G servers available: one is used to host Keysight RICtest platform and another is used to host coreSIM software.

Figure 7.22 shows the O-RU conformance test setup at AERPAW OTIC. Conducted RF output from O-RU can be connected to the Keysight VXT transceiver (Vector Signal Analyzer + Vector Signal Generator) to perform the O-RU conformance tests. The enhanced Common Public Radio Interface (eCPRI) port of the O-RU is connected to the Keysight O-DU emulator. Keysight O-RAN

Figure 7.21 AERPAW Fixed Node Deployments in the Research Triangle Park (AERPAW).

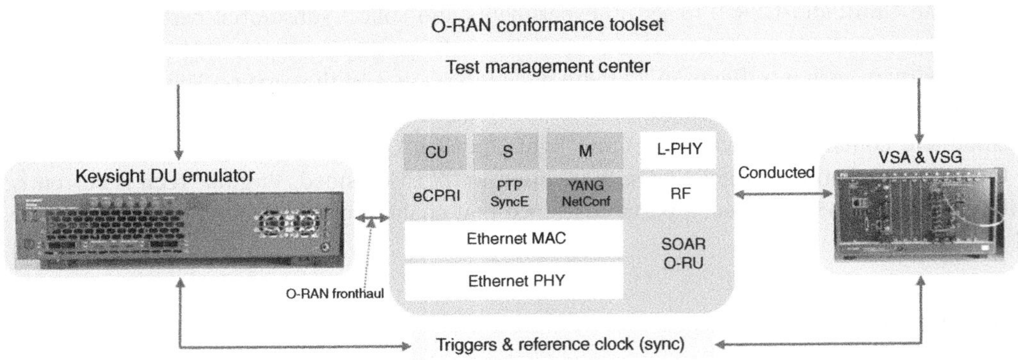

Figure 7.22 AERPAW OTIC O-RU Conformance Test Setup (AERPAW).

studio software and VSA, VSG software running on O-DU emulator S5040A, completes the conformance test setup. We have openTAP Software that binds all the software packages and automates the test case development and testing. This setup would be able to run standard O-RU conformance tests as defined in the ORAN ALLIANCE standard specifications. Upon successful completion of the conformance tests, AERPAW OTIC can issue appropriate conformance test certificates for the O-RU.

7.3.8 ARA Testbed

The Agriculture and Rural Communities (ARA) Testbed is a large-scale platform for advanced wireless research, being deployed across the Iowa State University (ISU) campus, the City of Ames, and nearby research and agricultural farms, as well as rural communities in central Iowa, covering a rural area with a diameter of over 60 km (approximately 37.28 miles) (ARA 2024). This testbed functions as a wireless living laboratory for smart and connected rural communities, facilitating the research and development of wireless technologies tailored to rural needs, aimed at providing affordable, high-capacity connectivity for rural areas and industries like agriculture.

At its core, ARA focuses on deploying advanced wireless platforms in real-world agricultural and rural environments, capturing the intricate systems, environmental characteristics, and community contexts that define rural broadband. For example, ARA showcases the future of precision agriculture across both crop and livestock farms, utilizing automated ground vehicles, cameras, and nanosensors. The project incorporates a diverse range of unique wireless technologies, including low-UHF mMIMO, mmWave wireless access, long-distance backhaul, free-space optical communications, and low-earth-orbit satellite systems. By utilizing both software-defined radio (SDR) and programmable COTS platforms, and leveraging open-source wireless software frameworks such as OpenAirInterface, srsRAN, and SD-RAN, ARA provides seamless coverage over hundreds of square miles of rural areas. This creates a highly programmable infrastructure for comprehensive wireless research in practical settings.

7.3.9 5GENESIS Consortium

In Europe, the 5GENESIS consortium is focused on implementing various 5G components and validating a range of use cases across its multiple testbeds (5GENESIS 2020). These testbeds incorporate edge-computing, NFV-enabled heterogeneous radio infrastructure, orchestration, and management frameworks, as well as terrestrial and satellite communications, and ultradense network deployments. Once completed, these testbeds will be fully compatible with the O-RAN ecosystem.

7.3.10 5G-EmPOWER

5G-EmPOWER, developed by Roberto Riggio at the Polytechnic University of Marche in Ancona, Italy (5G-EmPOWER-GitHub 2024), offers a near-RT RIC designed for heterogeneous RANs. In addition, it features nonstandard functionalities such as mobility management for both Wi-Fi and cellular networks, support for multi-tenancy, and the ability to deploy customized resource allocation schemes within network slices.

7.4 Major Takeaways of the Chapter

- There are several industry-led O-RAN research and development initiatives, notably O-RAN ALLIANCE and SD-RAN. Beyond that, several major vendors and operators are carrying out real-world O-RAN-based 4G and 5G deployments and trials across the world.
- Government agencies, such as NTIA and DoD, are accelerating O-RAN research, development, and deployment through various initiatives, such as the RIC Forum and Wireless Innovation Fund.
- Academia is playing a key role in O-RAN R&D by designing and developing various O-RAN capabilities, such as AI-driven xApps and rApps, as well as building open-source O-RAN testbeds for validation and experimentation.
- Notable academia-led O-RAN testbed resources include OAIC Platform, Colosseum, OpenRAN Gym, ColO-RAN, POWDER, CCI xG Testbed, AERPAW, ARA Testbed, 5GENESIS Consortium, and 5G-EmPOWER.
- Some of these academic testbeds, AERPAW, Colosseum, CCI xG Testbed, AERPAW, and ARA serve as OTIC centers.

References

5G Challenge 2023. https://5gchallenge.ntia.gov/ (accessed 24 July 2024).

5G-EmPOWER-GitHub 2024. https://github.com/5g-empower (accessed 24 July 2024).

5GENESIS 2020. https://5genesis.eu/ (accessed 24 July 2024).

ARA 2024. https://arawireless.org/about-ara/ (accessed 24 July 2024).

AWS O-RAN evolution 2024. https://docs.aws.amazon.com/whitepapers/latest/open-radio-access-network-architecture-on-aws/o-ran-evolution.html (accessed 24 July 2024).

CCI-xG-Testbed 2024. https://cyberinitiative.org/cci-xg-testbed.html (accessed 24 July 2024).

ColORAN 2024. https://openrangym.com/o-ran-frameworks/coloran (accessed 24 July 2024).

Colosseum 2024. https://www.northeastern.edu/colosseum/ (accessed 24 July 2024).

DoD Release 2022a. https://www.defense.gov/News/Releases/Release/Article/3114220/three-new-projects-for-dods-innovate-beyond-5g-program/ (accessed 24 July 2024).

DoD Release 2022b. https://www.defense.gov/News/Releases/Release/Article/3052013/dod-and-national-spectrum-consortium-team-for-open-ran-acceleration/ (accessed 24 July 2024).

DoD Release 2023. https://www.cto.mil/news/zts_battlefield_comms/ (accessed 24 July 2024).

Economic Times 2023. https://telecom.economictimes.indiatimes.com/news/telecom-equipment/know-all-about-open-ran-trials-deployments-happening-across-globally/101826682 (accessed 24 July 2024).
Fierce Networks 2023. https://www.fiercewireless.com/5g/dish-hails-5g-open-ran-provisions-defense-department (accessed 24 July 2024).
Johnson, D., Maas, D. and Van Der Merwe, J. (2022). NexRAN: Closed-loop RAN slicing in POWDER-A top-to-bottom open-source open-RAN use case. In *Proceedings of the 15th ACM Workshop on Wireless Network Testbeds, Experimental evaluation & Characterization* (pp. 17–23).
Keysight (2024). https://www.keysight.com/us/en/cmp/2020/5g-o-ran-software.html (accessed 24 July 2024).
Nasdaq 2023. https://www.nasdaq.com/articles/what-att-investors-need-to-know-about-the-$14-billion-open-ran-deal (accessed 24 July 2024).
National Instruments (2024) https://www.ni.com/en/solutions/semiconductor/wireless-infrastructure-development.html?%20cid=PSEA-7013q000001rC7YAAU-Consideration-GOGSE_156872434923&utm_keyword=open%20ran%20test&gad_source=1&gclid=EAIaIQobChMI0a6XsPz-gwMVEEpHAR1vxAZsEAAYASAAEgJxQfD_BwE (accessed 24 July 2024).
NTIA Innovation Fund 2024) https://www.ntia.gov/page/innovation-fund (accessed 24 July 2024).
OAIC 2024. https://www.openaicellular.org/ (accessed 24 July 2024).
OAIC-Documentation (2024). https://openaicellular.github.io/oaic/ (accessed 24 July 2024).
OAIC-GitHub (2024). https://github.com/openaicellular/oaic (accessed 24 July 2024).
OAIC-Testing-Documentation (2024). https://openaicellular.github.io/oaic/oaic_t.html (accessed 24 July 2024).
OAIC-Testing-GitHub (2024). https://github.com/openaicellular/oaic-t (accessed 24 July 2024).
ONF 2024. https://opennetworking.org/ (accessed 24 July 2024).
ONF Aether 2024. https://opennetworking.org/aether/ (accessed 24 July 2024).
ONF SD-RAN 2020. https://opennetworking.org/wp-content/uploads/2020/08/SD-RAN-v2.0.pdf (accessed 24 July 2024).
OpenRANGym 2024. https://openrangym.com/ (accessed 24 July 2024).
O-RAN ALLIANCE 2024. https://www.o-ran.org/what-we-do (accessed 24 July 2024).
O-RAN ALLIANCE OTIC 2024. https://www.o-ran.org/testing-integration#learn-otic (accessed 24 July 2024).
O-RAN Virtual Exhibition 2024. https://www.virtualexhibition.o-ran.org/classic/generation/2023 (accessed 24 July 2024).
O-RAN-White Paper (2023). https://assets-global.website-files.com/60b4ffd4ca081979751b5ed2/64561954d1802b0001099566_Overview%20of%20OTIC%20and%20O-RAN%20Certification%20and%20Badging%20Program_white%20paper_2023-04.pdf (accessed 24 July 2024).
OSC Software 2024. https://wiki.o-ran-sc.org/display/REL/Releases (accessed 24 July 2024).
Polese, M., Bonati, L., D'Oro, S. et al. (2022). ColO-RAN: Developing machine learning-based xApps for open RAN closed-loop control on programmable experimental platforms. *IEEE Transactions on Mobile Computing* 22 (10): 5787–5800.
POWDER 2024. https://powderwireless.net/ (accessed 24 July 2024).
RakutenMobile 2021. https://corp.mobile.rakuten.co.jp/english/news/press/2021/0322_01/ (accessed 24 July 2024).
Reus-Muns, G., et al. (2023). SenseORAN: O-RAN based radar detection in the CBRS band. *IEEE Journal on Selected Areas in Communications.* (accessed 24 July 2024).
RIC Forum 2024. https://its.ntia.gov/research-topics/ric-forum/ (accessed 24 July 2024).

Rohde-Schwarz (2024). https://www.rohde-schwarz.com/us/solutions/wireless-communications-testing/mobile-network-infrastructure-testing/testing-o-ran/testing-open-ran_255230.html (accessed 24 July 2024).

SD-RAN Documentation 2024. https://docs.sd-ran.org/master/introduction.html (accessed 24 July 2024).

Spirent (2024). https://www.spirent.com/products/open-ran-testing (accessed 24 July 2024).

Stephenson, N.H., Chiejina, A.J., Kabigting, N.B. and Shah, V.K. (2023). Demonstration of closed loop AI-driven RAN controllers using O-RAN SDR testbed. In *MILCOM 2023-2023 IEEE Military Communications Conference (MILCOM)*, pp. 241–242. IEEE.

TeckNexus 2021. https://tecknexus.com/5g-network/5g-magazine-open-ran-june-2021/the-top-openran-vendors-across-categories/ (accessed 24 July 2024).

Telecom Infra Project 2024. https://telecominfraproject.com/openran/ (accessed 24 July 2024).

TIP OpenRAN DU & CU Subgroup 2021. https://cdn.mediavalet.com/usva/telecominfraproject/Joo0LSJi10agbgWh7AWAUA/GI67cRgKi0uGj05nPlsbBA/Original/OpenRAN_ataglance_plus_Glossy_v08_2021_06_16-CU_DU.pdf (accessed 24 July 2024).

TIP OpenRAN Indoor Small Cell 2021. https://cdn.mediavalet.com/usva/telecominfraproject/0_JcBUzu1EuTC_tIJyR7lA/eMiL3w58vE6rJd0gpXSkPg/Original/OpenRAN_ataglance_plus_Glossy_v08_2021_06_16-Indoor_SC.pdf (accessed 24 July 2024).

TIP OpenRAN Outdoor Macro 2021. https://cdn.mediavalet.com/usva/telecominfraproject/VQzTtvMhOkmfgQkkl8w1MA/QWx_KLjVNkuU2jT6c2ziyA/Original/OpenRAN_ataglance_plus_Glossy_v08_2021_06_16-Outdoor_Macro.pdf (accessed 24 July 2024).

TIP OpenRAN RAN Intelligence and Automation 2021. https://cdn.mediavalet.com/usva/telecominfraproject/Xhv2dK0k4E6fLYIktAv0AQ/dHvZIwafWkiP53YBcphg6A/Original/RIA_OpenRAN_ataglance_Glossy_v08_2021_06_16.pdf (accessed 24 July 2024).

TIP OpenRAN RU Subgroup 2021. https://cdn.mediavalet.com/usva/telecominfraproject/qywkB9OlGEaIFul6KtmZ7g/1I9rZb5usUyx3X_RpFYlNg/Original/OpenRAN_ataglance_plus_Glossy_v08_2021_06_16-RU.pdf (accessed 24 July 2024).

Upadhyaya, P., Tripathi, N., Gaeddert, J. and Reed, J. (2023). Open AI cellular (OAIC): An open source 5G O-RAN testbed for design and testing of AI-based RAN management algorithms. *IEEE Network Magazine*. (accessed 24 July 2024).

Viavi 2024. https://www.viavisolutions.com/en-us/solutions/open-ran (accessed 24 July 2024).

8

O-RAN Challenges and Evolution of O-RAN Toward 6G

O-RAN is expected to be one of the prominent ways of designing and deploying a 5G Radio Access Network. However, with any new technologies, there are challenges to overcome. This chapter provides a glimpse of O-RAN challenges. Examples of potential O-RAN enhancements to address these challenges are also given. While O-RAN currently supports 5G and 4G radio interfaces, it is expected to support 6G in future. A brief overview of 6G is given, followed by potential enhancements to O-RAN to support 6G.

8.1 O-RAN Challenges and Solutions

Section 8.1.1 provides a brief overview of O-RAN challenges and solutions, while Section 8.1.2 takes a closer look at O-RAN challenges and solutions.

8.1.1 O-RAN Challenges and Solutions: An Overview

Figure 8.1 divides O-RAN challenges into the following categories: Orchestration and operation, administration, and maintenance (OAM); performance; security; product development; Total Cost of Ownership (TCO); interoperability; RAN product performance; and multivendor operational complexity. These challenges and their candidate solutions are discussed in detail in Section 8.1.2.

Orchestration and OAM. Locations of O-RAN components need to be determined properly. Furthermore, AI/ML models need to be orchestrated.

Performance. Overall performance aspects of O-RAN such as energy consumption, resource management, and quality of service (QoS) management need to be addressed.

Security. An increased threat surface of O-RAN compared to a traditional RAN requires a comprehensive solution.

Product Development. Increased virtualization and disaggregation in O-RAN make it difficult to outperform single-vendor custom solution that utilizes custom hardware.

TCO. Both Capital Expenditure (CapEx) and Operating Expenditure (OpEx) need to be considered to determine the TCO. An increased degree of disaggregation and the need for fiber transport network in O-RAN can increase O-RAN TCO.

Interoperability. While open interfaces enable O-RAN to utilize O-RAN components from different vendors, interoperability among components needs to be validated.

RAN Product Performance. Since O-RAN utilizes a high degree of virtualization, the time required to carry out certain complex physical layer processing may be longer compared to a traditional custom single-vendor RAN, requiring hardware accelerators.

Fundamentals of O-RAN, First Edition. Nishith D. Tripathi and Vijay K. Shah.

Companion website: www.wiley.com/go/tripathioran

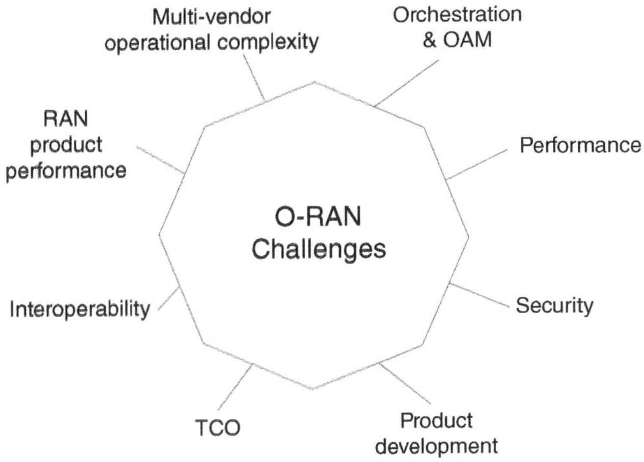

Figure 8.1 O-RAN Challenges.

Multivendor Operational Complexity. When O-RAN components from different vendors are combined to develop an integrated solution, operational complexity increases significantly.

The O-RAN ALLIANCE has developed comprehensive specifications including interoperability test specifications. However, the RAN is quite a complex portion of the cellular network. Numerous proprietary algorithms have been designed over multiple decades by RAN vendors to operate and optimize the RAN. Hence, switching from a vendor-specific RAN to the O-RAN with many open interfaces is a nontrivial task for a service provider. Widespread deployments of O-RAN have not occurred as of late 2024 due to the complexity of the O-RAN and challenges associated with comprehensive interoperability testing. While test specifications do define interoperability tests as described in Chapter 6, extensive field testing is needed to gain confidence in O-RAN.

As summarized in Chapter 1, O-RAN provides enhancements in two key areas—openness and intelligence. The openness aspect of O-RAN, where products from different vendors (e.g., an O-RAN Radio Unit (O-RU) from one vendor and an O-RAN Distributed Unit (O-DU) from another vendor) work together, is expected be realized earlier than the intelligence aspect of O-RAN (e.g., the use of RAN Intelligent Controllers [RICs], extended applications [xApps], and remote applications [rApps]). Potential conflicts among new xApps and rApps and conflicts between (i) O-RAN Apps (i.e., xApps and rApps) and (ii) traditional resource management algorithms developed by vendors in the O-DU and the O-RAN Central Unit (O-CU) need to be evaluated extensively prior to O-RAN Apps in commercial O-RAN deployments, especially in multivendor deployments.

8.1.2 O-RAN Challenges and Solutions: A Closer Look

Different categories of O-RAN challenges and associated candidate solutions are discussed in Sections 8.1.2.1 to 8.1.2.8.

8.1.2.1 Orchestration and OAM

Orchestration and OAM Challenges A single-vendor RAN utilizes vendor-specific orchestration to manage the RAN, simplifying the overall orchestration processes. However, since O-RAN is intended to utilize multiple vendors to provide flexibility to network operators, orchestration across components from different vendors becomes complicated. Deployment and administration of highly diverse hardware and software from different vendors is a nontrivial task.

Locations of O-RAN components for deployments need to be identified. These locations need to provide target capacity and throughput on the radio interface while minimizing intranetwork connectivity and bandwidth requirements. While cloud-based deployments provide scalability, provisioning suitable configurations of O-RAN components is an important challenge.

One of the key benefits of O-RAN is a flexible framework with well-defined interfaces to significantly enhance RAN performance. However, orchestrating artificial intelligence/machine learning (AI/ML) models is an important challenge (Aryal et al. 2023). Design of suitable AI/ML models, time scales of AI/ML model operations, defining and obtaining required inputs and data types, handling of fault prediction or drifting of AI/ML results, service requirements, and locations of AI/ML models are examples of key challenges related to intelligence management.

Candidate Solutions for Orchestration and OAM Challenges First of all, O-RAN specifications need to be fully complied with by vendors selected by the network operator. Individual compliance of vendor products to O-RAN specifications needs to be verified. Furthermore, the combinations of the vendor products need to be thoroughly tested. Any issues identified during individual product testing and integrated product testing need to be reflected in O-RAN specifications or deployment guidelines. AI-based techniques such as deep learning-based solutions and resource monitoring techniques may facilitate determination of ideal locations for implementing RAN services (Aryal et al. 2023). For example, locating AI/ML services near the edge is more suited for achieving low latency and handling user-sensitive data, and locating AI/ML services at a relatively centralized location is more suited for latency-tolerant services.

Several potential solutions for intelligence management are mentioned in Aryal et al. (2023). For example, an intelligent orchestration framework may be utilized to select an optimal location for the RAN function deployment. To reduce memory requirements while deploying AI/ML-based solutions, unused program modules and frameworks may be removed during the programming phase. Furthermore, a portion of data may be selected instead of all the available data to achieve faster convergence of AI/ML models and accurate results for AI/ML models. A method to determine whether to retrain an existing AI/ML model or switch to a new model can help address the drifting challenge. Drifting may be classified into sudden, gradual, incremental, or reoccurring. Retraining of an AI/ML model would be required for sudden and gradual drift types, and selection of a new model is needed for incremental and reoccurring drift types.

8.1.2.2 Performance

Performance Challenges Examples of performance-related O-RAN challenges include *energy consumption*, *resource management*, and *QoS* (Aryal et al. 2023).

Energy Consumption Challenges The RAN consumes a vast majority of the total energy consumption in the cellular network. For example, according to Brik et al. (2022), the RAN may consume more than 50% of the network energy consumption. Since the O-RAN utilizes more O-RAN components due to increased disaggregation of the RAN and introduction of additional components such as RICs compared to the non-O-RAN implementation of a 5G RAN, the O-RAN may consume more energy than traditional non-O-RAN implementation of the 5G RAN. Energy efficiency is quite important to network providers due to its impact on OpEx.

Candidate Performance Solutions for Energy Consumption Several approaches are mentioned in Aryal et al. (2023) to address the energy consumption in O-RAN. For example, certain RAN features may be turned off for small cells during high network load. Infrastructure sharing can be exploited to off-load some traffic to the cells that are relatively less congested. Another approach to address heavy traffic is to utilize an O-DU with a higher processing capacity. Additionally, information sharing among network functions (NFs) can also increase energy efficiency. AI/ML-based techniques such as sequential multiagent deep reinforcement learning, concurrent multiagent deep reinforcement learning, and team multiagent deep reinforcement learning may be exploited to determine an optimal solution that achieves optimal network performance from the perspectives of power and resource allocation. Joint selection of optimal O-DU for NF placement and resource block allocation may reduce energy consumption in O-RAN.

Resource Management and QoS Challenges Efficient resource management has always been a significant challenge for the RAN due to rising popularity of wireless communications and relative scarcity of radio resources. With increasing complexity of newer generations of cellular technologies such as 5G, resource management has become quite complex. Furthermore, O-RAN introduces the frameworks of RICs, which further increase the overall resource management complexity, because RICs need to interact with traditional 5G RAN components such as the next-generation Node B (gNB)-CU and the gNB-DU.[1] Resource management will continue to be an important topic for O-RAN just like it has been a challenge for a traditional RAN. Indeed, O-RAN aims to optimize RAN operations via openness and intelligence. Resource management is central to RAN optimization.

The QoS is an important aspect of in a wireless communication system. Optimal QoS management optimizes network performance and user experience. While QoS management is not unique to O-RAN, it is more complex for O-RAN-like resource management due to the increased degree of disaggregation compared to traditional O-RAN. Note that QoS management is coupled with resource management. Indeed, resource management needs to consider QoS requirements of users and multiple services of each user. Due to increased disaggregation of the RAN in O-RAN, identifying fronthaul open interface requirements to achieve ultra-low latency is a unique challenge in O-RAN. Furthermore, placement of functionalities in the network deployment and achieving scalability for enhanced performance are additional challenges faced by O-RAN.

Candidate Performance Solutions for Resource and QoS Management As mentioned earlier in this section, the resource management challenge is not unique to O-RAN. The good news is that O-RAN is inherently poised to tackle this challenge through its architecture with open interfaces and RICs. Suitable planning and execution to manage specific RAN functions are a key to addressing the resource management challenges.

Several potential approaches for O-RAN-based resource management are highlighted in Aryal et al. (2023). For example, xApps of the Near-Real-Time (RT) RIC can be used to address the challenges of resource management by implementing resource allocation policies. A cell-splitting strategy can be used to reduce traffic congestion in a cell. A resource allocation xApp can allocate a suitable number of resource blocks to configured network slices. A dynamic traffic forecasting scheme can be used to forecast future traffic demand in federated O-RAN, which can facilitate resource management in a proactive and predictive manner.

Several candidate mechanisms for O-RAN-based QoS management are mentioned in Aryal et al. (2023). Network operations management, proper hardware–software selection, and continuous

1 The O-CU includes the functions typically performed by the 3GPP-defined gNB-CU, and the O-DU and the O-RU together include the functions typically performed by the 3GPP-defined gNB-DU.

monitoring of app-level services can help enhance QoS. The deployment of network functionalities at the edge of the network can potentially alleviate traffic congestion at the O-RAN interfaces. RIC applications can be exploited to analyze and optimize key network performance metrics. Power regulation may be helpful in maximizing overall throughput. Continuous monitoring of deployed AI/ML models may avoid any decrease in performance. Enhanced O-RAN-based cell selection procedure that chooses a cell capable of providing the best performance rather than just the best radio connectivity can enhance overall performance.

8.1.2.3 Security

Security Challenges O-RAN has more interfaces and network elements compared to a traditional RAN. Hence, the threat surface is much larger in O-RAN, leading to new security challenges emerge in O-RAN. Chapter 4 on O-RAN Security provides a detailed view of O-RAN security. The O-RAN ALLIANCE has carried out an extensive analysis of security vulnerabilities and developed threat models. Note that openness and disaggregation of O-RAN lead to increased potential for threats on the network as well as user data (GCC Open RAN Consortium 2024).

In one approach, security challenges in O-RAN can be viewed from the perspectives of confidentiality and availability (Aryal et al. 2023). Confidentiality could become a challenge in O-RAN, because the information present or generated at the entities such as the O-DU contains user-sensitive information and may get exposed when shared with RICs. Furthermore, compared to a traditional closed-system RAN solution from a single vendor, the multivendor O-RAN solution poses confidentiality challenges regarding user-sensitive data. The availability of the system is more challenging in O-RAN compared to a traditional RAN due to an increased degree of disaggregation and the existence of products from multiple vendors in a multivendor O-RAN system. The location of RAN functionalities influences the availability of O-RAN. The RIC applications also influence availability. Since xApps and rApps manage control loops as discussed in Chapter 3, their availability affects the availability of O-RAN.[2]

Candidate Security Solutions To address the confidentiality challenge, an intelligent Zero Trust Architecture (i-ZTA) that utilizes AI/ML for information security can be utilized (Aryal et al. 2023). Key elements of i-ZTA, the policy enforcement point and policy decision point, exploit intelligence to make access decisions.

Several approaches to enhance location-related availability of O-RAN are mentioned in Aryal et al. (2023). A Binary Integer Programming optimization model that selects the deployment location for O-RAN functions in the O-Cloud. Backup servers for virtual RAN functions can perform functions in case primary servers fail. Furthermore, a RIC Fault-Tolerant framework can be developed to make the applications fault-tolerant during programming.

Note that the O-RAN ALLIANCE has developed specifications to strengthen security as discussed in Chapter 4. Several candidate solutions to enhance O-RAN security are suggested in GCC Open RAN Consortium (2024). For example, O-RAN monitoring solutions can provide advanced features that can enable detection, trigger warnings and prevention, and act against these threats. O-RAN operators can exploit virtualization to deploy tested security patches and software updates quickly and remotely. Furthermore, due to a comprehensive O-RAN ecosystem, network operators and suppliers can coordinate with one another by sharing incident information about threats,

---

2 O-RAN is intended to be operational even without xApps and rApps. However, intelligence aspects of O-RAN are no longer available when all xApps and rApps (that have utilized intelligence to optimize RAN functions) are unavailable.

vulnerabilities, and exploits and providing greater access to data about network performance to enable accelerated development and deployment of security mitigations. Furthermore, open interfaces in O-RAN can be used to distribute security analytics throughout the network and move RAN monitoring to the edge. Additionally, automation enables zero-touch management that eliminates the inherent security risks in human access to NFs. Closed-loop management can quickly detect and fix degraded configurations or anomalous network activities. Finally, compliance of O-RAN components to various standards and certifications such as Third Generation Partnership Project (3GPP) Security Assurance Specifications, European Telecommunication Standards Institute, and GSMA Network Equipment Security Assurance Scheme (NESAS) and certification schemes for cloud services can enhance security in O-RAN.

8.1.2.4 Product Development

Product Development Challenges Challenges related to deployment and operations are summarized below in two areas—*testbed availability* and *data management and acquisition* (Aryal et al. 2023). An end-to-end (E2E) testbed for the O-RAN system is important for implementing and testing various use cases and product combinations. The O-RAN Software Community (OSC) releases software for O-RAN functions such as RICs and Service Management and Orchestrations (SMOs). Integrating OSC-specified software with these functionalities with implementations from other open-source communities on an ongoing basis is a non-trivial challenge. Since O-RAN introduces intelligence-based framework in the form of RICs, data management is essential for all AI/ML-based processing in O-RAN. One data management challenge is how to store, categorize, and select the RAN data per use case requirement. A data acquisition challenge is how to access proprietary data handled by the RAN equipment such as the O-RU, the O-DU, and the O-CU due to privacy and competition concerns. Another data acquisition challenge is the lack of open-source data for research.

Candidate Development Solutions To address testbed-related challenge, E2E testbeds that enable testing of intelligent solutions for next-generation O-RAN systems can be very helpful. Examples of these testbeds include srsRAN-based frameworks and Colosseum-based frameworks (Aryal et al. 2023). Open RAN Gym utilizes srsRAN for RAN software. A framework that utilizes Colosseum as the data factory and ColO-RAN to implement O-RAN control architecture has been developed. In another variation, CellOS is used to extend srsRAN implementation, while OSC-based RIC functionalities are utilized.

To help address data acquisition challenges, frameworks such as Colosseum and Arena could be used to extract realistic datasets (Aryal et al. 2023). Diverse data sources such as access points, cameras, and Wi-Fi access points may be used to extract location-specific contextual information. While challenging, it may be possible in some cases to obtain datasets from commercial service providers.

8.1.2.5 Total Cost of Ownership

Total Cost of Ownership Challenges The TCO includes CapEx and OpEx. O-RAN is expected to reduce the TCO for service providers due to flexibility of vendor selection for different O-RAN components and xApps and rApps due to a broad supply chain. However, the TCO significantly depends on the selected O-RAN topology and the resource and transport costs dictated by the chosen topology with a hyperconverged cloud platform, zero-touch automation, and lower processing latency (GCC Open RAN Consortium 2024). Furthermore, while virtualized Distributed Unit (vDU)/virtualized Central Unit (vCU) pooling can provide substantial TCO reduction over years, it may not be a feasible architecture for many operators due to the need for an expensive fiber transport network and the limited availability of far-edge data centers to host vDU functions. Maintaining the virtualized environment and radio frequency (RF) part of O-RAN affects OpEx. Increased need for integration testing with multivendor products increases the overall costs. Relatively shorter product life cycles due to virtualized products also incur hardware and software upgrade costs including additional integration testing costs. The use of general-purpose servers in O-RAN instead of dedicated hardware in traditional RAN may consume more power, increasing energy costs, and hence OpEx.

Candidate Solutions for the TCO Reduction Several solutions to reduce the TCO are mentioned in GCC Open RAN Consortium (2024). For example, the increase in Commercial Off The Shelf (COTS) hardware capacity, CU/DU capacity, and remote radio unit/antenna multibands approach could significantly reduce CapEx. To reduce OpEx related to energy costs, cloud data centers in O-RAN can leverage centralized cooling, lighting, and electricity purchasing agreements. Furthermore, OpEx related to the tower space rentals can be reduced due to relative centralization of the O-RAN equipment such as O-DUs, O-CUs, and RICs.

8.1.2.6 Interoperability

Interoperability Challenges Increased disaggregation of the gNB in O-RAN and the support for O-RAN components from different vendors significantly increase flexibility but make interoperability a significant challenge. It is much easier to test and manage RAN components from a single vendor. However, when O-RAN components from different vendors are integrated to develop an E2E solution, interoperability of these components needs to be verified. Furthermore, interoperability verification needs to be repeated with every introduction of new hardware (GCC Open RAN Consortium 2024).

Candidate Interoperability Solutions As baseline, O-RAN test specifications can be used as baseline. OTICs can be utilized to facilitate testing. Databases containing interoperability-verified O-RAN components with fully defined hardware and software configurations and software versions can be used to facilitate integration. The RAN integrator (e.g., the network operator) would need the right resources to perform interoperability verification (GCC Open RAN Consortium 2024). Suitable planning and fallback strategies can be developed to minimize any downtimes.

8.1.2.7 RAN Product Performance

RAN Product Performance Challenges Initial multivendor O-RAN may not have performance or security parity compared to conventional single-vendor network deployments (GCC Open RAN Consortium 2024). Single-vendor RAN includes vendor-specific optimized algorithms to manage various aspects of RAN. Furthermore, single-vendor RAN benefits from decades of experience of RAN vendors. When O-RAN components from new vendors are utilized or O-RAN components from new vendors are combined with O-RAN components of traditional vendors, it may take time and efforts to make the performance of such integrated O-RAN exceed the performance of a conventional single-vendor network.

Candidate Solutions for RAN Product Performance Challenges To address performance challenges, several approaches are mentioned in GCC Open RAN Consortium (2024). The multivendor O-RAN can be optimized by exploiting intelligent algorithms and the associated parameters. Elaborate AI-based mechanisms based on xApps and rApps can facilitate such optimization. Comprehensive testing with suitable use cases can be carried out for the intended combination(s) of O-RAN components from different vendors. Note that the use of AI and ML in the Near-RT RIC's xApps, the Non-RT RIC's rApps, and the SMO orchestration adds intelligence to the O-RAN, enhancing the Key Performance Indicators (KPIs) and customer experience. Furthermore, network slicing and centralized policy-driven user data analytics can enhance the overall network KPIs and performance despite the occurrence of unplanned events and changes in customer behavior (GCC Open RAN Consortium 2024).

8.1.2.8 Multivendor Operational Complexity

Challenges of Multivendor Operational Complexity While O-RAN characteristics such as virtualization, disaggregation, and open interfaces are quite beneficial, they create several operational challenges (GCC Open RAN Consortium 2024). For example, life cycle management of each O-RAN component is needed including software and hardware considerations. There is a need to host image registries and regular patch upgrades with historical records for Containerized Network Functions (CNFs). Unit/block/interoperability testing is needed for initial deployment as well as each software upgrade. More extensive E2E global IP schema modeling with dual-stack is needed compared to a traditional single-vendor RAN. Testing of end of service or sale monitoring and replacement of software components is needed. Firmware upgrades for hardware components and corresponding software requirements are more extensive in O-RAN. Maintaining SLAs, issue tracking, and Level 3 support coordination become more complex due to O-RAN components from multiple vendors. Indeed, inter-vendor coordination for technical issues including patches or upgrades becomes complicated. E2E application and service reliability need to be ensured. Agility for service development delivery is another challenge. In case of a brownfield deployment, interoperability and interfacing with existing network elements need to be ensured.

Operational challenges also depend on deployment scenarios (GCC Open RAN Consortium 2024). For example, certain radio processing needs to be completed in few microseconds. Hence, hardware accelerators are typically required to off-load certain L1 processing from generic processors. Introduction of hardware accelerators increases complexity. In one implementation approach, a given COTS platform or server performs functions of multiple RAN and core entities such as O-CU, O-DU, RICs, User Plane Function (UPF), and edge computing applications. Such approach reduces costs and provides flexibility but increases complexity. Similarly, leveraging the containerization of O-CU and O-DU, capacity scaling of CNFs, redundancy, replacement or reboot, and reduced latency techniques can enhance the overall reliability of applications but increase overall complexity.

Candidate Solutions for Multivendor Operational Complexity To address significant operational challenges, the role of a System Integrator (SI) becomes very significant (GCC Open RAN Consortium 2024). The SI can integrate various hardware and software components of O-RAN. The SI can carry out extensive interoperability testing before commercially launching the system. The network operator can act as an SI, or the network operator can obtain service of an SI. The SI would continue to play an important role beyond the initial commercial launch, because of the need for ongoing updates. Human resources can be provided suitable training to develop relevant skills to operate a multivendor O-RAN.

8.2 6G: A Concise Introduction

While 5G networks are being deployed commercially around the globe in the 2020s, research on 6G is also being performed. The initial focus of 6G-related research is to define a vision and requirements for 6G and identify technologies that could be part of 6G. Several organizations, global and regional,[3] are expected to play an important role in 6G. Examples of global 6G organizations are the International Telecommunication Union (ITU), the 3GPP, and the Next-Generation Mobile Network (NGMN) Alliance. Examples of regional 6G organizations are the Next G Alliance (NGA) in North America, 6G Smart Networks and Services Industry Association (6G-IA) in Europe, 6G Forum in South Korea, Beyond 5G Promotion Consortium in Japan, and Bharat 6G Alliance in India. For example, the NGA initially published a 6G vision paper containing a roadmap for 6G along with six audacious goals for 6G (NGA 2022) and has since published multiple white papers covering different aspects of 6G (NGA 2024). The 3GPP is expected to develop specifications for 6G. The actual specification work on 6G by the 3GPP is likely to start in Release 21 (Qualcomm 2022).

8.2.1 6G Usage Scenarios

The ITU has typically defined target use cases and performance requirements for different generations of wireless technologies such as International Mobile Telecommunication-2020 (IMT-2020) for 5G. The ITU has identified use cases and potential performance goals for 6G in ITU (2023). Figure 8.2 illustrates a wheel diagram showing six usage scenarios envisioned by the ITU for 6G (ITU 2023).

In Figure 8.2, three scenarios are extensions of IMT-2020 or 5G scenarios, while three scenarios are new relative to 5G. In the three extension scenarios, immersive communication is the extension of the 5G's enhanced Mobile Broadband usage scenario, massive communication is the extension

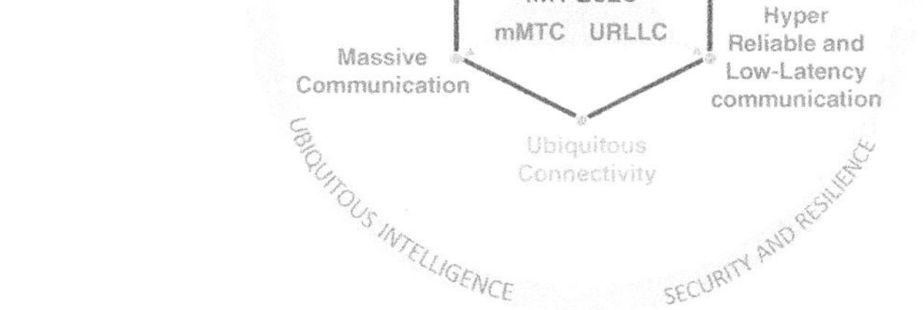

Figure 8.2 IMT-2030 Usage Scenarios and Overarching Aspects. Source: The ITU-R Framework for IMT-2030, July 2023/Reprinted with permission from ITU.

3 In this book, global organizations are the organization that have significant participation from multiple continents and countries, and regional organizations are the organizations that have significant participation from a specific region or country.

of the 5G's mMTC usage scenario, and Hyper Reliable and Low-Latency Communication (HRLLC) is the extension of the 5G's Ultra-Reliable and Low-Latency Communication (URLLC) usage scenario. Three new usage scenarios are Ubiquitous Connectivity, Integrated AI and Communication, and Integrated Sensing and Communication. Additionally, four overarching aspects have been defined; they represent design principles common to all six usage scenarios. These overarching aspects are sustainability, connecting the unconnected, ubiquitous intelligence, and security/privacy/resilience.

8.2.1.1 Immersive Communications

This usage scenario involves a rich and interactive video experience or immersive experience for users. Interactions with machine interfaces are included. Example use cases associated with this usage scenario are communication for immersive eXtended Reality (XR), remote multisensory telepresence, and holographic communications.

8.2.1.2 Massive Communication

This usage scenario involves connections for a massive number of devices or sensors. Example use cases associated with this usage scenario are expanded and new applications in smart cities, transportation, logistics, health, energy, environmental monitoring, agriculture, and areas requiring diverse Internet of Things (IoT) devices.

8.2.1.3 HRLLC

This usage scenario involves time-synchronized operations with highly reliable and low-latency communications. Example use cases are communications in an industrial environment, emergency services, tele medicine, and monitoring of electrical power transmission and distribution.

8.2.1.4 Ubiquitous Connectivity

This usage scenario involves connectivity to bridge the digital divide and serves users and IoT devices in uncovered or scarcely covered geographic areas including rural, remote, and sparsely populated areas.

8.2.1.5 Integrated AI and Communication

This usage scenario involves distributed computing and AI applications. Example use cases associated with this usage scenario are automated driving, autonomous collaboration between devices for medical assistance applications, off-loading of heavy computation operations across devices and networks, and creation of and prediction with digital twins.

8.2.1.6 Integrated Sensing and Communication

This usage scenario involves applications and services that require sensing capabilities and offers wide-area multidimensional sensing that provides spatial information about unconnected and connected objects including their movements and surroundings. Example use cases associated with this usage scenarios are navigation, activity detection and movement tracking, environmental monitoring, and provisioning of sensing data/information on surroundings for AI, XR, and digital twin applications.

8.2.2 6G Capabilities and Performance Targets

Figure 8.3 specifies capabilities and associated performance targets defined by the ITU for 6G (ITU 2023).

Peak Data Rate is the maximum achievable data rate under ideal conditions per device with example target values being 50, 100, and 200 Gbps.

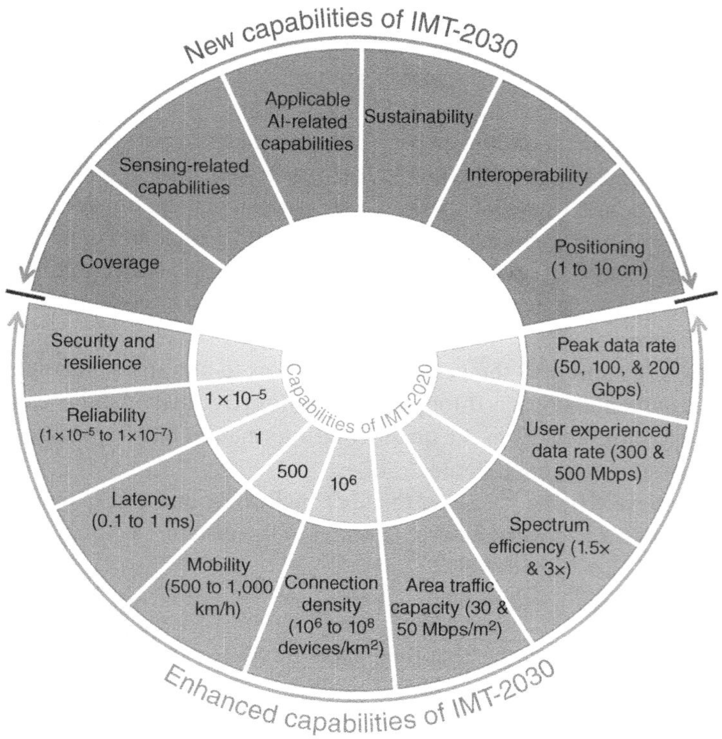

Figure 8.3 Performance Targets for IMT-2030 Capabilities. Source: The ITU-R Framework for IMT-2030, July 2023/Reprinted from permission from ITU.

User-Experienced Data Rate is the data rate that is achievable ubiquitously across the coverage area to a mobile device with example target values being 300 and 500 Mbps.

Spectrum Efficiency is the average data throughput per unit of spectrum resource and per cell with example target values being 1.5 and 3 times greater than that of IMT-2020.

Area Traffic Capacity is the total traffic throughput served per geographic area with example target values being 30 and 50 $Mbps/m^2$.

Connection Density is the total number of connected and/or accessible devices per unit area with example target values being 10^6–10^8 devices/km^2.

Mobility is expressed as the maximum speed at which a defined QoS and seamless transfer between radio nodes (which may belong to different layers and/or radio access technologies) can be achieved with example target values being 500–1,000 km/h.

Latency is the latency or delay over the air interface and refers to the contribution by the radio network to the time from when the source sends a packet of a certain size to when the destination receives it with example target values being 0.1–1 ms.

Reliability is the capability of successfully transmitting a predefined amount of data within a predetermined time duration with a given probability with example target values being 1×10^{-5} to 1×10^{-7}.

Coverage is the ability to provide access to communication services for users in a desired service area and is specified to be the cell edge distance of a single cell through link budget analysis.

Positioning is the ability to calculate the approximate position of connected devices. Positioning accuracy is the difference between the calculated horizontal/vertical position and the actual horizontal/vertical position of a device with example target values being 1–10 cm.

Sensing-Related Capabilities are the abilities to provide functionalities in the radio interface such as range/velocity/angle estimation, object detection, localization, imaging, and mapping. Metrics such as accuracy, resolution, detection rate, and false alarm rate can be used to measure these capabilities.

Applicable AI-Related Capabilities are the abilities to provide certain functionalities throughout IMT-2030 to support AI-enabled applications. Distributed data processing, distributed learning, AI computing, AI model execution, and AI model inference are examples of these functionalities.

While **security** refers to the preservation of confidentiality, integrity, and availability of information and protection of networks, devices, and systems against cyberattacks, **resilience** refers to capabilities of the networks and systems to continue operating correctly during and after a natural or man-made disturbance.

Environmental **sustainability** is the ability of the network and devices to minimize greenhouse gas emissions and other environmental impacts throughout their life cycle. Sustainability can be quantified by energy efficiency, which refers to the quantity of information bits transmitted or received per unit of energy consumption in bit/Joule.

Interoperability is the radio interface being based on member inclusivity and transparency to enable functionality or functionalities between different entities of the system.

8.2.3 6G Technology Enablers

6G would need to have certain capabilities to be able to support the usage scenarios discussed above. Figure 8.4 lists candidate 6G technology enablers identified by the NGA (NGA Alliance 2023; NGMN Alliance 2021).

Five major categories of technology enablers for 6G identified by the NGA are component technologies, network O&M and service enablement technologies, radio technologies, system and network architectures, and trustworthiness technologies.

8.2.3.1 Component Technologies

Examples of key component technologies for 6G include semiconductor technologies, RF subsystems for diverse frequencies, devices, and modems, and holographic displays.

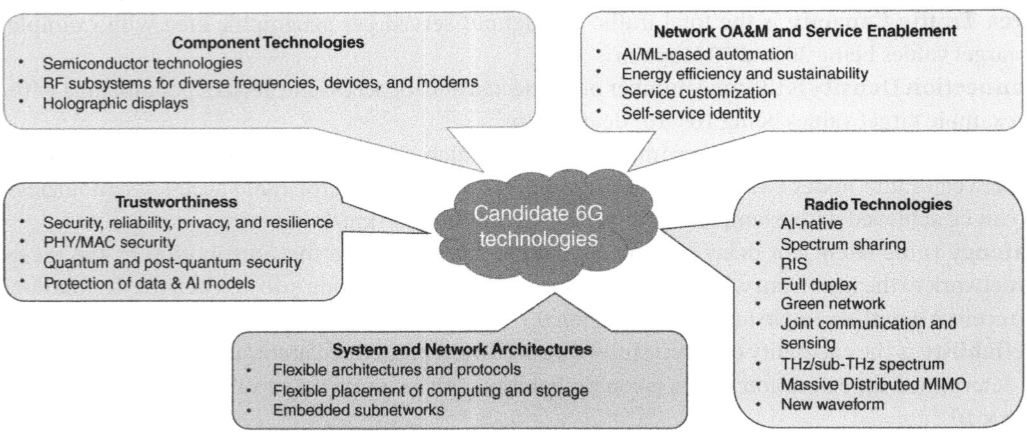

Figure 8.4 Potential 6G Technology Enablers.

8.2.3.2 Radio Technologies

Examples of key radio technologies for 6G include AI, spectrum sharing, Reconfigurable/Reflective Intelligence Surface, full duplex, green network, joint communication and sensing, THz/sub-THz spectrum, massively distributed MIMO, and a new waveform.

8.2.3.3 System and Network Architectures

6G may use flexible architectures and protocols, flexible placement of computing and storage, and embedded subnetworks.

8.2.3.4 OAM and Service Enablement

OAM and service enablement in 6G may make use of AI/ML-based automation, energy efficiency and sustainability, service customization, and self-service identity.

8.2.3.5 Trustworthiness

Examples of key trustworthiness technologies for 6G include PHY/MAC security on the radio interface, quantum and postquantum security, and protection of data and AI models.

8.3 O-RAN in 6G

O-RAN is expected to play an important role in 6G as well. The O-RAN specifications defined till Fall 2024 utilize 3GPP-defined 5G and 4G specifications. As mentioned in Section 8.2, the 3GPP's first 6G specifications are planned for Release 21. Based on the planned functional freeze in March 2027 and the planned end date of June 2027 for Release 20, Release 21 normative specifications work can be expected to start in 2027 (3GPP 2024). As the 3GPP works on 6G specifications in Release 21, the O-RAN ALLIANCE may start evaluating the impact of 6G on O-RAN. *The exact impact of 6G on O-RAN will not be known for at least few years from Fall 2024*. However, the O-RAN ALLIANCE has already established a group called the next Generation Research Group (nGRG) that focuses on research of open and intelligent RAN principles in 6G and future network standards to facilitate O-RAN evolution (O-RAN ALLIANCE 2024).

The nGRG charter includes (along with other areas) (i) providing a forum to facilitate O-RAN-related 6G research efforts and publish research findings and (ii) leveraging industry and academic 6G research efforts and determining how O-RAN may evolve to support 6G and beyond by considering regional research efforts, ITU-R, and 3GPP development.

The overall nGRG work scope is divided into three time horizons: near-term, mid-term, and long-term. In the near-term horizon spanning 2022 to 2025, the work scope involves (i) aligning with industry partners, (ii) performing a high-level impact analysis of potential 6G technology trends and the ITU-R Vision for the next IMT on the O-RAN architecture, and (iii) establishing research items based on defined research priorities. The near-term work scope aims to create outputs such as white papers and research reports. In the mid-term horizon spanning 2025 to 2027, the nGRG aims to provide inputs to O-RAN work groups/focus groups to prepare for O-RAN 6G standards studies and coordinate O-RAN 6G collaborations with other Standards Development Organizations (SDOs) based on the research findings. Finally, the nGRG's long-term work scope beyond 2027 aims to align with other 6G research organizations, SDOs, and programs

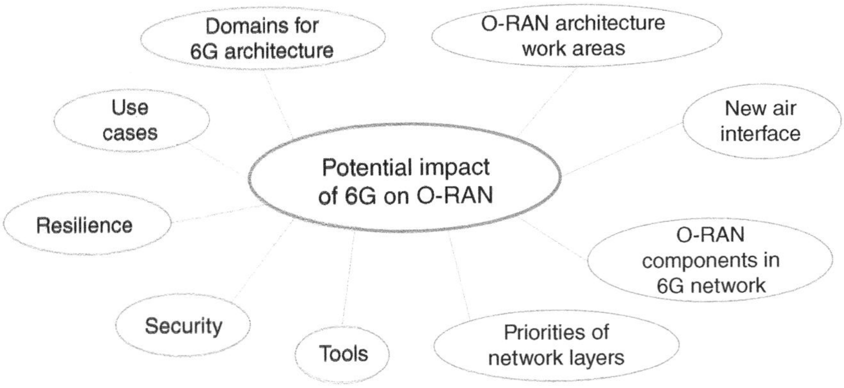

Figure 8.5 Impact of 6G on O-RAN.

through liaisons via O-RAN Standard Development Focus Group and collaborate with O-RAN Testing and Integration Focus Group, IEEE, and National Science Foundation (NSF) on potential 6G testbeds.

A brief summary of the work performed by the nGRG as of Fall 2024 is given below.

The nGRG carried out a survey to steer the O-RAN research toward 6G, find the gaps in the current O-RAN standards, and enhance O-RAN standards to close the gaps (O-RAN ALLIANCE 2023). The survey asked questions related to (i) 6G use cases and requirements that may impact existing and evolving O-RAN, (ii) wanted functional and nonfunctional capabilities in 6G, and (iii) gaps between requirements and wanted system capabilities in current O-RAN architecture and standards. Figure 8.5 classifies the findings of this survey in nine categories;[4] see O-RAN ALLIANCE (2023) for details of the survey's findings. These categories include use cases, domain for 6G architecture, O-RAN architecture work areas, new air interface, O-RAN components in the 6G network, priorities of network layers, tools, security, and resilience. These categories are briefly discussed below.

8.3.1 Use Cases

The nGRG survey has identified 13 categories of use cases. These categories include quantum systems, privacy and security, new public safety and government applications, specialized vertical industry sectors, RT analytics, limits to Moore's law, resilience by design, automation or autonomics, sensory and brain experiences, AI/ML in every part of the network, beyond 5G, extreme-edge or edge centric networks, softwarization and disaggregation, and 6G spectrum toward sub-terahertz (THz).

Among these use case categories, the AI/ML in every part of the network was found to be of most interest, followed by the RT analytics, softwarization and disaggregation, and beyond 5G edge-centric network. Privacy and security, automation/autonomics, and 6G spectrum toward THz communication of relatively less interest. Interest in THz communication appears to have reduced significantly since the early days of 6G research. New categories identified by survey participants beyond the originally envisioned 13 categories include sustainability and energy efficiency, ease of network operation, coverage extension technologies, and robotics.

4 These categories are created in the book to facilitate discussions of 6G-related topics for O-RAN. The nGRG survey itself does not specify these categories.

Seven prioritized use cases identified by ITU-T include Holographic-Type Communications (HTC), Tactile Internet for Remote Operations (TIRO), Intelligent Operation Network (ION), Network and Computing Convergence (NCC), digital twin (DT), Space Terrestrial Integrated Network (STIN), and Industrial IoT (IIoT) with Cloudification (O-RAN ALLIANCE 2023). The ITU-T use cases with the highest priority are found to be XR/HTC, DT, and NTN/STIN, followed by NCC, ION, IIoT, and TIRO. Compared to these ITU-T use cases, additional use cases mentioned by survey participants include AI, massive communication, and robotics.

The NGA has identified four main categories of use cases, which are network-enabled robotic and autonomous systems, multisensory extended reality, distributed sensing and communications, and personalized user experiences (O-RAN ALLIANCE 2023). Prioritized NGA use cases in the descending priority order are network-enabled robotic and autonomous systems, multisensory extended reality, XR/HTC, personalized user experiences, and distributed sensing and communications.

Additional overall use cases of interest identified by the nGRG include autonomous vehicles/systems, AI, massive communication, sensing, context awareness and localization, network for sensors, IP breakout (UPF) and RAN control (Near-RT RIC and RT RIC) interactions, URLLC and Time-Sensitive Networking (TSN) for private networks and industrial applications, integration of cellular and Wi-Fi, user-centric network, global connectivity, extreme communications, versatile integrations and intelligent customization, Urban Air Mobility, energy efficiency, scheduler profile adaptation, spectrum sharing in lower and mid bands, RAN slicing, extreme communications, versatile integrations and intelligent customization, metaverse-based remote management, metaverse, and Intent-driven autonomous network management (O-RAN ALLIANCE 2023).

8.3.2 Domains for 6G Architecture

The nGRG survey explored interest in domains such as (i) transport; (ii) core; (iii) RAN; (iv) Management, Orchestration, and SMO; (v) services; and (vi) others (e.g., DT, AI, security and privacy, service-based architecture [SBA], and energy efficiency). Among these domains, RAN is found to be the domain where significant improvements are needed in 6G.

8.3.3 O-RAN Architecture Work Areas

The nGRG survey explored interest in the following O-RAN architecture work areas: (i) Automation/Autonomics on RAN interacting with other domains, (ii) xApps and rApps only, (iii) new functions/interfaces, and (iv) others. The work area of xApps and rApps was of most interest to members. The work area of automation on RAN interaction with other domains was found to be the next area of interest, followed by the area of new functions and interfaces. Additional areas of interest suggested by members include common SMO for full automation and cross-domain optimizations and customization of each node to enhance coordination between Apps and RAN nodes, AI/ML-native architecture, and identification of the features for software or hardware implementation.

8.3.4 New Air Interface

A new air interface would directly affect O-RAN. Additional software or specialized hardware may be needed to support a new 6G air interface. The need for new hardware was found to be the majority view of survey respondents. The need for edge computing to reduce the TCO was found to be important for energy efficiency and sustainability. The new air interface may support joint communications and sensing. Hardware accelerators and xHaul interfaces and transport protocols of O-RAN may need to be enhanced to support a new 6G air interface supporting extreme throughput.

8.3.5 O-RAN Components in the 6G Network

A 6G is expected to be a mix of O-RAN components with new 3GPP-based 6G and legacy (e.g., 5G and 4G) deployments based on the nGRG survey findings. Note that the SMO and the RICs would need to communicate with legacy components for network monitoring and visibility. There would be a need to support user-centric communication, where connectivity is provided irrespective of communication technology and sustainability through control and management of 3GPP-based 6G systems by RICs. Additional open interfaces and components related to AI-native RAN architecture such as SMO and RIC may be needed to support combined or mixed deployments. Research and experiments would be needed to verify feasibility of combined deployments.

8.3.6 Priorities of Network Layers

The nGRG survey explored the following network layers: integrated models across multiple layers (E2E), service layer, Ethernet/switching, physical layer, and IP. Among these options, integrated models across multiple layers (E2E) were found to be of the highest priority. O-RAN would need to strike a balance between the level of required integration and the achievable benefits.

8.3.7 Tools

The nGRG survey explored if E2E tools, E2E service modeling, RAN design tools, or some other tools will be needed in 6G O-RAN. Among these options, the need for RAN design tools in 6G was found to be of the highest priority. The need for specific tools would likely be clear after 6G specifications are defined. Since 6G may use newer technologies such as joint sensing and communications, user positioning, and reconfigurable intelligent surfaces, new tools that can be useful for these technologies would be needed. Ray tracing may play a more prominent role in 6G due to new radio technologies. Tools that can provide enhanced channel modeling for sub-THz and THz communications and large antenna arrays may also be needed.

8.3.8 Security

The nGRG survey found that a vast majority of respondents considered the need to handle security directly in the design phase to be important. The survey found the "Security by Design" principle to be relevant in 6G. Adding security features after an initial design may result in a suboptimal solution. Furthermore, cross-domain security approaches are considered to be necessary in the design phase.

8.3.9 Resilience

6G needs to be resilient and needs to support service delivery at the target quality levels with sustainability. 6G aims for resilience that maintains service quality levels on the network even when adverse events occur. Additionally, resilience is also the ability to recover service levels after an adverse event occurs. The nGRG survey identified the need for resilience to be directly incorporated into the 6G networks at the design stage, ensuring that functional safety in use cases may require architectural design features such as statelessness. Furthermore, network and consumer services may require different resiliency levels. 6G software and hardware need to be flexible enough to adapt to diverging demands. In 6G, AI/ML can be exploited to monitor network KPIs and modify network settings to enhance resilience. The overall network architecture design may use redundancies in data centers and key components to further increase resilience.

8.4 Major Takeaways of the Chapter

O-RAN faces several challenges before it is widely deployed around the globe. However, a comprehensive ecosystem, flexibility, openness, intelligence, and comprehensive integration and testing can address such challenges. Here are the major takeaways from this chapter.

- O-RAN challenges can be classified into the following categories: Orchestration and OAM, performance, security, product development, TCO, interoperability, RAN product performance, and multivendor operational complexity.
- Locations of O-RAN components need to be determined properly. Furthermore, AI/ML models need to be orchestrated.
- Overall performance aspects of O-RAN such as energy consumption, resource management, and QoS management need to be addressed.
- An increased threat surface of O-RAN compared to a traditional RAN requires a comprehensive solution.
- Increased virtualization and disaggregation in O-RAN make it difficult to outperform single-vendor custom solution that utilizes custom hardware.
- Both CapEx and OpEx need to be considered to determine the TCO. An increased degree of disaggregation and the need for fiber transport network in O-RAN can increase O-RAN TCO.
- While open interfaces enable O-RAN to utilize O-RAN components from different vendors, interoperability among components needs to be validated.
- Since O-RAN utilizes a high degree of virtualization, the time required to carry out certain complex physical layer processing may be longer compared to a traditional custom single-vendor RAN, requiring hardware accelerators.
- Multivendor Operational Complexity. When O-RAN components from different vendors are combined to develop an integrated solution, operational complexity increases significantly.
- In the ITU's vision of 6G usage scenarios, three scenarios are extensions of IMT-2020 or 5G scenarios, while three scenarios are new relative to 5G. Three extension scenarios are immersive communication, massive communication, and HRLLC. Three new usage scenarios are Ubiquitous Connectivity, Integrated AI and Communication, and Integrated Sensing and Communication.
- The ITU has specified a variety of performance targets for 6G including traditional metrics such as peak data rate, user-experienced data rate, spectrum efficiency, area traffic capacity, connection density, mobility, coverage, and latency and new metrics such as reliability, positioning, sensing capabilities, AI capabilities, sustainability, and interoperability.
- Five major categories of technology enablers for 6G identified by the NGA are component technologies, network O&M and service enablement technologies, radio technologies, system and network architectures, and trustworthiness technologies.

- The nGRG of the O-RAN ALLIANCE focuses on research of open and intelligent RAN principles in 6G and future network standards.
- Potential impact of 6G on O-RAN can be classified into categories such as use cases, domain for 6G architecture, O-RAN architecture work areas, new air interface, O-RAN components in the 6G network, priorities of network layers, tools, security, and resilience.

References

3GPP (2024). *Releases*. https://www.3gpp.org/specifications-technologies/releases (accessed 21 July 2024).

Aryal, N. et al. (2023). Open radio access network challenges for next generation mobile network. *26th Conference on Innovation in Clouds, Internet and Networks and Workshops (ICIN)*. March 2023, https://hal.science/hal-03967401 (accessed 21 July 2024).

Brik, B. et al. (2022). Deep learning for b5g open radio access network: Evolution, survey, case studies, and challenges. *IEEE Open Journal of the Communications Society* 3: 228–250.

GCC Open RAN Consortium (2024). Open RAN for Brownfield Operators Challenges and Opportunities. White Paper, October 2022, https://telecominfraproject.com/wp-content/uploads/Open-RAN-for-Brownfield-Operators-Challenges-and-Opportunities.pdf (accessed 21 July 2024).

ITU (2023). *The ITU-R Framework for IMT-2030*. July 2023, https://www.itu.int/en/ITU-R/study-groups/rsg5/rwp5d/imt-2030/Documents/IMT-2030%20Framework_WP%205D%20Management%20Team.pdf (accessed 21 July 2024).

NGA (2022). *Next G Alliance Report: Roadmap to 6G*. February 2022, https://nextgalliance.org/white_papers/roadmap-to-6g/ (accessed 21 July 2024).

NGA (2024). *6G Library*. https://nextgalliance.org/6g-library/ (accessed 21 July 2024).

NGA Alliance (2023). *6G Radio Technology Part I: Basic Radio Technologies*. November 2023, https://nextgalliance.org/white_papers/6g-radio-technology-part-i-basic-radio-technologies/ (accessed 21 July 2024).

NGMN Alliance (2021). *NGMN 6G Drivers and Vision*. April 2021, https://www.ngmn.org/work-programme/ngmn-6g-drivers-and-vision.html (accessed 21 July 2024).

O-RAN ALLIANCE (2023). *O-RAN Towards 6G. Report ID: RR-2023-01*. October 2023, https://mediastorage.o-ran.org/ngrg-rr/nGRG-RR-2023-01-O-RAN-Towards-6G-v1_3.pdf (accessed 21 July 2024).

O-RAN ALLIANCE (2024). *Introduction (NGRG—Next Generation Research Group)*. https://public.o-ran.org/display/NGRG/Introduction (accessed 21 July 2024).

Qualcomm (2022). *Vision, Market Drivers, and Research Directions on the Path to 6G*. December 2022, https://www.qualcomm.com/content/dam/qcomm-martech/dm-assets/documents/Qualcomm-Whitepaper-Vision-market-drivers-and-research-directions-on-the-path-to-6G.pdf (accessed 21 July 2024).

Appendix A

Energy Efficiency in Open Radio Access Networks

This Appendix discusses energy consumption in Open Radio Access Networks (O-RAN). Traditional methods to increase energy efficiency (EE) are described, followed by EE methods for O-RAN.

A.1 Introduction

Fifth-generation (5G) wireless networks achieve high throughput through advanced techniques like massive MIMO (M-MIMO) and increased network densification, but these approaches often require additional base stations (BSs) or hardware, leading to higher power consumption, increased costs, and a larger carbon footprint. To improve EE, the Third Generation Partnership Project (3GPP) has incorporated energy-saving mechanisms in the 5G New Radio (NR) standards, with the O-RAN providing a framework for standardized interfaces to monitor and optimize network efficiency using machine learning (ML) and other intelligent methods.

Cellular network design has historically prioritized performance without much focus on energy efficiency (Wu et al. 2015). This has led to Radio Access Networks (RANs) consuming over half of the total energy in cellular networks (Brik et al. 2022). Disaggregating O-RAN functionalities into separate hardware components and operating them individually, as mentioned in Mollahasani et al. (2022) and Wu et al. (2015), has contributed to increased energy use. EE is now a significant concern for network providers because energy consumption directly affects operational costs (Mollahasani et al. 2021; Pamuklu et al. 2021; Wu et al. 2015).

One way to address this energy usage issue, as suggested in Niknam et al. (2022), is to turn off certain features on small cells during peak network load. Another approach is infrastructure sharing, where excess traffic is rerouted to less congested cells. The authors in Mollahasani et al. (2022) propose adding an extra O-RAN distributed unit (O-DU) with higher processing capacity as another solution. Information sharing among network functions, highlighted in Iturria-Rivera et al. (2022), can also improve EE. The authors used various multiagent deep reinforcement learning (RL) strategies to find the optimal approach for balancing power use and resource allocation to maximize network performance. According to Pamuklu et al. (2021), reducing energy consumption in O-RAN architecture requires a combined strategy of selecting the optimal O-DU for placing network functions and allocating resource blocks effectively.

In other words, the key pillars for enhancing energy efficiency in the O-RAN framework are as follows: (i) power-efficient hardware, (ii) monitoring and reporting energy efficiency key performance indicators (KPIs) for both hardware and software, (iii) incorporating energy-efficient

Fundamentals of O-RAN, First Edition. Nishith D. Tripathi and Vijay K. Shah.

Companion website: www.wiley.com/go/tripathioran

features at a functional level, and (iv) automating energy efficiency through artificial intelligence and orchestration. Furthermore, the TIP Open RAN MoU Group (TIP Open RAN MoU Group 2024) released a Technical Priority Document, which establishes specific energy efficiency targets for various network components, outlining acceptable power consumption limits in different scenarios depending on user traffic load and capacity requirements.

A.2 EE Improvement Methods in Traditional RAN Environment

In this section, we summarize some traditional EE methods that have been followed in the industry (Hoffmann and Dryjanski 2023). There are various strategies to enhance EE in traditional mobile networks and RANs, and we can apply them at multiple layers (Feng et al. 2012).

A.2.1 Turning On/Off the BS

At the top level, EE can be improved by turning off the entire BS, effectively putting it into the sleep mode. Consider Figure A.1 (Hoffmann and Dryjanski 2023).

A macro-BS is always active to provide coverage for a given area, while several pico-BSs can be dynamically switched on or off based on current traffic demand. For instance, during the daytime in a busy city center where many people are at work, the pico-BSs may need to be active, but at night, some can be put into sleep mode as most people return to their suburban homes. This process is governed by RL, a method that involves learning through interaction. The agent's goal is to determine which configuration of active pico-BSs achieves the best energy efficiency given a specific distribution of users.

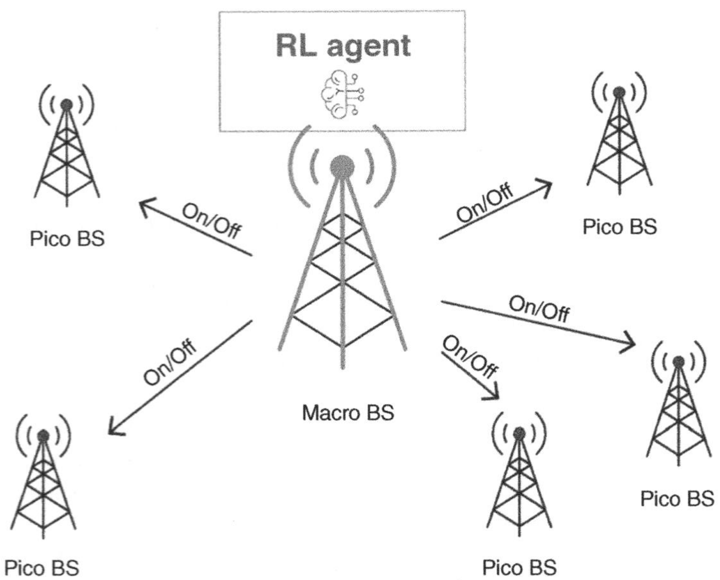

Figure A.1 BS On/Off Technique. (Source: Hoffmann and Dryjanski 2023, Copyrighted by Rimedo Labs. Reproduced with permission).

A.2.2 Antenna Selection

At a more granular level, EE optimization involves specific components within the BS, such as enhancing the efficiency of power amplifiers or reducing the number of active antennas. This scenario involves optimizing the number of active antennas in an M-MIMO BS as shown in Figure A.2. M-MIMO arrays can contain hundreds of antenna elements, with each element affecting both the overall power consumption and user throughput. The proposed algorithm in Hoffmann and Kryszkiewicz (2021) dynamically adjusts the number of active antennas based on the spatial distribution of users within a cell. Because the M-MIMO system is complex, with nontrivial dependencies involving interference and spatial channel correlations, the RL agent determines the optimal number of active antennas to maximize energy efficiency.

A.2.3 Other Methods

EE improvements can also focus on protocol-specific elements, like dynamic power allocation, adjusting discontinuous reception (DRX) parameters, or blanking resource blocks strategically under particular network conditions. Moreover, energy savings (ES) can be achieved by deactivating individual component carriers.

The aforementioned methods can be categorized as *direct approaches* to improve EE. To summarize, the following elements can be switched off (or set to sleep mode), from the highest to the lowest level: the entire BS, individual sectors/cells, the transceiver chain linked to a specific antenna, single component carriers, specific resource blocks, or even resource elements like pilot signals that are not needed when there is no/minimal traffic.

Additionally, the EE of wireless networks can be boosted indirectly. *Indirect approaches* to EE optimization include load balancing across neighboring cells or frequency bands within a

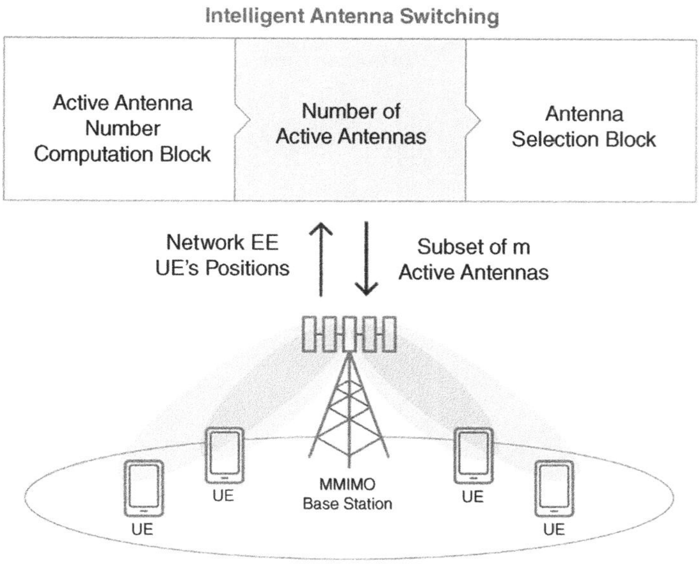

Figure A.2 Antenna Selection Technique. (Source: Hoffmann and Dryjanski 2023, Copyrighted by Rimedo Labs. Reproduced with permission).

single cell. Energy-efficient load balancing is achieved through intelligent traffic steering, dynamic switching of users between cells and frequency bands, which enables turning off of underutilized segments, along with adaptive power allocation. Furthermore, to manage and analyze the substantial amount of data collected from the network, ML algorithms play a crucial role. These algorithms support effective decision-making, such as when to turn off power-hungry hardware component.

A.3 EE Improvement Methods in O-RAN

Now, we will discuss some of the most common yet effective methods that have been proposed in the literature.

A.3.1 Cell Switch On/Off

The traffic load in mobile networks fluctuates throughout the day and between weekdays and weekends. This reflects users' mobility and behavior. For example, during the weekdays, from Monday to Friday, BSs in the cities experience high load and traffic, because major big workplaces in the United States are typically in urban areas (Monitor 2020) (and 88% of jobs are based in metropolitan statistical areas [MSAs]), with peak loads from morning to afternoon, while suburban BSs see low activity until later in the day when people return home. Here, we can see a pattern that presents us with an opportunity to save energy by shutting down entire cells or specific operating frequencies (carriers) during periods of low usage. The O-RAN ALLIANCE introduced a comprehensive framework for EE optimization over extended periods (minutes, hours, or longer) in the non-real-time RAN Intelligent Controller (non-RT RIC) by intelligently turning cells on or off. The communication process between different O-RAN components to enable on/off switching of cells based on ML inference in the non-RT RIC is described in O-RAN ALLIANCE (2024). Figure A.3 illustrates

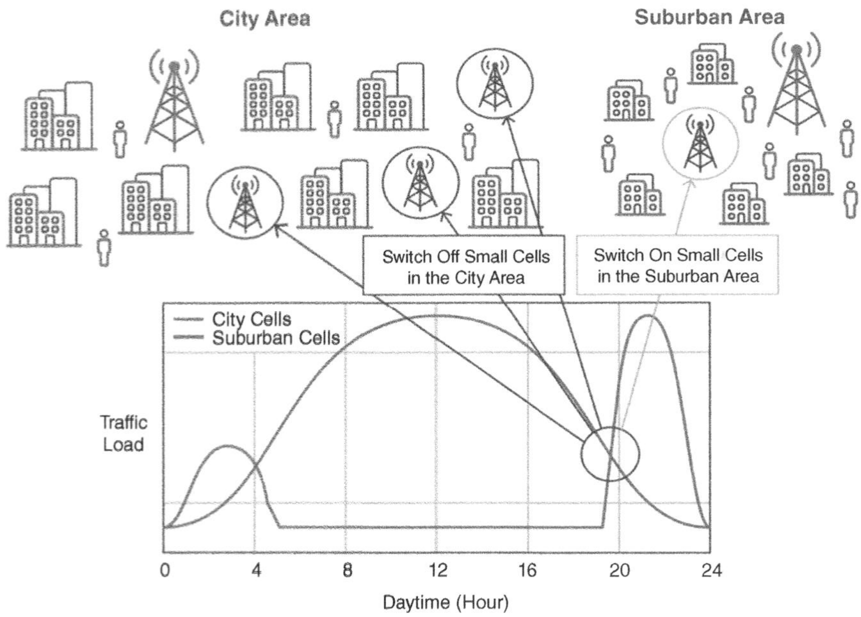

Figure A.3 Antenna Selection Technique. (Source: Hoffmann and Dryjanski 2023, Copyrighted by Rimedo Labs. Reproduced with permission).

how toggling cells on or off in various sections of the designated area, depending on the daily traffic pattern can be advantageous toward EE (Hoffmann and Dryjanski 2023).

A.3.2 Radio Frequency (RF) Channel Switching

As discussed in the previous section, the traffic load in mobile networks fluctuates throughout the day. Typically, it is lower at night when most people are asleep and increases during the day as people use data for work and other activities. Taking these variations into account, the number of active radio frequency (RF) channels can be adjusted to achieve ES while still meeting the subscribers' quality of service (QoS) demands. As shown in Figure A.4 (Hoffmann and Dryjanski 2023), during peak hours, all RF channels of a BS are active. When the traffic load decreases to a medium level (corresponding to morning or afternoon times), 40% of RF channels may be turned off to conserve energy. Finally, during periods of low traffic (corresponding to early morning or late evening hours), only 20% of the RF channels need to be active to maintain subscribers' QoS, while the remaining channels can be switched off to further maximize ES. It is to be noted that turning off RF channels requires re-configuring and re-tuning the BS; for instance, with fewer active antennas, the number of supported beams decreases, and the beams might lose their functional shapes.

A.3.3 Traffic Steering

Hoffmann and Dryjanski (2023) discusses the implementation of a policy-based traffic steering extended application (xApp) within O-RAN. The xApp is a microservice designed to optimize and manage network traffic, which enables more flexible and intelligent traffic steering within the O-RAN framework. This technique allows mobile network operators to dynamically route traffic based on predefined policies, which enhances network efficiency as well as user experience. They explore the benefits of this xApp, such as improving resource utilization and providing mobile network operators with better control over traffic flows, leading to a more efficient and adaptive network environment. They also evaluate technical considerations in implementing the xApp and how it integrates with other O-RAN components to support advanced traffic management strategies.

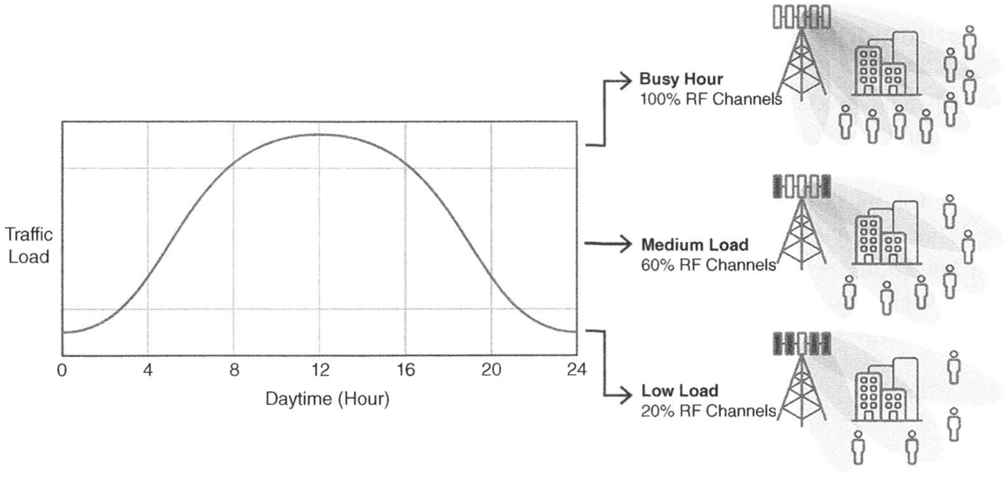

Figure A.4 RF Channel Switching On/Off Technique. (Source: Hoffmann and Dryjanski 2023, Copyrighted by Rimedo Labs. Reproduced with permission).

A.3.4 Leveraging Silicon Compute

O-RAN hardware is one variable of the energy efficiency equation in network design. Even when hardware components are optimized, there is a baseline level of energy consumption required to maintain network connectivity. The other variable is meticulously designed monitoring and optimizing software, tailored to O-RAN hardware. To achieve significant energy savings, it is crucial to monitor, analyze, and optimize network operations in real time, ensuring that the hardware operates as efficiently as possible. Ericsson (Ericsson 2023) has been a key player in collaborative efforts to improve EE within the Cloud-RAN ecosystem. The company has partnered with others to develop software features for the Cloud-RAN DU that significantly reduces energy consumption without compromising performance. Ericsson claims that this collaboration is part of a broader initiative to establish a software ecosystem that can greatly enhance EE, which in turn provides a blueprint for future network designs.

One notable example of this collaboration is the integration of Intel's Real Time Dynamic scaling of CPU cores with C-state technology into the RAN market, which significantly improves EE. C-states are power-saving modes in Intel CPUs, which allows individual cores to reduce energy usage when idle. The deeper the C-state, the more power is saved, with C0 being the active state and higher-numbered C-states representing progressively lower power modes. For most mobile networks, over 80% of energy is consumed in the RAN, and the remainder in the core network (Ericsson 2023). Out of the estimated 80%, around 80% of energy is used to power the radios, while the remaining 20% goes to the DU (Ericsson 2023). Even this 20% energy is significant in terms of ES and carefully optimized software algorithms designed to leverage the built-in C-states in Intel CPUs can prove to offer major gains in EE. Through collaboration with Intel and the application of their C-state management technology, Ericsson has developed algorithms that intelligently manage C-states in Cloud-RAN DUs. These algorithms allow CPUs to enter sleep modes during off-peak hours when traffic is lower, thereby reducing energy consumption. However, the key to this approach is ensuring that CPUs can quickly wake up when network demand increases, avoiding disruptions in service or data loss. The combination of intelligent C-state management and real-time network optimization enables a more energy-efficient RAN while maintaining the flexibility and responsiveness required for dynamic traffic patterns.

A.4 Summary

Energy efficiency is a crucial focus in 5G and future mobile networks, defined as the ratio between average user throughput and average power consumption. As these networks pursue higher throughput, more BSs and hardware are deployed, leading to increased energy demands. To tackle this challenge, the O-RAN ALLIANCE is prioritizing energy-saving mechanisms within the RAN. Examples of energy-saving techniques within O-RAN include turning off hardware components during low traffic periods. Collaboration and software innovation are important in achieving energy efficiency. Adopting energy-efficient practices in the broader O-RAN ecosystem can yield significant energy-saving benefits.

References

Brik, B., Boutiba, K., and Ksentini, A. (2022). Deep learning for B5G open radio access network: Evolution, survey, case studies, and challenges. *IEEE Open Journal of the Communications Society* 3: 228–250.

Ericsson. (2023). *Open Ran Architecture: Embracing Energy Efficiency*. https://www.ericsson.com/en/blog/2023/7/open-ran-architecture-embracing-energy-efficiency (accessed 18 August 2024).

Feng, D., Jiang, C., Lim, G. et al. (2012). A survey of energy-efficient wireless communications. *IEEE Communications Surveys & Tutorials* 15 (1): 167–178.

Hoffmann, M. and Dryjanski, M. (2023), *"The O-RAN Whitepaper 2023 – Energy Efficiency in O-RAN"*, Whitepaper, Rimedo Labs, https://mailchi.mp/0afa75992d20/the-o-ran-whitepaper-2023-energy-efficiency (accessed 18 August 2024).

Hoffmann, M. and Kryszkiewicz, P. (2021). Reinforcement learning for energy-efficient 5G massive MIMO: Intelligent antenna switching. *IEEE Access* 9: 130329–130339.

Iturria-Rivera, P.E., Zhang, H., Zhou, H. et al. (2022). Multi-agent team learning in virtualized open radio access networks (O-RAN). *Sensors* 22 (14): 5375.

Mollahasani, S., Erol-Kantarci, M. and Wilson, R. (2021, December). Dynamic CU-DU selection for resource allocation in O-RAN using actor-critic learning. *2021 IEEE Global Communications Conference (GLOBECOM)*, pp. 1–6. IEEE.

Mollahasani, S., Pamuklu, T., Wilson, R., and Erol-Kantarci, M. (2022). Energy-aware dynamic DU selection and NF relocation in O-RAN using actor–critic learning. *Sensors* 22 (13): 5029.

Monitor, W.F. (2020). *Where the Jobs Are: Urban, Suburban, Rural Migrations, and the Hollowing Out of Middle-Skill Jobs*. https://wfmonitor.com/2020/06/12/where-the-jobs-are-urban-suburban-rural-migrations-and-the-hollowing-out-of-middle-skill-jobs/ (accessed 23 April 2024).

Niknam, S., Roy, A., Dhillon, H.S., Singh, S., Banerji, R., Reed, J.H., et al. (2022, December). Intelligent O-RAN for beyond 5G and 6G wireless networks. *2022 IEEE Globecom Workshops (GC Wkshps)*, pp. 215–220. IEEE

O-RAN ALLIANCE. (2024). *O-RAN Use Cases Detailed Specification 14.0*. O-RAN.WG1.Use-Cases-Detailed-Specification-R003-v14.00 https://specifications.o-ran.org/specifications (accessed 18 August 2024).

Pamuklu, T., Mollahasani, S. and Erol-Kantarci, M. (2021, October). Energy-efficient and delay-guaranteed joint resource allocation and du selection in o-ran. *2021 IEEE 4th 5G World Forum (5GWF)*, pp. 99–104. IEEE.

TIP Open RAN MoU Group. (2024). https://telecominfraproject.com/openran-mou-group/ (accessed 18 August 2024).

Wu, J., Zhang, Y., Zukerman, M., and Yung, E.K.N. (2015). Energy-efficient base-stations sleep-mode techniques in green cellular networks: A survey. *IEEE Communications Surveys & Tutorials* 17 (2): 803–826.

Index

Fundamentals of O-RAN, First Edition. Nishith D. Tripathi and Vijay K. Shah.

Companion website: www.wiley.com/go/tripathioran

S

Printed and bound by CPI Group (UK) Ltd, Croydon, CR0 4YY
17/09/2026
14964382-0001